Encyclopedia

of Organic, Sustainable, and Local Food

Edited by
Leslie A. Duram

UNIVERSITY OF NEBRASKA PRESS • LINCOLN AND LONDON

Encyclopedia of Organic, Sustainable, and Local Food, edited by Leslie A. Duram was originally published in hardcover by Greenwood Press, an imprint of ABC-CLIO, LLC, Santa Barbara, CA. Copyright © 2010 by Leslie A. Duram. Paperback edition by arrangement with ABC-CLIO, LLC, Santa Barbara, CA. All rights reserved.

Manufactured in the United States of America

∞

First Nebraska paperback printing: 2011

Disclaimer: For the Marketing and USDA Economic Research Service entries by Carolyn Dimitri, Barry Krissoff, and Fred Kuchler, the views expressed are those of the author and not necessarily those of the Economic Research Service or the U.S. Department of Agriculture.

Library of Congress Cataloging-in-Publication Data
Encyclopedia of organic, sustainable, and local food / edited by Leslie A. Duram.—
Paperback ed.
p.   cm.
Includes bibliographical references and index.
Reprint. Originally published in hardcover: Santa Barbara, CA: Greenwood Press, 2010.
ISBN 978-0-8032-3625-7 (pbk.: alk. paper)
1. Food supply—United States. 2. Local foods—United States. 3. Natural foods—United States.
4. Sustainable agriculture—United States. I. Duram, Leslie A.
HD9005.E645 2011
338.1'97303—dc22   2010043210

To Jon, Kyle, and Maggie; we are a wonderful team.
And to all the women who have encouraged me and
believed in me: Pearl, Annjeanette, Rita, Ria, Zita, and, of
course, Mom, who taught me to laugh often and loudly.

Contents

Preface

This encyclopedia introduces students and general readers to topics in organic, sustainable, and local food and farming. At the same time, it provides in-depth information on a variety of specialized topics that will inform both general readers as well as experts. The volume introduction helps readers to understand the current relevance and importance of the organic, sustainable, and local food movements in contemporary U.S. society.

To start, users can quickly find topics of interest in the list of entries and topical list of entries and the index. The more than 140 entries, written by internationally respected scholars, government experts, and dedicated activists, cover a wide range of relevant topics in organic, sustainable, and local food including definitions of farmers markets and the like, and also discuss health issues, economic terms, organic big business, United States Department of Agriculture (USDA) organic standards, food labels, and numerous organizations that are so important to the organic food movement. Topics that seem familiar because of recent media attention, such as organic milk, Whole Foods, or climate change, are described here objectively. Other topics may seem surprising, such as Walmart or food deserts, and thus require some explanation and grounding.

Whereas certified organic agriculture has specifically defined methods of production, "sustainable agriculture" is a more general concept that implies long-term environmental and economic viability. Some entries describe the opposite form of production: chemical (or conventional) agriculture. This helps the reader understand the uniqueness of organic farming in the United States in sharp contrast to the vast dominance of chemical agriculture practiced since World War II. Local food initiatives have seen significant growth in recent years and play an increasingly important role in the North American food system.

Each contribution draws on current sources of information, including reports, books, journal articles, and Web sites, some of which are listed in the entries' further readings sections. Cross-references to related entries and appendixes are also included. The four appendixes provide primary documents giving key legislation, recommendations, and commentary related to organic food and farming.

In addition, a chronology provides the background and context through which to understand these current food issues. This historical foundation helps us understand why current topics are important for today's society and environment. After looking back, readers are encouraged to look toward the future—read more about key topics and stay informed about pressing issues. A selected bibliography has been included to aid the reader in this endeavor.

Introduction

The popularity of organic food is growing. What began as a fringe "hippie" movement has now blossomed to over $23 billion in sales each year. Organic food includes vegetables, fruits, grains, and any agricultural crop or food that has been grown or produced using organic methods, which are based on building healthy soil naturally without the use of synthetic chemical fertilizers and pesticides. In addition, no genetically modified organisms (GMOs), no sewage sludge, and no irradiation are used in organic agriculture. There is an organic certification system that many farmers opt for, although smaller organic farmers may grow using organic methods and not seek certification. This is particularly the case if farmers know their customers and a relationship of trust has been established. Indeed, the local food movement is a popular trend that builds closer relationships between farmers and consumers by selling crops closer to where they are produced. "Sustainable food" is a general term that encompasses the notion of food production that maintains economic and ecological viability for the long-term.

Some consumers know that certified organic food means chemical-free crops or crops grown in a field that has gone at least three years without the use of prohibited synthetic chemicals. Other consumers have the general notion that organic is healthier. Still others, perhaps, are just caught up in the trend. For whatever reason, organic food, and thus organic farming, is becoming mainstream. Consumers may be most interested in the health benefits of their organic broccoli or the popularity of their fair-trade, shade-grown, organic latte, but there are broader social and ecological benefits as well.

First, in terms of social sustainability and sense of community, organic farming represents an opportunity for farm families to stay on the farm and remain in agriculture. Over the last one hundred years, small farms have been gobbled up by their larger neighbors; average farm size has increased significantly and the number of farmers has decreased dramatically. It is hard to survive in agriculture without being a part of the "bigger is better" mentality that relies on genetically engineered crops, chemical fertilizers, and toxic pesticides to produce commodities for livestock, biofuels, and export. While U.S. agriculture has been "biggering" (to quote the ever-wise Dr. Seuss), the alternative agricultural movement has quietly taken root. Organic farming is a true alternative that allows some farmers to stay on their land and produce real food for their communities. This, in turn, allows a small town to stay afloat, rather than board up its Main Street.

The local food movement is a cultural phenomenon that is closely related to organic and sustainable farming. By building local food networks, people are getting reconnected to their community through their food. Indeed, farmers markets, community-supported agriculture, and community gardens all lead to social connectedness that flies in the face of the impersonal global alienation that treats food like any other industrial commodity. In other words, organic farming can provide cultural, economic, and social opportunities for both rural and urban America.

Second, organic agriculture has notable ecological benefits. Its diverse cropping and farming techniques mimic nature. This supports more wildlife, as buffer strips of native grasses and windrows of trees increase the habitat availability for birds and other native wildlife. Increased biological diversity is seen on and around organic farms. Organic farming techniques are not reliant on fossil fuel-based agrichemicals. These farms likely produce food for their local communities, which further decreases their fossil fuel use. Local food greatly reduces food miles, the distance food travels from farm to dining table in the United States, which currently averages 1,800 miles. Thus, organic farms help reduce the huge contribution that agriculture makes to global warming. Soil building is a key component of organic methods, and deeper, healthy topsoil leads to less erosion. In addition, organic farming methods do not rely on synthetic, petroleum-based chemicals, which run off the farm fields and poison streams, lakes, and rivers. Therefore, the ecological benefits of organic farming methods are as pronounced as the social benefits.

Although the potential benefits of organic food are great, so too are the possible negative trends. Indeed, organic food is becoming a victim of its own success; in some places its popularity has caused profit-seeking corporations to overshadow small farms. As the USDA National Organic Program (NOP) now regulates organic farming and food, some small-scale organic farmers are left out. Most organic brands are now owned by large multinational corporations. So, as organics shift to the mainstream and big business takes over organic food, local food has become the "new organic" or, rather, the new alternative.

This encyclopedia seeks to stimulate readers to question the current food system and think about what type of food system they want in the future. It informs readers about the importance of growing and eating organic food and encourages people to get involved in their local food community.

Chronology

1862 U.S. Department of Agriculture (USDA) is set up without Cabinet status.

1862 President Abraham Lincoln appoints a chemist, Charles M. Wetherill, to serve in the new Department of Agriculture. This begins the Bureau of Chemistry, the predecessor of the Food and Drug Administration (FDA).

1883 Harvey W. Wiley, chief chemist of the USDA, expands studies of food adulteration. He campaigns for a federal law and is called the "Crusading Chemist."

1889 USDA is raised to Cabinet status.

1894 USDA develops the first list of dietary standards for Americans.

1902 Congress appropriates $5,000 to the Bureau of Chemistry to study chemical preservatives and colors and their effects on digestion and health, because of increasing public awareness.

1906 Pure Food and Drug Act is passed by Congress and signed by President Theodore Roosevelt. It enables the federal government to remove a food or drug product from circulation if the government proves it is unsafe. Food processors are not required to prove their products are safe prior to being sold.

1906   Meat Inspection Act is passed to set up the first system to address filthy meat-packing plants.

1907   First Certified Color Regulations, requested by manufacturers and users, list seven colors found suitable for use in foods.

1910   Federal Insecticide Act is passed. It is the first regulation of agricultural chemicals, focused only on protecting farmers from deceptive manufacturers.

1914   In *U.S. v. Lexington Mill and Elevator Company* the U.S. Supreme Court rules that for a food with dangerous chemical residues to be banned from foods, the government must prove the harm it caused in humans. The mere presence of such an ingredient is not sufficient to ban the food.

1916   USDA publishes first daily food guides, consisting of five food groups: milk and meat, cereals, vegetables and fruits, fats and fat foods, and sugars and sugary foods.

1920   Sir Albert Howard, early researcher on organic farming methods, carries out his work on composts in India and informs sustainable agriculture groups in the United States and the United Kingdom.

1921   General Mills Corporation creates a character named Betty Crocker for advertising to convince generations of Americans to use processed foods.

1924   Educator, philosopher, and lecturer Rudolph Steiner publishes *Agriculture*, which describes the basics of biodynamic agriculture.

1927   Biodynamic produce marketed for the first time. It is inspired by Rudolf Steiner's lectures on biodynamic farming.

1928   Demeter symbol is established as quality control symbol for biodynamic agriculture.

1933   Agricultural Adjustment Act (AAA) initiates crop and marketing controls.

1933   FDA recommends a complete revision of the obsolete 1906 Pure Food and Drug Act. The first bill is introduced into the Senate, launching a five-year legislative battle.

1935   AAA is amended to provide marketing orders and continuing funds for removal of agricultural surpluses.

1935   Resettlement Administration is created to combat rural poverty (in 1946 it becomes the Farmers Home Administration).

1938 Federal Food, Drug and Cosmetic Act, which gives the FDA the authority to oversee the safety of food, drugs, and cosmetics, is enacted.

1938 USDA Yearbook of Agriculture is titled "Soils and Men" and provides an excellent account of organic farming that is still used today by modern organic farmers.

1938 A powerful new pesticide called DDT is discovered by a Swiss chemist.

1939 Food stamp plan begins to assist low-income Americans obtain minimum food access.

1940 Lord Northbourne uses the term "organic" to describe a dynamic, holistic, sustainable agricultural system in his book *Look to the Land*.

1940 Sir Albert Howard's book *An Agricultural Testament* is published. It describes many of the principles of organic agriculture.

1940 FDA moves out from under the supervision of the Department of Agriculture; is placed in the Federal Services Administration.

1941 The first Recommended Dietary Allowances, including recommendations for calories and nine nutrients, are released by the Food and Nutrition Board of the National Academy of Sciences.

1941–45 Victory Gardens (vegetable gardens) are planted during World War II to ensure an adequate food supply for civilians and troops. During World War II, nearly twenty million Americans plant Victory Gardens.

1942 First use of DDT pesticide on U.S. farms.

1942 J. I. Rodale begins publishing *Organic Gardening* magazine.

1942–49 Price controls and food rationing occur during wartime emergency.

1945 Nerve gas research conducted during the war results in the development of chemicals toxic to insects, producing an explosion in the production of pesticides.

1946 Walnut Acres in Pennsylvania, owned by visionary Paul Keene, begins producing organic foods and becomes the country's first organic food brand.

1946 Lady Eve Balfour's book *The Living Soil* is published, setting out the early vision of the organic movement and leading to the establishment of the Soil Association in the United Kingdom, which remains the primary certification and education organization in that country.

1946 First cases of DDT resistance in houseflies (after just four years of use.)

1946   Congress passes legislation to fund permanent National School Lunch program to provide meals for low-income students through federal subsidies.

1947   The Federal Fungicide, Insecticide, and Rodenticide Act (FFIRA) replaces the 1910 Pesticide Act. The law, like its predecessor, regulates the synthetic organic pesticides and does not address environmental or public health.

1947   Sex hormones are first introduced into livestock production to increase fat and weight of animals. One of those hormones, DES, is hailed as the most important development in the history of food production. Several decades later, DES is found to cause cancer.

1948   U.S. food industry begins increasing the amount of monosodium glutamate (MSG) added to processed foods for flavor enhancement, doubling the amount each decade. Researchers later discover that MSG can trigger toxic reactions in some people.

1949   FDA publishes first guidance to food industry, "Procedures for the Appraisal of the Toxicity of Chemicals in Food," which becomes known as the "black book."

1950   Houseflies show almost 100 percent resistance to DDT in U.S. dairies.

1950   Oleomargarine Act requires prominent labeling of colored oleomargarine, to distinguish it from butter.

1950s   J. I. Rodale, founder of Rodale Press, popularizes the term "organic."

1953   Factory Inspection Amendment clarifies previous law and requires FDA to give manufacturers written reports of conditions observed during inspections and analyses of factory samples.

1954   FDA carries out first radiological examination of food because of tuna imported from Japan following atomic blasts in the Pacific. Around-the-clock monitoring is conducted to deal with the emergency.

1954   Miller Pesticide Amendment spells out procedures for setting safety limits for pesticide residues on raw agricultural commodities.

1954   Assistant Secretary of Agriculture Earl Butz claims that agriculture "is now a big business" and farmers must "adapt or die."

1956   "Basic Four" food groups are outlined in USDA's "Essentials of an Adequate Diet." Recommendations include the following minimum number of servings: two servings of milk and milk products; two servings of meat, fish, poultry, eggs, dry beans, and nuts; four servings of fruits and vegetables; and four servings of grain products.

1958   Food Additives Amendment is enacted, requiring manufacturers of new food additives to establish safety. The Delaney Clause prohibits the approval of any food additive shown to induce cancer in humans or animals.

1958   FDA publishes in the Federal Register, the first list of substances Generally Recognized As Safe (GRAS). This evolving list is still influential in food industry.

1959   U.S. cranberry crop is recalled three weeks before Thanksgiving for FDA tests to check for aminotriazole, a pesticide found to cause cancer in laboratory animals. Cleared berries are labeled safe by FDA inspection—the only such FDA endorsement ever allowed on a food product.

1959   Four University of California scientists advise farmers to use farming practices that later become known as integrated pest management (IPM), a major pest-control tool for modern organic and sustainable farmers.

1960   Color Additive Amendment enacted, requiring manufacturers to establish the safety of color additives in foods, drugs, and cosmetics. The Delaney Clause still prohibits the approval of any color additive shown to induce cancer in humans or animals.

1960   Federal Hazardous Substances Labeling Act, enforced by FDA, requires prominent label warnings on hazardous household chemical products.

1961   Codex Alimentarius, or food code, sets global food standards for consumers, food producers, and processors, national food control agencies, and the international food trade.

1962   The Committee for Economic Development publishes "An Adaptive Program for Agriculture," which promotes huge corporate farms at the expense of family farms. This report guides government policy for more than a decade.

1962   Consumer Bill of Rights is proclaimed by President John F. Kennedy in a message to Congress. Included are the right to safety, the right to be informed, the right to choose, and the right to be heard.

1962   Publication of *Silent Spring* by scientist Rachel Carson. It documents negative consequences of agricultural chemicals. In response, chemical companies, the American Medical Association, and large-scale farmers begin an immediate and aggressive counterattack to Carson's views. Still, the book influences the environmental movement and renews public interest in organic agriculture.

1963   The President's Science Advisory Committee publishes a report vindicating Rachel Carson's research and assertions.

1963   The National Agricultural Chemicals Association doubles its public relations budget.

1964   Food Stamp Act sets up federal assistance program that provides food vouchers to low income Americans. States distribute funds, but the USDA administers the program.

1965   Cesar Chavez and Delores Huerta organize the United Farmworkers Union in Delano, California. Pesticide abuses are a major focus of their organizing work.

1965   The President's Science Advisory Committee decries the excessive use of agricultural chemicals and states: "The corporation's convenience has been allowed to rule national policy."

1966   Fair Packaging and Labeling Act requires all consumer products in interstate commerce to be honestly and informatively labeled, with FDA enforcing provisions on foods, drugs, cosmetics, and medical devices.

1968   *Science* magazine publishes an article linking bird population declines with reproductive failure caused by pesticide accumulations in their tissues.

1969   White House Conference on Food, Nutrition, and Health recommends systematic review of GRAS substances in light of FDA's ban of the artificial sweetener cyclamate. President Richard M. Nixon orders FDA to review its GRAS list.

1969   White House Conference on Food, Nutrition, and Health brings public attention to the importance of nutrition; later it influences legislation for expanding the food stamp program, food labeling, and the school lunch program.

1969   National Environmental Policy Act (NEPA) is the foundation of a new era of environmental protection laws.

1970   Environmental Protection Agency (EPA) is established; takes over FDA and USDA duties for pesticide regulations.

1970   *Acres USA* magazine begins publication. It is now one of North America's oldest, largest magazines, covering commercial-scale organic and sustainable farming.

1970   Term "agroecology" is used commonly in research to connote the interconnectedness of agricultural production and ecosystem sustainability.

1971   Northeast Organic Farming Association of Vermont establishes organic certification standards, making it one of the oldest organic farming associations in the US.

1971   *Diet for a Small Planet* is published by Francis Moore Lappé by Food First; it influences generations of vegetarians and environmentalists.

1971   Japanese food scientists synthesize in a laboratory a cheaper sweetener called high-fructose corn syrup. It lengthens shelf-life and thus is useful in baked goods, candy, and even frozen foods.

1972   The registration for DDT is suspended in the United States due to concerns about long-term environmental and health effects.

1972   International Federation of Organic Agricultural Movements (IFOAM), a nongovernmental organization, publishes the first standards for organic agriculture.

1972   EPA bans the pesticide DDT for its cancer-causing potential in humans, but it continues to be sold to other countries that export food to the United States. More than 250 pests worldwide are resistant to DDT.

1972   Author and commentator Jim Hightower publishes the book *Hard Times, Hard Tomatoes* that details the agribusiness bias of the land grant college system.

1972   Cascadian Farm is established on a few acres in Washington state near the Cascade Mountains. It becomes a powerhouse in organic foods and is acquired by General Mills.

1973   California Certified Organic Farmers (CCOF) is established. It is one of the first organizations to certify organic farms in the United States. CCOF now provides certification services to all stages of the organic food chain from farms to processors, restaurants, and retailers.

Mid-   Organic farming grows because of DDT ban, consumer opposition
1970s to chemical pesticides, and increasing environmentalism.

1974   First genetic engineering research is proposed.

1974   Oregon Tilth organic certification agency is established. It is a nonprofit organization supporting and promoting biologically sound and socially equitable agriculture through education, research, advocacy, and certification.

1975   Federal Insecticide, Fungicide, Rodenticide Act (FIFRA) is amended to require USDA and EPA coordination. It establishes Scientific Advisory Panel (SAP) review of proposed actions and improves protection of pesticide manufacturers' trade data.

1975   World Conference on Animal Production report estimates that factory farmed animals contain up to thirty times more saturated fat than animals raised just three decades earlier.

1976   Director of National Cancer Institute, Arthur Upton, tells a congressional committee that half of all cancers are caused by diet.

1976   Frontier Natural Products Co-op is established. It specializes in natural and organic products, particularly organic herbs and spices and aromatherapy products.

1977   National Institutes of Health issues the first of three warnings that an epidemic of obesity is looming in the United States.

1978   Bob's Red Mill Natural Foods opens in Milwaukie, Oregon. Now produces over four hundred whole grain products, many of which are certified organic. Slogan: "Whole grain foods for every meal of the day."

1979   Antibiotic- and hormone-treated meat from the United States is banned in Europe.

1979   American Community Gardening Association (ACGA) is established to advocate for and educate about community gardens in the United States and Canada.

1980   First Whole Foods Market opens in Austin, Texas, with a staff of nineteen; Whole Foods is valued at $4.7 billion by 2005.

1980   Total documented U.S. sales of organic food are under $1 million.

1980   Washington State implements the first state-run organic certification program.

1980   Testing by the FDA finds that 38 percent of all grocery foods sampled contain pesticide residue.

1980   USDA Agriculture Study Team on Organic Farming publishes the first federal *Report and Recommendations on Organic Farming*. Provides a general overview of the status of organic agriculture in the United States: methods, implications for environmental and food quality, economic assessment, and research recommendations.

1981   Article "Organic Farming in the Corn Belt" in *Science* by William Lockeretz, Georgia Shearer, and Daniel H. Kohl is one of the earliest scientific studies of modern organic farming.

1982   Demeter USA is established; this is the brand for products with biodynamic agricultural methods.

1982   Chemical manufacturers push for support of genetic engineering to make crops resistant to their herbicides (thus allowing farmers to spray more). Opponents question why they did not simply engineer weed-resistant crops.

1982   FDA publishes first Red Book (successor to 1949 "black book"), officially known as *Toxicological Principles for the Safety Assessment of Direct Food Additives and Color Additives Used in Food*.

1983   Austria becomes the first country in the world to set official guidelines for organic farming.

1984   Community-supported agriculture (CSA) is brought to the United States by Jan VanderTuin from Switzerland. These are subscription farms, where consumers pay dues to be members and receive weekly produce.

1984   Michigan State University becomes the first land-grant college to sponsor a major conference on sustainable agriculture.

1984   Some 447 species of insects and mites are known to be resistant to one or more pesticides; fourteen weed species are resistant to one or more herbicides.

1984   Earthbound Farms begins selling organic berries from its two-and-a-half acre farm in Carmel Valley, California. It later becomes Natural Selection Foods, which is the world's largest grower of organic produce, operating 24,000 organic farms in the United States, Mexico, and New Zealand.

1985   Organic Crop Improvement Association (OCIA) is established as farmers group and organic farm certifier.

1985   Organic Trade Association (OTA) is established and plays a key role in policy and marketing issues for organic products.

1986   Initial CSA projects begin in the United States—delivering harvest shares from Robyn Van En's Indian Line Farm in Massachusetts and the Temple-Wilton Community Farm in New Hampshire.

1986   Cattle in Britain begin to suffer from bovine spongiform encephalopathy (BSE), commonly called "mad cow disease" due to the behavior of the sick cows. It is caused by feeding animals the blood and meat of infected animals, a common practice in large cattle feedlots, and a practice that is never allowed in organic livestock production.

1987   Demeter International is established. It now represents over 4,000 producers in nearly fifty countries who use biodynamic agricultural methods.

1988   FDA is officially an agency of the Department of Health and Human Services, with a Commissioner of Food and Drugs appointed by the president with responsibilities for research, enforcement, education, and information.

1988   Organic Valley Cooperative is established. Today it has over 1,200 members who produce award-winning, certified organic dairy, produce, and meats.

1988   USDA Cooperative State Research, Education and Extension Service establishes the Sustainable Agriculture Research and Education (SARE) program to promote sustainable farming systems through nationwide research and education grants program.

1988   Washington is the first state to implement an organic certification program. Today, thirteen states have such programs: Colorado, Idaho, Iowa, Indiana, Kentucky, Maryland, New Hampshire, New Mexico, Nevada, Oklahoma, Rhode Island, Texas, and Washington.

1989   The European Union (EU) bans the importation of U.S. meat treated with any growth hormone.

1989   Quality Assurance International is established. The QAI symbol is seen on many products. It offers cutting-edge certification programs.

1989   World Health Organization makes a conservative estimate that there are a million pesticide poisonings in the world each year, resulting in 20,000 deaths.

1989   Alar, a chemical widely used in apple production, is banned as a known carcinogen. Consumers are scared and apple purchases drop for several months. Interest in organic apples increases rapidly.

1989   Britain bans human consumption of certain organ meats, including brain and spinal cord. The United States prohibits import of live cattle, sheep, bison, and goats from countries where BSE (mad cow disease) is known to exist in cattle.

1990   Number of CSA farms in the United States totals only fifty.

1990   Two out of five people are expected to get cancer in their lifetime.

1990   Organic retail sales reach $1 billion in the United States.

1990   Congress passes the Organic Foods Production Act of 1990, mandating that the USDA develop national standards and regulations for organically grown crops.

1990   EPA report "Citizens Guide to Ground-Water Protection" states that seventy-four different pesticides have been found in the groundwater of thirty-eight agricultural states.

1990   Sales of organic products top $1 billion.

1990   Organic Farming Research Foundation (OFRF) is founded as a non-profit organization to foster improvements and adoption of organic farming methods through organic farming research grants and policy/networking programs.

1990   Nutrition Labeling and Education Act requires all packaged foods to display easy-to-read nutrition label.

1990   There are 2,140,000 farms in the United States on about 1.1 billion acres, of which 600,000,000 acres is cropland. Farmers spray more than 800,000,000 pounds of pesticides on U.S. crops.

1991   The European Union adopts organic regulations.

1991   Term "food miles" is created to describe the distance food travels from farm field to dining room table.

1991   Austria grants subsidies to encourage growers to convert to organic production and also sets organic livestock standards, another first.

1992   Tissue analysis demonstrates a substantial link between pesticides and breast cancer.

1992   European Union approves the first government-enforced standards for organic production.

1992   National Organic Standards Board created to define national standards for organic certification.

1992   USDA introduces Food Guide Pyramid, which encourages the consumption of grains, vegetables, and fruits, but continues to recommend two to three servings each day of foods from a dairy product group and from a meat group.

1993   The National Research Council warns that children are at particular risk from pesticide residues because standards were created to measure the effect of pesticides on adult males, ignoring children's and female's lower body weight.

1993   National Academy of Sciences calls for greater regulation of pesticides and stricter tolerance standards.

1993   BSE (mad cow disease) diagnosed in 120,000 cattle in Britain.

1994   Sales of organic food top two billion dollars with sales increasing 20 percent annually, making it the fastest growing segment of grocery markets.

1994   USDA Agricultural Marketing Service begins to gather data on farmers markets in the United States. It discovers that there are 1,700 in operation now.

1994   Studies link organochlorine chemicals (such as DDT, dieldrin, and other agrichemicals) with male reproductive problems and breast cancer. Problems include a 50 percent drop in average male sperm counts in the last forty years.

1994   FDA approves genetically engineered recombinant bovine growth hormone (rBGH or rBST) after a long activist fight. The hormone was developed to increase milk production despite the huge surplus of milk in the United States. It is banned in Canada, Europe, Australia, and elsewhere.

1994 The genetically engineered Flavr-Savr tomato is approved for sale. It tastes terrible and few buy it.

1994 Monsanto and Delta-Pine Land get federal permission to grow genetically modified Roundup Ready cotton.

1996 Food Quality Protection Act amends the Food, Drug, and Cosmetic Act, eliminating application of the Delaney Clause, which had prohibited food additives that cause cancer in humans and animals. In exchange the EPA begins to evaluate the combined toxic effects of all pesticide exposures, including those from water and food. Consideration of body weight and possible impacts on children must be included.

1996 British government changes its position and admits that BSE (mad cow disease) can be transmitted to humans. Four-and-a-half million cattle are destroyed.

1996 Dolly the sheep, the first animal cloned from an adult animal, is born at the Roslin Institute in Edinburgh, Scotland.

1996 World Food Summit concludes. Delegates promised full efforts to ease world hunger.

1997 More than 600 insects and mites are resistant to one or more pesticides. Approximately 120 weeds are resistant to one or more herbicides. Approximately 115 disease organisms are resistant to pesticides.

1997 Four million acres of genetically engineered cotton, soybeans, and corn are grown in the United States.

1997 FDA approves food irradiation as a way of killing bacteria such as *E. coli* in beef. *E. coli* results from dirty slaughterhouse conditions. Activists claim the process destroys nutrients and creates chemicals that may be mutagenic and carcinogenic.

1997 FDA bans protein made from cows, sheep, deer, and ruminants in feed for other ruminants.

1997 USDA proposes and publishes the National Organic Program (NOP), a rule detailing organic production, handling and labeling requirements as an amendment to the Organic Foods Production Act. USDA receives a record 275,000 public comments on numerous issues (mainly opposition to proposed use of genetic engineering, irradiation, and sewage sludge in organic production).

1998 FDA reports that 55 percent of all grocery foods sampled contain pesticides.

1999   US Geological Survey Circular 1225 reports on a comprehensive research program that found pesticides in every fish and stream sample in the United States.

1999   In the United States, 45 percent of the cotton acreage, 45 percent of the corn acreage, and 57 percent of the soybean acreage is planted with genetically modified seed.

1999   Export ban of British beef is lifted after three and a half years.

1999   Minnesota passes the Organic Agriculture Promotion and Education Act to encourage organic agriculture, the first U.S. legislation to authorize an organic certification cost-share program that assists farmers in paying for certification.

1999   Codex Alimentarius, a joint commission of the Food and Agriculture Organization and the World Health Organization (WHO) of the United Nations, approves international guidelines for the production, processing, labeling, and marketing of organic foods.

1999   Small Planet Foods, parent company of Cascadian Farm (produces over one hundred certified organic products) acquired for $11.4 million by General Mills. Realizing typical consumer perceptions are anti-big-business, the name General Mills never appears on any Cascadian Farms products.

1999   Restaurant Nora in Washington, D.C., becomes one of North America's first certified organic restaurants, meaning that 95 percent or more of everything at the restaurant has been produced by certified organic growers and farmers.

1999   USDA's Community Food Security Initiative begins. It is aimed at aiding grassroots efforts to reduce hunger and improve nutrition.

2000   Sales of organic products in the United States reach $8 billion.

2000   After receiving more than 275,000 public comments, the USDA releases second proposed National Organic Program (NOP) that removes the controversial components introduced in 1997 (use of sludge, irradiation, or genetic engineering). All agricultural products sold as organic must be in full compliance by October 2002. Official organic seal and labeling rules also introduced.

2001   USDA Economic Research Service (ERS) Report by Catherine Greene titled "U.S. Organic Farming in the 1990s: Adoption of Certified Systems" *Agriculture Information Bulletin* No. 770 is released. It shows that certified organic crop land more than doubled during the 1990s.

2001   Organic Farming Research Foundation (OFRF) releases a report by Jane Sooby titled "Organic Farming Systems Research at Land

Grant Institutions 2000-2001" which documents the lack of publicly funded university research into organic farming.

2001   Corporate food giant Dole begins marketing organic bananas in the United States.

2002   Article by Paul Mäder, Andreas Fliessbach, David Dubois, Lucie Gunst, Padrout Fried, and Urs Niggli appears in *Science*, based on twenty-one years of organic agricultural experiments. It finds that organic methods can produce similar yields as chemical agriculture with significantly lower inputs.

2002   Key USDA ERS report by Carolyn Dimitri and Catherine Green, "Recent Growth Patterns in the U.S. Organic Foods Market," is published in Agriculture Information Bulletin Number 777. It describes the increase in production and consumption of organic products in the United States.

2002   For the first time the USDA's National Agriculture Statistics Service, Agriculture Census collects and reports data on organic farming.

2002   General Mills begins to market organic "O's" cereal under the well-known organic brand Cascadian Farms (but does not indicate on the box the true corporate ownership).

2002   Movie *My Father's Garden* depicts the transition to chemical agriculture through historical footage and the personal story of a farmer (the filmmaker's father) who dies of cancer before she could get to know him.

2003   FDA requires food labels to include trans fat content, the first substantive change to the nutrition facts panel on foods since 1993.

2003   A cow in Washington state tests positive for mad cow disease (BSE).

2003   Five major weeds develop resistance to the widely used Roundup herbicide; harsher chemicals must be used to control resistant weeds.

2003   Eighty-four percent of U.S. canola acreage is planted with GMO seed.

2003   Catherine Greene and Amy Kremen author "U.S. Organic Farming in 2000-2001: Adoption of Certified Systems" USDA-ERS Agricultural Information Bulletin No. 780, which updates earlier data on organic farming.

2004   FDA bans feeding cow blood, chicken waste, and restaurant scraps to cattle in an effort to reduce the risk of mad cow disease (BSE) in the United States.

2004   Seventy-six percent of cotton, 45 percent of corn, and 85 percent of soybean acreage in the United States is planted with GMO crops. Many farmers have few options for obtaining other seed, since Monsanto and other corporations do not offer anything but genetically engineered (GE) seed.

2004   USDA Cooperative State Research, Education, and Extension Service (CSREES) initiates the Integrated Organic Program that, for the first time, offers grants to fund research projects that address organic agriculture issues, priorities, or problems.

2004   Passage of the Food Allergy Labeling and Consumer Protection Act requires the labeling of the following foods : peanuts, soybeans, cow's milk, eggs, fish, crustacean shellfish, tree nuts, and wheat. Genetically modified foods continue to go unlabeled because FDA considers them "nutritionally equivalent" to non-GMO foods.

2005   More than 9,000 farmers are accused by Monsanto of violating its patent rules. Most settle out of court, but more than 190 farmers and businesses end up in legal proceedings.

2005   Award-winning documentary film *The Real Dirt on Farmer John* opens and details the huge community-supported agriculture (CSA) farm "Angelic Organics" located in northern Illinois.

2006   Forty-four hundred farmers' markets are active in the United States.

2006   Walmart is the largest seller of organic milk in the United States.

2006   USDA National Agricultural Library publishes "Organic Farming and Marketing: Publications for the United States Department of Agriculture, 1977-2006." It is compiled by Mary V. Gold and Rebecca S. Thompson who work at the Alternative Farming Systems Information Center.

2007   Funding is reduced for farm programs like conservation security, sustainable practices, and water quality. Billions of dollars in farm subsidies continue to be paid to farmers based on the number of bushels of commodity crops they produce, thus underscoring the federal emphasis on high yields over environmental sustainability.

2007   *King Corn* movie is released, which uses humor to depict the illogical agricultural policies that support industrial corn production that is bankrupting farmers and making Americans obese from high-fructose corn syrup.

2007   Twelve weeds have developed resistance to the herbicide Roundup, at the same time that genetically modified Roundup Ready Soybeans comprises 85 percent of the U.S. soybean crop.

2008   Organic farmers, processors, and trade groups warn that there is widespread contamination of U.S. corn, soybeans, and other crops by genetically engineered varieties, which is threatening the purity of organic and natural food products.

2008   USDA ERS Report by Carolyn Dimitri and Lydia Oberholtzer "The U.S. Organic Handling Sector in 2004: Baseline Findings of the Nationwide Survey of Organic Manufacturers, Processors, and Distributors" (EIB-36) finds that most organic firms are small to mid-size.

2008   There are an estimated 2,300 community-supported farms (CSAs) in the United States.

2008   Angelic Organics, an organic/biodynamic farm serving the Chicago area, is a CSA farm with more than 1,400 families as shareholders (feeding over five thousand individuals)—one of the largest in the Midwest.

2008   "Farm Fresh to You" is an organic CSA that serves over four thousand families in northern California, making it one of the largest CSA farms.

2008   According to the American Community Gardening Association, there are five thousand community gardens in the United States. According to New York community gardener Adam Honigman, "Community gardening is 50 percent gardening and 100 percent local political organizing."

2008   The USDA Agricultural Marketing Service states that there are 4,685 farmers markets in the United States, up 8 percent in just the last two years.

2008   The Organic Trade Association and Natural Foods Merchandiser estimates that global sales of organic products topped $23 billion, growing 20 to 25 percent annually. The United States is 45 percent of this total. The market research firm Packaged Facts publishes "Natural and Organic Food and Beverage Trends in the U.S.," which estimates that organic products will actually top $32.9 billion.

2008   U.S. Congress mandates $7.5 billion spending on farm subsidies to conventional agriculture compared with $15 million for programs for organic and local foods.

2009   Global economic recession hits organic food, too; expected growth in organic food sales is 6 percent rather than continued double-digit growth.

2009   Film *Food Inc.* documents how multinational food corporations keep food consumers ignorant about industrial food production and what shoppers can do about it—buy local food.

2009 Obama administration names Kathleen Merrigan, a long-time proponent of organic and sustainable agriculture, Deputy Secretary of Agriculture, the number two position in the USDA.

2009 New Secretary of Agriculture Tom Vilsack creates the first organic "people's garden" outside USDA headquarters in Washington, D.C.

2009 First lady Michelle Obama establishes an organic vegetable garden at the White House. The last time there was such a garden was when Eleanor Roosevelt planted a Victory Garden during World War II.

List of Entries

Topical List of Entries

1990 Farm Bill, Title XXI

The Organic Foods Production Act (OFPA) established the National Organic Program (NOP) and the National Organic Standards Board as Title XXI of the 1990 Farm Bill (Food Agriculture Conservation and Trade Act of 1990, Pub. L. No. 101-624, Nov. 28, 1990, 104 Stat. 3359). This law established the guidelines for federal regulations governing organic production and handling, including the authority to institute the regulations and the administrative structure for regulation. In 1997, after seven years of discussions with agricultural interests, the U.S. Department of Agriculture proposed organic regulations (National Organic Program; Proposed Rule, 7 CFR Part 205). A record 250,000 public comments were received during the open comment period, mostly in opposition to use of sewage sludge, irradiation, and genetically modified organisms that had been permitted under the proposed rules but conflicted with internationally recognized standards. The USDA revised the program description and promulgated it for full implementation by October 2002.

Prior to passage of the OFPA and promulgation of federal organic rules, private and state government agencies handled standards setting, certification, and enforcement. The patchwork of nonuniform rules that emerged became unwieldy for producers and consumers as the industry grew and the face-to-face interaction of local farmers markets was expanded to interstate and international commerce. Both industry and consumer groups supported a uniform definition of "organic" that could be enforced by law to discourage mislabeling and fraud. Federal action was justified by the growing share of organic sales comprised by interstate commerce, which is regulated by the federal government. Federal regulatory authority was also required for standing in negotiating trade disputes and for the management of international trade in emerging world organic markets.

With multiple certifiers and jurisdictions offering different standards, some much more stringent than others, and some (mostly grower groups) performing internal certifications of members, it became clear that a single federal regulation had to be established. OFPA required that a National List of approved and prohibited substances for use in production, processing, and handling be developed. Furthermore, a farm plan or handling plan for maintaining compliance with organic rules would be required. The lengthy public discussion that preceded the proposed standards between 1990 and 1997 mostly addressed these materials and procedures and required specific activities, such as recordkeeping and farm planning, to establish certification requirements. The impetus for developing the final rules became urgent during 1998 and 1999 as major food processing companies began large-scale acquisitions of independent organic firms.

To a large extent, the final regulations relating to materials and procedures relied on the European organic standards passed in 1992, the Codex Alimentarius Commission organic food standards established in 1991, and the American Organic Standards issued in 1999 by the Organic Trade Association (OTA), the main industry trade organization. The OTA's executive director from 1990 to 2007, Katherine DiMatteo, is typically credited with guiding the development of these voluntary standards and with OTA's coordinated efforts to move the USDA regulatory process forward. Kathleen Merrigan, administrative director of the USDA Agricultural Marketing Service (AMS) from 1999 to 2001, oversaw the successful revision of the national organic standards and is commonly referred to as the "midwife of the National Organic Program" for her proactive role in managing the process. Bob Scowcroft, since 1990 the executive director of the Organic Farming Research Foundation (OFRF), the primary organization supporting organic farm research and education, was instrumental in mobilizing farm organizations to petition their legislative representatives and take an active role in the regulation comment process.

OFPA also established the federal role of accreditor. The NOP, administered by the USDA AMS, checks certifier rules and procedures and accredits the certifying organizations, which in turn administer federal, state, or private certification standards for their clients and issue certifications. This approach provides for maximum choice by producers and handlers in choosing a certifier and encourages a mix of state and private certification options. Standards more stringent than the federal regulations may be used for certifications by accredited organizations, but states must permit entry and sale of any certified product that meets federal standards, regardless of the certifier. Proponents of the national law demanded third-party ("arm's length") inspections by organizations with no financial interest in whether farmers and handlers being inspected passed certification.

In recognizing the dynamic nature of the organic industry, the bill's sponsor, Senator Patrick J. Leahy (D-VT), crafted OFPA to account for innovations in the industry through two provisions. First, an orderly review process for deleting and adding substances to the National List was established. Second, it was explicitly recognized that synthetic materials might be required when acceptable organic alternatives are not available. This provision was designed to allow for orderly industry transition to the ideal all-organic input mix.

The implementation of the NOP was not without controversy. Some farmers and consumers did not think the regulations were strong enough. In January 2005, the U.S. Court of Appeals for the First Circuit, based in Boston, ruled in favor of two of seven issues and asked for clarification of a third issue raised in a lawsuit challenging the NOP. The lawsuit was brought by Arthur Harvey, an organic blueberry farmer in Maine, against then-Secretary of Agriculture Ann Veneman. Consequently, thirty-six synthetic materials previously permitted in limited quantities in organic processed foods were prohibited in food labeled "organic" or "100% organic" and future use of any synthetic materials in processing was prohibited. Also, dairy herds transitioning to organic production were required to use 100 percent organic feed for the first nine months compared with 80 percent organic feed prior to the decision. The USDA AMS was also charged by the court with clarifying the process for permitting nonorganic agricultural ingredients when the organic form of the product was not commercially available.

The OTA commissioned a study of the economic effect of this court decision and presented their results to the USDA. Following a USDA AMS report indicating that the changes would have a devastating effect on the organic food processing industry, Congress amended the OFPA in November 2005 as a rider to the 2006 Agricultural Appropriations Bill. It restored the NOP regulation allowing the use of synthetic ingredients in processed products labeled as organic provided they appear on the National List. The amendment also provided that dairy operators could feed transitional dairy herds crops and forage from organically managed land in the final year of transition to organic production. Subsequent to the amendment, the NOP tightened the applicability of the commercial availability clause, which made it possible to substitute nonorganic versions of inputs in production and processing.

This was not the end of legal challenges and proposed legislation to formulate an organic program consistent with the varying views of organic in the industry. In 2006, Arthur Harvey filed another lawsuit, focusing on synthetic processing aids and food contact substances. This lawsuit was subsequently dismissed by the U.S. District Court of Maine. In March 2007, H.R.1396 was introduced by Representative Lynn Woolsey (D-CA) to prohibit the labeling of cloned livestock and products derived from cloned livestock (such as milk, cheese, meat, skins, wool, lard) as organic. The bill was not passed, but is indicative of the continued interest in the evolution of the organic industry and the government's role in ensuring a standard for certification that meets the requirements of the groups who produce and consume organic foods and products. The OFPA does not govern body care products, textiles, wild-caught or wild-collected products, and other categories of items that consumers want in organic form. The organic sector promises to be an active area for legislation and regulatory action for at least the next ten years.

Further Reading

Endres, A. B. "An Awkward Adolescence in the Organics Industry: Coming to Terms with Big Organics and Other Legal Challenges for the Industry's Next Ten Years." *Drake Journal of Agricultural Law* 12(17) 2007: 1–36.

The National Agricultural Law Center. University of Arkansas. Food Agriculture Conservation and Trade Act of 1990, Pub. L. No. 101-624, Nov. 28, 1990, 104 Stat. 3359, part 9 of 11. Text of the Organic Foods Production Act Title XXI-Organic Certification, 3935–51. www.nationalaglawcenter.org/assets/farmbills/1990-9.pdf.

Organic Trade Association. "Organic Foods Production Act Backgrounder." www.ota.com/pp/legislation/backgrounder.html.

See Also: Policy, Agricultural; Policy, Food; APPENDIX 2: Organic Foods Production Act of 1990

Luanne Lohr

2008 Farm Bill

The colossal U.S. federal legislation known as the 2008 Farm Bill reauthorized and replaced hundreds of federal policies and programs related to food and agriculture. The 673 pages of small print, technically called the "Food, Conservation and Energy Act of 2008" (HR 6124), was vetoed by President George W. Bush, and his veto was overridden by Congress; it became law (Public Law 110–234) on June 18, 2008.

Most of the provisions are in effect through 2012, continuing the established five-year cycle for this omnibus legislation. The Congressional Budget Office estimates that the bill will cost $307 billion over five years. The Farm Bill is most widely known for the controversial provisions covering federal supports for commodity crops (e.g., corn, wheat, soybeans, cotton, sugar). Title I of the bill covers the subsidies, loan programs, crop insurance, and government purchases for these crops. The commodity support programs are estimated to cost $35 billion. The largest part of the bill's expenditures, however, goes to benefits for nutrition assistance (e.g., food stamps, food banks, etc.), totaling $209 billion. The bill also covers federal programs for conservation practices on U.S. agricultural land, agricultural research and education, the U.S. Forest Service and the National Forests, biofuels and agricultural energy usage, rural development programs, farm credit, and international trade.

Hearings held on the bill by the U.S. Department of Agriculture as well as various congressional subcommittees began in 2005. This began three years of enormously complex political maneuvering. The biggest controversies centered on crop subsidies and the images of millionaire farmland owners collecting checks from the U.S. Treasury for doing nothing. The ferocious debate over limiting subsidy payments became even more embroiled as the price of oil rose sharply and the demand for corn-based ethanol pushed up the price of food across the globe. In the end, the bill made modest reforms by lowering the income limits for support program eligibility and lowering the actual payment ceiling for some programs. For example, individuals are now ineligible for most commodity supports if they have more than $500,000 in nonfarm adjusted gross annual income.

A significant change from previous farm bills was the focus on so-called specialty crops (i.e., fruits, vegetables, nuts, and herbs). U.S. farm and food policies had been almost exclusively concerned with midwestern and southern commodity crops. Fruit and vegetable producing states were not well represented on the Agriculture Committees and were largely overlooked in federal policy. The 2008 bill changed this dramatically, allocating nearly $2.5 billion to research, promotion, and government purchases of specialty crops. This change was clearly related to the explosion of diet-related health problems in the United States (e.g., obesity, diabetes), and the admonitions by health professionals to increase consumption of fresh fruits and vegetables.

For organic agriculture, the 2008 Farm Bill was an important watershed. The organic sector built a considerable amount of political momentum and organization around the bill and achieved some historic (if still incremental) gains. Almost twenty years after the Organic Foods Production Act was included with great struggle in the 1990 Farm Bill, organic agriculture was embraced in a largely noncontroversial fashion as part of the U.S. agricultural mainstream. About $107 million in mandatory funding was allocated directly to programs specifically for organic agriculture, with tens of millions more authorized for future discretionary spending. By contrast, the 2002 bill provided about $20 million.

ORGANIC AGRICULTURAL RESEARCH AND EDUCATION

The lack of production and market information is a crucial limiting factor for growth and improvement of U.S. organic agriculture. Therefore, a primary goal in the farm bill was to increase mandatory funding for USDA research, education, and data collection. The main policy arguments were anchored by a call for a "fair share" of USDA research funding and supported with specific evidence about the environmental, economic, and health benefits of organic agriculture.

The biggest winner was the Organic Agricultural Research and Extension Initiative (OREI). This program provides competitive grants to projects that combine research, extension, and education for organic agriculture. In 2002, this program received the first mandatory funding ever for organic research grants, totaling $15 million over five years. In the 2008 bill, OREI was allocated $78 million in mandatory funding over four years, plus additional discretionary authority for congressional appropriations of up to $25 million per year. Two new purposes were added to the existing language for the program: to develop seed varieties specifically suited to organic systems and to study the environmental and conservation outcomes of organic production.

Funding for the Organic Production and Market Data Initiative was another high priority. This program for collecting and analyzing data about the organic sector was created in the 2002 Farm Bill but received no mandatory funding. The 2008 bill allocated $5 million in mandatory funding for Organic Data Initiatives, plus additional authority for appropriations up to $5 million per year. This program will address the collection of price information, which is important for crop insurance and other issues, and allow for studies to follow-up the 2007 Census of Agriculture regarding the organic producer community.

Language was sought to also increase organic research within USDA's Agricultural Research Service (ARS). Although both House and Senate bills were completed with language encouraging USDA to allocate a fair share of in-house ARS funding for organic objectives, this language was not included in the final bill. Though it would have been helpful in arguing for increased ARS organic activity in the appropriations process, it would have been nonbinding in any case.

Additionally, support for classic plant and animal breeding within USDA's main research grant programs (the "Seeds and Breeds Initiative") was included as a purpose of USDA's primary competitive grants division, known now as the Agriculture and Food Research Initiative (formerly the National Research Initiative). This provision is expected to benefit organic producers who need better seed breeding, without the use of transgenic technologies.

CONSERVATION PROGRAMS AND ORGANIC CONVERSION ASSISTANCE

Under the "Conservation Title" of the farm bill are a number of programs that provide producers and landowners with financial incentives, cost-sharing, and technical assistance for maintaining and improving water, soil, and air quality. A core advantage of organic agriculture is its beneficial effects on natural resources. Yet organic agriculture has received very little attention from the USDA conservation programs. A key goal for this farm bill was integration of organic farming and ranching into the purposes of USDA conservation programs and streamlining the administrative "crosswalk" between organic certification and qualification for conservation payments.

The Conservation Stewardship Program (CSP), formerly known as the Conservation Security Program, appeared to hold the most promise for rewarding the benefits generated by organic practices. Since its inception in 2002, however, the CSP program has been crippled by funding issues, so expansion of CSP funding was also a primary goal in this farm bill. In the final 2008 bill, CSP funding was successfully increased by $1.1 billion. This was projected to be sufficient for stewardship payments on 13 million acres per year. Also, the CSP application process is now specifically required to be coordinated with organic certification.

Because of the environmental benefits expected from organic agriculture, the conservation programs were also identified as the logical place to create a specific program to support conversion of conventional operations to organic. The rate of conversion of conventional U.S. farms and ranches to organic practices is lagging well behind the demand for most organic products (dairy farms may be an exception). Although there are numerous obstacles for producers who wish to make the transition, from unfair crop insurance policies to the ongoing dearth of research, providing financial and technical support within the conservation programs could at least offset some of the increased production costs during the conversion period. In addition, extra emphasis on training and technical support for transitional producers help ensure that conversions actually succeed. Without good technical support and training, much of the financial assistance funds for conversion will be ineffective. Several different approaches to conversion support appeared during construction of the bill. In the end, payments for organic conversion were authorized under the Environmental Quality Incentives Program (EQIP). A specific dollar amount was not set aside for this purpose, but will be included within the overall allocation for the program. The bill allows for up to $20,000 per year in conversion support, with a maximum of $80,000 over six years. The Secretary of Agriculture is also directed to ensure that technical assistance is available for producers who receive conversion support.

The bill also provides for prioritizing USDA conservation loans to transitional producers, and organic production initiatives are mentioned as examples for additional programs such as the Cooperative Conservation Innovations Program.

ORGANIC CERTIFICATION COST-SHARE PROGRAM

The 2002 Farm Bill created a new program to help offset the costs of organic certification for producers and processors. Only $5 million was allocated to the National Organic Certification Cost-Share program for 2002 to 2008. That funding was slow to get out through the states to the users, but once the system was in place the money was used up quickly. By 2006, most states had exhausted their funding for this program. In the new legislation, the program was renewed with $22 million in mandatory funding. It will provide annual support of up to $750 for up to 75 percent of certification costs per operation. As well, additional funding for Certification

Cost-Share in sixteen states is provided as part of the Agricultural Management Assistance program.

CROP INSURANCE EQUITY FOR ORGANIC PRODUCERS

Organic farmers have been treated unfairly on both ends (premiums and payouts) of crop insurance. Due to the perception of increased risk, they have had an automatic 5 percent surcharge on premiums. Further, organic producers have not been able to insure organic crops for their actual organic market value but instead have received payouts based on conventional prices. The goal for this farm bill was to eliminate these disparities, and partial success was achieved. USDA is required to develop improvements in crop insurance policies for organic producers, including a review of the necessity for the premium surcharge. The burden of proof is on USDA to show that there is clear and systemic evidence of a need for any surcharge. USDA is also required to develop and implement options for organic payouts with the goal of offering the payout for all organic crops within five years as sufficient data become available.

Further Reading

American Farmland Trust. www.aft.org.
Center for Rural Affairs. www.cfra.org.
Organic Farming Research Foundation. www.ofrf.org.
National Sustainable Agriculture Coalition. www.sustainableagriculture.net.

See Also: Policy, Agricultural; Policy, Food

Mark Lipson

Agribusiness

"Agribusiness" is the common term for the economic activities and structures surrounding large-scale food production, processing, and distribution. Because it is a significant economic, political, social, and ecological influence in U.S. life, there is concern that it exerts too much power in food production. The movement toward organic and local food is partly a statement of opposition to the current agribusiness system.

Agribusiness accounts for about one-fifth of the U.S. economy and employs roughly one-fourth of U.S. laborers. Its strength emerged from transitions in agricultural production methods and processes that began in the early part of the last century; its current growth and development rests in organizational structures of vertical and horizontal integration, which support agribusiness corporations in the era of globalization. Although some question the inevitability of agribusiness consolidations, in many ways, large-scale organic production mimics these same trends.

The rise of agribusiness accompanies increasing economic consolidation at all levels of the food system. Despite the growing presence of small specialty companies (particularly in the area of organics), multinational corporations account for the vast majority of wealth and economic activity in this sector. Agribusinesses exist throughout the food system, from production to processing to retail. Future market trends for agribusiness suggest expansion of genetic patenting and ownership at the international level through patenting of intellectual property.

Agribusiness firms are some of the most well-known businesses in the world. For example, ConAgra, Cargill, and Monsanto provide seed and other inputs for grain cropping on six continents. In seed oils production and processing, Archer Daniels Midland, Bunge, and Cargill are leading the development of new markets in ethanol and biodiesel. Iowa Beef Processors (IBP), Tyson, and Novartis are names known to meat and poultry consumers internationally.

The rise of corporate multinational agribusiness both paralleled and drove other changes in the structure of agriculture and food production in the United States. Most notable in these changes was the industrialization of agriculture, which began in the early 1920s and accelerated through the postwar years of the 1950s. During this time, rural populations rapidly depleted and farm size began to increase. Both trends continue unabated today. The modern transition of the 1960s through the 1980s to increasingly sophisticated machinery and specialized seed stocks changed the economic balance of farming and moved farming towards "economies of scale," or a system that prioritizes increasing production quantity to offset increasing input costs. The emphasis on greater efficiency in quantity of production allows crop and commodity prices to be maintained at levels lower than the cost of production, that is, lower than input costs.

Input costs are the expenses required to raise and harvest food crops. They include annual necessities, such as hybrid seed purchases, fertilizers, herbicides, and insecticides. However, input costs also include the cost of machinery, energy (whether fuel or electricity) and, in the case of confinement livestock operations, specialty buildings, feed, and medicines. In modern agriculture, these costs are often so prohibitive that short-term operating loans to farmers are common. Operating loans provide funds for annual inputs, so that the crop can be planted in the spring with the expectation that the operating loan will be paid off from fall profits.

The balance of input costs and commodity prices has disrupted farm economies through what is referred to as the "production/price cycle." This phrase describes a common condition for modern farmers: production quantity needs to increase to economically compensate for lower per unit return, yet as all farmers produce more, the general market price drops accordingly, and the cycle repeats. This cycle has promoted the growth and expansion of agribusiness. Hence, U.S. production agriculture continues to consolidate within the economies of scale model, and the need for products and services promoting production quantity continues.

The suppliers of these products and services have shifted from local independent companies to the large agribusiness firms. As industrialization of agriculture and its accompanying inputs increased, specialty firms extended their operations to include products and services that reflect different inputs for multiple stages of the production process. This diversity in operation was often achieved through strategic merger and buyout, creating vertical and horizontal integration in the agribusiness industry.

Vertical integration describes a corporate organization drawing income from, or otherwise owning or controlling, each phase of the production cycle. For example, under vertical integration a single agribusiness firm sells the seed for planting, the fertilizer to ensure its growth, the herbicide to prevent challenge from weeds, and the insecticide to protect from loss or damage by bugs. In addition, agribusiness firms are also the purchasers of raw commodities for food processing and distribution. In this way, vertical integration allows agribusiness firms to be both buyer and seller throughout the production cycle, and to exponentially increase economic returns.

Horizontal integration also has changed the form and structure of agribusiness firms, particularly in relation to industrialized livestock production. Horizontal integration describes expansion of market control or ownership by an agribusiness firm through increasing shares of a single product or market. Again, mergers, cluster agreements, and buyouts of firms serve to consolidate market power at new levels. For example, only two agribusiness firms, IBP and Cargill, now control roughly 58 percent of the beef market, over 80 percent of the market is controlled by only four entities.

Market consolidation is also apparent in food processing and retail. Three global companies have overwhelming control in global food processing—Nestle, Unilever, and Phillip Morris. Similarly, food retail worldwide is dominated by only four companies—Tesco (UK), Ahold (Netherlands), Carrefour (France), and Walmart (USA). The impact of this consolidation is felt by farmers and consumers alike. Farm values and profits have steadily fallen with the rise of agribusiness, and consumer prices for food have remained steady or risen during the same period. The process of market clustering also continues here, as retailers and processors sign exclusive contracts. For example, Walmart has signed contracts with Smithfield, Tyson/IBP, and Farmland for all meat and poultry supplies. These forms of market agreements reinforce agribusiness domination of the food system.

Currently, the biotechnological age in agriculture has continued to expand the reach of agribusiness under globalization. With the rise of genetic patenting and genetic modification in seed development, agribusiness multinational firms have been ever more attentive to purchasing and patenting genetic stock for production crops around the world. Genetic patenting refers to the practice of designing specific gene sequences for use in production, and also to the process of identifying specific gene sequences in formerly unidentified samples. This latter practice is raising multiple questions, particularly for peasant and indigenous small-scale and subsistence farmers, who have built seed stock for generations or who find previously available stocks are now patented. At the same time, the establishment of intellectual property rights laws in developing countries is being touted as the best protection of future food safety and development. The issue becomes one of intellectual property, and the legal status of commercial discovery. Intellectual property is defined as the legal possession of creations of the mind, particularly those which have non-exclusive use. For example, anyone can reproduce a seed with a patented gene sequence, but the right to commercial use of that seed is limited to the intellectual property holder. It becomes an issue of ability rather than of proprietorship. In this manner, definitions of market and product ownership and control are expanding internationally beyond the material items themselves, to the protected knowledge behind them.

Further Reading

Danbom, David. *Born in the Country: A History of Rural America*. Baltimore: John Hopkins University Press, 2006.

Hendrickson, Mary, and William Heffernan. "Concentration of Agricultural Markets." National Farmers Union 2007. www.nfu.org/wp-content/2007-heffernanreport .pdf.

International Federation of Agricultural Producers. *Industrial Concentration in the Agri-food Sector. Paris, France*: May, 2002.

Tripp, R., D. Eaton, and N. Louwaars. *Intellectual Property Rights: Designing Regimes to Support Plant Breeding in Developing Countries*. The International Bank for Reconstruction and Development/The World Bank: 2006.

USDA-NASS. *Increasing Production Costs*. USDA Web site. March 29, 2006. www.usda.gov.

Watal, J. *Intellectual Property Rights in Indian Agriculture*. Indian Council for Research on International Economic Relations. July, 1998. www.icrier.org/pdf/ jayashreeW.PDF.

See Also: Agriculture, Conventional; Brands; Factory Farming; Genetically Engineered Crops

Meredith Redlin

Agrichemicals

The word "agrichemical" is a contraction of "agricultural" and "chemical." It is a generic term for a broad range of materials used to provide essential mineral compounds (fertilizers), control pests (pesticides), and improve yield and quality of crops and animals (drugs, hormones, and growth regulators). Agrichemicals are critical to the productivity of agricultural systems. The use of synthetic agrichemicals along with improved crop varieties and mechanization contributed to large increases in crop yields during the twentieth century. Some leading agrichemical companies are Bayer Crop Sciences, Syngenta, BASF, Monsanto, and Dow Agrosciences. Organic production is seen as a more traditional form of agriculture because it prohibits the use of modern synthetic agrichemicals.

Use of chemicals in agriculture dates to ancient times. Romans burned sulfur to kill insects and controlled weeds with salt. They used honey and

Tractor spraying agrichemicals in orchard. Courtesy of Leslie Duram.

arsenic to control ants. Dead bodies, animal blood, bones, seaweed, fish, manure, and crop rotations were used to fertilize crops. Ashes and ground limestone (lime) were used to adjust soil pH. During the nineteenth and early twentieth centuries, farmers used sulfur, nicotine sulfate, sulfur copper acetoarsenite (Paris green), and calcium arsenate to help control insects. In 1890, copper sulfate and lime were first used to control fungal diseases on grapes. These pesticides poisoned fruit and vegetables and accumulated in orchards and crop fields. In the early nineteenth century, superphosphate became the first fertilizer produced by a chemical process. It was created when bones were treated with sulfuric acid. In the 1860s, synthetic potassium fertilizers were first available in Europe.

Just before World War I, the Haber-Bosch process for converting atmospheric nitrogen into ammonia was developed in Germany. At first, ammonia produced using this method was too expensive for use in agriculture but was used for manufacturing explosives. World War II increased demand and led to improved, more affordable methods of synthesizing ammonia and nitrates. After the war, use of synthetic fertilizers became widespread, increasing over 400 percent in next sixty years. Synthetic fertilizers now supply half of the nitrogen that reaches the world's croplands.

Synthetic, carbon-based pesticides are also a result of World War II. Dichlorodiphenyltrichloroethane (DDT) was introduced during the war and is credited with preventing insect-borne human diseases, such as a typhus epidemic in Naples, Italy. Some early synthetic pesticides were harmful to humans and the environment. Pesticides such as DDT, dieldrin, and chlordane, can persist in the environment for long periods of time. These pesticides bioaccumulate, build up to toxic concentrations, as they pass through the food chain. Predators, high on the food chain, accumulated elevated levels of these pesticides in their bodies. One effect of bioaccumulation in birds was thinning of eggshells. Peregrine falcon, brown pelican, osprey, double-crested cormorant, and bald eagle populations all declined due to thinner eggshells and reduced hatching success. Widespread environmental problems from persistent pesticides led Rachel Carson to write *Silent Spring* (1962). This book brought attention to the effect of agrichemicals on ecosystems and marked the start of the environmental movement. Environmental concerns about highly persistent pesticides led to their ban during the 1970s. Recently, antimalaria programs in Africa have brought DDT back for use in treating mosquito netting and interiors of huts. Production of synthetic carbon-based pesticides more than quadrupled from 400 million pounds in 1950 to 1.7 billion pounds in 2000.

Pesticides can be classified according to the pests they kill. Fumigants act as sterilants, killing a wide range of organisms. They are used to sterilize the upper level of soil. Also, fumigants are important to stop introduction of pests into new areas during international shipments of agricultural products and to control pests of stored grains. Methyl bromide, a widely used fumigant, is being phased out under the Montreal Protocol because it depletes stratospheric ozone. Systems replacing methyl bromide with other fumigants, plastic mulches, or biofumigants crops have been developed. Fungicides control fungus in crops and animals. They are essential for production of high-value fruits, vegetables, and ornamental crops, especially in humid areas. Fungicides allow blemish-free fresh produce to survive long distance shipping. Herbicides kill weeds and are the most commonly used group of pesticides in North America. They allow large-scale agricultural production with minimal labor. Since the 1980s, crops have been genetically modified for resistance to broad-spectrum herbicides, such as glyphosate (the active ingredient in Roundup). In North America, over two-thirds of soybeans and the majority of cotton, maize, and canola crops are genetically modified. The last major category of pesticides is insecticides, which are used to control insects. Insecticides are widely used in production of crops to protect against insect vectors of animal and human diseases and to control insects in structures. Maize, cotton, and other crops have been genetically modified to contain Bacillus thuringiensis (Bt) endotoxins, controlling caterpillar insects.

Problems continue to plague pesticide use in agriculture. Pesticides can contaminate water through runoff and leaching. Over twenty pesticides have been detected in wells and over eighty pesticides have the potential to move into groundwater. Long Island, New York, is particularly susceptible to groundwater contamination because of shallow water tables, sandy soils, and intensive agriculture. Aldicarb, a carbamate insecticide, and atrazine, a triazine herbicide, are examples of pesticides found in well water. After prolonged or multiple exposures to a pesticide, some pest organisms are no longer controlled. It is estimated that over a thousand pest species have developed resistance to pesticides. Toxicity is the harmfulness of a pesticide to humans. Symptoms of acute toxicity, such as skin burns, headaches, and loss of consciousness, appear within twenty-four hours of pesticide exposure. Repeated exposure to small amounts of pesticides in the environment, diet, and drinking water can cause chronic diseases. The U.S. Environmental Protection Agency (EPA) defines the minimal acceptable risk of a pesticide as an increased cancer rate of one person in one million people. Pesticides can also affect the endocrine (hormone) system. The Food Quality Protection Act (1996) requires the EPA to consider the greater risk to infants and children from dietary exposure to pesticides.

Fertilizers contain mineral nutrients (primarily nitrogen, phosphorus, and potassium) that improve crop yields. They can be in the form of dry powders, pellets, gases, or liquids. Fertilizers are the most widely used agrichemicals. In some regions, fertilizers are applied to almost all land used to produce crops (even ornamentals and turf). Nitrogen is the most common nutrient in fertilizers. In 2000, world consumption of nitrogen fertilizer was 81.7 million metric tons (Mt). China consumes 23 million Mt of nitrogen fertilizer per year. It is the largest consumer and manufacturer of nitrogen-based fertilizer. Western Europe, India, and the United States each consume about 12 million Mt of nitrogen fertilizer per year. The most rapid growth in fertilizer use is expected in Asia. Most of commercial fertilizer nitrogen is derived from ammonia in energy-intensive reactions that rely on natural gas. Higher energy prices are causing fertilizer costs to increase rapidly. Urea is the world's most widely traded nitrogen fertilizer because it is more stable and easier to transport than ammonia. In North America, the greatest nitrogen fertilizer use takes place during fall applications of anhydrous ammonia before spring planting of corn. Nitrogen in the form of urea is widely used to fertilize home lawns. Nitrogen-use and land-use efficiency of agriculture needs to be improved, and the risk of negative effects to the environment need to be minimized.

There are problems with the use of fertilizers. Nitrogen from fertilizers and manures are converted into nitrates by soil bacteria. Nitrates can get into groundwater and surface waters, contaminating drinking water. High levels of nitrates in drinking water are a threat to human health. Atmospheric ammonia is generated from decomposing manure, losses during production, and application of fertilizers and biomass burning. Atmospheric ammonia contributes to soil acidification and global warming through formation of aerosols and nitrous oxide. Marine dead zones occur in the Gulf of Mexico, the Baltic Sea, the Black Sea, off the Oregon coast, and the Chesapeake Bay. Nitrogen and phosphorus enter rivers primarily through soil erosion and runoff of fertilizers. The nitrogen and phosphorus in marine ecosystems cause algal blooms. When the algae die and decompose, dissolved oxygen in water is reduced (hypoxia), killing other organisms, including fish. Nitrogen fertilizer used in the farming states of the Mississippi River Basin is one cause of the Gulf of Mexico dead zone. Phosphate runoff also causes overgrowth of weeds and algae in lakes and streams. Because of these environmental concerns, some areas, such as Florida, are restricting the use of fertilizers.

Further Reading

Brand, Charles. "Some Fertilizer History Connected with World War I." *Agricultural History* (19) 1945: 104–13.

Carlisle, Elizabeth. "The Gulf of Mexico Dead Zone and Red Tides." Tulane University Environmental Biology Web Site. www.tulane.edu/~bfleury/envirobio/enviroweb/DeadZone.htm.

Ching, Lim Li. "GM Crops Increase Pesticide Use." Institute of Science in Society Web Site. www.i-sis.org.uk/GMCIPU.php.

Delaplane, Keith S. "Pesticide Usage in the United States: History, Benefits, Risks, and Trends." University of Georgia College of Agricultural and Environmental Sciences Web Site. http://pubs.caes.uga.edu/caespubs/pubcd/B1121.htm.

Environmental Protection Agency. "Pesticide Issues." Environmental Protection Agency Pesticides Web Site. www.epa.gov/pesticides.

Fixen, Paul, and Ford West. "Nitrogen Fertilizers: Meeting Contemporary Challenges." *Ambio: A Journal of the Human Environment* (31) 2001: 169–76.

See Also: Antibiotics and Livestock Production; Fungicides; Growth Hormones and Cattle; Health Concerns; Herbicides; Methyl Bromide; Pesticides

Lauren Flowers and John Masiunas

Agricultural Subsidies

Few topics related to the farming sector generate as much controversy as agricultural subsidies and farm programs. The complexities of programs lumped into the public's concept of subsidies are mind boggling. "Subsidy" is a catch-all word often used to represent the various types of direct and indirect farm revenue support and risk management programs authorized approximately every five years by federal agricultural legislation (the Farm Bill). Farm support dates to 1933, when income support programs were introduced. Each successive Farm Bill was structured to reorient farm support programs to conform to economic, political, and environmental goals current at the time of passage. With growing real farm household incomes, productivity gains, and low debt-to-asset ratios for commercial-scale farms, justification for support programs has received significant challenges in recent years.

In 2006, U.S. farmers received $15.8 billion from all types of programs, of which approximately 19 percent of payments were for conservation programs. Approximately 44 percent of U.S. farmers participated in one or more farm programs in 2006, receiving an average payment of $12,687, or 39 percent of net cash income. However, payments are not equally distributed among farmers. The largest 11 percent of farms in terms of gross receipts received 56 percent of all government payments in 2006.

Payments are not equally distributed across commodities, either. Commodities accounting for 58 percent of the value of U.S. farm output in these years received no payments at all. Many crops—fruit and vegetable crops, most animal products, livestock, aquaculture, and tree nuts—are not eligible for the majority of direct payment programs. Corn, wheat, cotton, and rice, which accounted for 18 percent of U.S. farm output value, topped the list, receiving 79 percent of all commodity payments in 2005.

Direct payments and loan programs are what most people think of when they use the term "subsidies." These programs are administered by the Commodity Credit Corporation (CCC), a federally owned and operated corporation within the U.S. Department of Agriculture (USDA) that handles all money transactions for agricultural price and income support programs.

Other types of payments are made to farmers to support income and protect against low prices. Disaster payments are made for farmers suffering crop failure due to extreme weather conditions that effectively destroy crop yields for a region ($527.2 million paid in 2007). Payments to farmers are also made for agreeing to engage in conservation activities on operating farmland and for establishing easements to retire land from

Commodity Programs Administered by the Commodity Credit Corporation— Explanation of Programs and Amounts Paid in 2007

FIXED DIRECT PAYMENTS

Calculated as payment rate times yield times payment acres, the latter being a percentage of the historical acreage planted to that crop on the farm. Payment rate is commodity specific, set in the Farm Bill. Available to producers with eligible historical production of wheat, feed grains, cotton, rice, and oilseeds. Limit: $40,000 per person per year. Payout: $5 billion.

COUNTER-CYCLICAL PAYMENTS

Payments accrue whenever the effective price for a commodity is less than the target price set in the Farm Bill. The effective price is the direct payment rate (see fixed direct payments) plus the higher of the loan rate set by the CCC or the national market price for the commodity. If the effective price is higher than the target price, as when market price is high, counter-cyclical payments are zero. Calculated as payment rate (target price minus effective price) times yield times payment acres. Available to producers of wheat, feed grains, cotton, rice, oilseeds, pulse crops, and peanuts. Limit: $65,000 per person per year. Payout: $1.1 billion.

LOAN DEFICIENCY PAYMENTS AND MARKETING LOAN GAINS

All or part of the crop harvest serves as collateral for a loan from the CCC at commodity-specific loan rates set by the

continued

Farm Bill. Loan is repaid at the lower of a rate based on average market prices for the commodity during the preceding 30 days or a rate that minimizes the accumulation of stocks and storage costs of the commodity. When the repayment rate is less than the loan rate, the difference is a marketing loan gain and is a gain to the farmer. A loan deficiency payment allows the producer to receive the value of a marketing loan gain without borrowing and subsequently repaying a commodity loan. Available to producers of wheat, feed grains, cotton, rice, oilseeds, wool and mohair, honey, pulse crops, and peanuts. Limits on LDP and MLG: $75,000 per person per year. Payout LDP: $54.4 million. MLG: $271.8 million.

COMMODITY CERTIFICATES

Commodity certificates issued by the CCC are purchased by farmers at posted county price for wheat, feed grains, and oilseeds or at the effective adjusted world price for rice or upland cotton. Farmers use the certificates to acquire crop collateral they pledged to the CCC to obtain a commodity loan. When the posted county price or effective adjusted world price paid for the certificates is below the loan rate paid by the CCC on the collateral purchased with the certificates, producers realize the difference as a gain. Limit: None, but resulting loan gain is subject to MLG limits. Net value of CC: $818.4 million.

Source: www.ers.usda.gov/Briefing/FarmPolicy

production or prevent land from coming into production ($3.1 billion in 2007). The milk income loss program, which cost $73.4 million in 2007, pays farmers a percentage of the differential between a market reference price and the target price. Phase out of government issued supply-constraining quotas for peanuts and tobacco cost $901.2 million in 2007.

Besides payments, the federal government also supports U.S. farmers through commodity purchases that are used in school and senior meal programs and international food aid or are stored at government expense. Exports are promoted through spending on export credit guarantees, market access programs, export enhancements, and foreign market development. Research and information dissemination on domestic production and international markets is paid for by the federal government as well.

One of the most market-oriented federal supports is crop insurance. Farmers can purchase subsidized multiperil insurance programs, administered by the USDA's Risk Management Agency, to protect against the financial effects of crop loss due to natural hazards. Available to most farmers, this program is not as widely used as might be expected because disaster payments have been relatively common when significant losses due to weather have occurred.

Analyses conducted since 2006 indicate that current farm programs fail to benefit the majority of farmers or commodities, are subject to fraud and abuse, do not increase food security or reduce consumer food costs significantly, and are not doing enough to protect vital water quality and soil resources. Added to these complaints, the organic agriculture sector has argued that although they are eligible to participate, most organic farmers find farm programs biased against them either explicitly or by not recognizing the full value of ecological farming systems. For example, crop insurance payment rates were not calculated on the basis of organic prices, which tend to be higher than conventional prices. The benefits to water quality of the added organic matter generated by soil-conserving practices were not considered justification for inclusion of organic farming as a reimbursable conservation practice in most states.

The USDA has never supported subsidies and cost sharing for organic conversion, except for cost sharing for certification costs in some states authorized by the 2002 Farm Bill. Despite the heavy reliance on subsidies in the conventional sector and the significant barrier imposed by the cost of transitioning to organic farming, the USDA has consistently held that organic farmers do not require special programs. It has been suggested that this position was taken because funding would have been diverted from other programs paid out through the CCC, which angered conventional commodity groups. It has also been suggested that growth in the sector was assumed to be an indicator that government support was not needed.

Prior to the passage of the 2002 Farm Bill and the implementation of the National Organic Standards in 2002, the Organic Farming Research Foundation (OFRF) reported that 47 percent of organic farmers did not participate in any government programs between 1997 and 2002. Commodity programs (direct payments, counter-cyclical payments, and loan deficiency program) accounted for the majority of organic farmer participation (34 percent between 1997 and 2002), followed by crop insurance (21 percent), disaster payments (19 percent), and various conservation programs. These low percentages reflected the misalignment between the programs and the needs of organic farmers. Organic farmers were facing greater income risk, especially during transition from conventional to organic, and were competing with heavily subsidized European producers.

The OFRF and the Organic Trade Association (OTA) lobbied for changes in subsequent Farm Bills. In the 2002 Farm Bill, certification cost sharing was introduced on a limited basis and organic producers were exempted from producer assessments used to support conventional agriculture research and promotion. Organic proponents won their biggest gains in the 2008 Farm Bill, which included a fourfold increase in certification cost sharing to $22 million, mandated a review of underwriting risk and loss history for organic crops that could raise the payment levels and reduce premiums, qualified organic farming as a conservation practice eligible for payments, and prioritized organic transition for guaranteed loan programs. These changes make it possible for farmers to receive the financial support needed to make the transition to organic agriculture. However, percentage of acreage devoted to organic farming in the United States lags significantly behind that in Europe, where environmental and economic benefits are documented and recognized by governments.

European countries have established targets for percentage of land to be converted to organic production and market subsidies, conversion support, and direct payments by individual countries and the European Union through the EU organic legislation. Under the agri-environmental support programs, farmers enter into a voluntary five-year contract with a government agency that commits the farmer to specific farming practices perceived as beneficial to the environment in return for payment. In 2001, almost €500 million were spent on organic lands under the two main EU agri-environmental regulations. The average organic farm payment of €183 to €186 per hectare was double the €89 per hectare paid to support conventional farms. As a consequence, acreage enrollment rates are significant, ranging from 33 percent of acreage in France to 94 percent in Denmark.

There is little chance of such an extensive support program being undertaken in the United States unless organic farming becomes recognized as a significantly more productive system in terms of both commercial scale output and environmental goods such as cleaner water. However, the tone set by the 2008 Farm Bill makes clear that policy makers have acknowledged the major barriers to transitioning to organic farming and are using the politically acceptable tools at hand to address them.

Further Reading

Dimitri, C., and L. Oberholtzer. *Market-Led Growth vs. Government-Facilitated Growth: Development of the U.S. and EU Organic Agricultural Sectors.* Outlook Report WRS0505. Washington, D.C.: U.S. Department of Agriculture, 2005.

Gardner, B. L., and D. A. Sumner. *AEI Agricultural Policy Series: The 2007 Farm Bill and Beyond.* Washington, D.C.: American Enterprise Institute, 2007. www.aei.org/research/farmbill/publications/pageID.1476,projectID.28/default.asp.

Lohr. L. "The Conservation Security Program: Leveling the Playing Field for U.S. Organic Farmers." *Organic Farming Research Foundation Information Bulletin* (11) Fall 2002: 1, 6–8.

U.S. Department of Agriculture. Data on government payments. USDA Web site. www.ers.usda.gov/data/FarmIncome/finfidmuxls.htm#payments.

U.S. Department of Agriculture. "2008 Farm Bill Side-by-Side." USDA Web site. www.ers.usda.gov/FarmBill/2008/2008FarmBillSideBySide112508.pdf.

U.S. Department of Agriculture. "Economic Research Service Briefing Room: Farm and Commodity Policy." www.ers.usda.gov/Briefing/FarmPolicy/.

See Also: Agriculture, Conventional; Policy, Agricultural; Policy, Food

Luanne Lohr

Agriculture, Alternative

Alternative agriculture is a movement that started as part of the "back to the land" era of the 1960s and gained momentum in the 1970s as an alternative approach to farming and food production. The movement started as a way to counter the growing strength of agribusiness corporations, agrichemical usage, and the disconnection between food producers and consumers. The alternative agriculture movement is particularly concerned with the ecological aspects of production. There are broad aims to protect farmland, water supplies, and soil from degradation, pollution, and overuse.

A separate but important goal is to protect farmers who are considered valuable stewards of land and producers of healthy, safe food. Alternative agriculture implies a production system different than the conventional system of agriculture. There is no one system or approach that encompasses the meaning of alternative agriculture. Rather, alternative agriculture is a range of approaches or possibilities that are different in some way from the conventional approach. Out of the alternative agriculture movement came many alternative approaches to producing and selling food.

Although alternative agriculture can be difficult to clearly define, it can be characterized by agricultural production that emphasizes decentralization, independence, community, diversity, and working with natural nutrient cycles. Alternative agriculture aims to be economically and environmentally sustainable over the long term. At the most basic level, alternative agriculture seeks to understand the ecological limitations of production and work within those limitations to improve production quality rather than quantity. On a practical level, this translates to reduced usage of off-farm inputs, especially synthetic chemicals, and at least partial disengagement from the industrial food system. Farmers using alternative agriculture may not be entirely disengaged from the conventional food system; instead they may participate in some aspects of the conventional food system while incorporating some elements of alternative agriculture into their farming or marketing system. There are a range of technological and management options to reduce off-farm inputs, decrease environmental damage, and increase price premiums. Alternative agricultural approaches include organic, sustainable, and low-input farming practices and often utilize local or direct marketing channels. These approaches all employ the concept of integration or whole farm planning. Successful and sustainable agriculture needs to be about more than just production. The concept of integration recognizes that farming incorporates economic, social, and ecological issues. All three must receive equal attention for alternative agriculture to be successful. Often, alternative agriculture focuses too heavily on input substitution. A focus on input substitution limits the benefits to the environment and does not extend to the social and economic situations of producers and consumers. When there is no consideration of economic or social aspects of alternative agriculture, the lack of integration in the farming system will mean the system does not have long-term sustainability.

BENEFITS

The economic, social, and ecological benefits of alternative agriculture are broad ranging and have the potential to reach farmers, consumers, and local communities. Smaller farms using alternative agricultural approaches require proportionally higher amounts of human labor, which create more meaningful farm jobs and bolster the local community. There is also a higher value placed on the profession of farming, which encourages more people to get involved. Local knowledge of a farming system is passed on through apprenticeships and can be very valuable in creating a viable production system and improving local social capital. This leads to higher job satisfaction and quality of life, which in turn attracts more young people to the profession of farming and creates more long-term farming opportunities. Farms that utilize an alternative agriculture approach often supply a local food system, which has a ripple effect throughout the community. It brings the decision-making back to the farmers and local consumers. This creates more equity among farmers and connects consumers with the source of their food. This connection is important in keeping consumers interested in the safety and sustainability of the food system. Rural communities that are composed of mainly small to mid-scale farms with local direct markets have more economic prosperity than communities surrounded by large corporate farms. When corporate farms are present, the profits are taken out of the community to larger cities, but small to mid-scale farms circulate more of their income in the local community. A strong local economy based on alternative agriculture initiatives contributes to economically and socially viable communities.

Farms using alternative production methods provide ecosystem services that include enhanced biodiversity, soil and water quality, reduced greenhouse gases, and better wildlife habitat. The farms are less dependent on fossil fuels, and they help maintain the rural landscape while releasing fewer chemicals into the environment. The farms also help maintain genetic information and regulate biotic and climatic regions. Alternative approaches encourage more diversity of crops and livestock, which reduces the vulnerability of the farm. This means farmers are less likely to have problems with pests, depleted soils, extreme weather events, or unstable market prices. This is achieved through crop rotation, incorporating livestock and crops for nutrient cycling, and integrated pest management. Alternative agriculture incorporates multiple ways of increasing nutrients in the soil. Therefore, farmers increase the health and productivity of the soil and reduce the need for expensive inputs. Currently these ecosystem services are not reflected in farmers' incomes through premium prices or subsidy payments. These ecosystem services do benefit the farmer through increased resilience, better quality production, and reduced costs for external inputs.

CHALLENGES

Working outside the conventional food system can be challenging, because there has traditionally been little support for alternative agriculture. Alternative agriculture requires more information, labor, and knowledge than other types of production. It requires a shift in management, beliefs, and practices. Much of the process requires working independently with little support from traditional agricultural information services. There are five main obstacles faced by farmers using alternative production practices: financial obstacles, lack of technical information and assistance, a lack of research to improve farming methods that do not rely on expensive inputs, marketing difficulties, and policies that focus on commodity production instead of quality food production. As more farmers and consumers begin to see the benefits of alternative agriculture, support has increased in many aspects. Research on alternative production strategies using agroecology and whole farm approaches has increased significantly as well as information services to disseminate the research information. Financially there is still very little support, as farm policy in North America still pro-

> ## World Wide Opportunities on Organic Farms (WWOOF)
>
> Do you want to WWOOF? You can volunteer on an organic farm in numerous locations across the globe. In exchange, you get free room and board. WWOOF organizations have lists of organic farmers and gardeners who want volunteer help during some seasons. There is a wide variety of tasks and experiences. "WWOOFers" are volunteers who select the hosts that match their interests and then contact them to arrange a stay. These volunteers often live on the family farm.
>
> *Source:* www.wwoof.org

motes a commodity subsidy approach. One of the problems facing industrial agriculture is the heavy use of credit to purchase capital for the farm. Farmers using alternative farming approaches generally make smaller investments and remain outside the conventional capital investment cycle. The drawback is that the farmers are generally not eligible for subsidy payments or insurance.

Consumer support has grown, and now farmers markets, farm stores, and local independent food supply chains are increasing. If farmers are strategic, they can emphasize their alternative production approaches in the marketing of their products. When products are sold into the conventional food system, they do not receive a price premium. It can be challenging to find alternative outlets for their products, but if they are successful there are many opportunities for farmers to receive a fair price for their products. Food production practices are unlikely to change without changing social, economic, and political structures. Alternative agriculture can be supported by consumers who demand food products that are produced from alternative methods of production, policies that support alternative production, and education for both producers and consumers on the benefits of alternative agriculture. The challenges to alternative agriculture can be overcome by changing the current structures of research, government funding, and agribusinesses.

Further Reading

Alteriri, Miguel. "Ecological Impacts of Industrial Agriculture and the Possibilities for Truly Sustainable Farming." In *Hungry for Profit: The Agribusiness Threat to Farmers, Food and the Environment*, Fred Magdoff, John Bellamy Foster, and Frederick Buttel, eds., 77–92. New York: Monthly Review Press, 2000.

Beus, Curtis E., and Riley E. Dunlop. "Conventional Versus Alternative Agriculture: The Paradigmatic Roots of the Debate" *Rural Sociology* 55(4) 1990: 590–616.

Bowler, I. R. "Sustainable Agriculture as an Alternative Path of Farm Business Development." In *Contemporary Rural Systems in Transition*, I. R. Bowler, C. R. Bryant, and M. D. Nellis, eds., 237–53. Wallingford: C.A.B. International, 1992.

Ilbery, B., and D. Maye. "Alternative or Conventional? An Examination of Specialist Livestock Production Systems in the Scottish-English Borders." In *Rural Change and Sustainability: Agriculture, the Environment, and Communities*, S. Essex, A. Gilg, R. Yarwood, J. Smithers, and R. Wilson, eds., 95–106. Oxfordshire: C.A.B. International, 2005.

Lapping, Mark. "Toward the Recovery of the Local in the Globalizing Food System: The Role of Alternative Agricultural and Food Models in the U.S." *Ethics, Place & Environment* 7(3) 2004: 141–50.

Maye, Damian, Lewis Halloway, and Moya Kneafsey. *Alternative Food Geographies: Representation and Practice*. Oxford: Elsevier, 2007.

Vandermeer, John. "The Ecological Basis of Alternative Agriculture." *Annual Review of Ecology and Systematics* 26, 1995: 201–2.

See Also: Agriculture, Biodynamic; Agriculture, Organic; Community-Supported Agriculture; Family Farms; Farmers Markets; Local Food

Shauna M. Bloom

Agriculture, Biodynamic

Biodynamic agriculture is an ecological agricultural production system that relies on balance for success: balance between soil-improving and soil-exhausting crops, companion planting that relies on the interdependence in plants, a manuring system that recycles animal and farm wastes, and use of preparations that foster a healthy soil life and plant growth. Biodynamic farms are self-sufficient, and, ideally, do not bring in soil, manure, or other additions. Thus, this system can be similar to organic farming methods, but the two have distinct certification systems.

Biodynamic agriculture was inspired by a series of eight lectures given by the Austrian philosopher Rudolf Steiner in 1924 in Germany. Steiner delivered the lectures after a group of farmers asked him why seeds and cultivated plants were degenerating and how the farmers might improve seed and plant quality. The lectures, reprinted as *Agriculture: A Course of Eight Lectures*, contain esoteric discussions on the relationships among the cycles of the moon, sun, water, soil, and agriculture. As part of these discussions, Steiner laid out methods for optimally creating and applying compost to soil.

Steiner provided the foundation for biodynamic agriculture, leaving it to others to bring the tenets of his lectures to practical agriculture. Ehrenfried Pfeiffer, who attended these lectures, was one of the main forces behind the spread of biodynamic farming in Europe in the late 1920s and the United States in the 1930s. He was part of the Experimental Agricultural Circle, which consisted of the participants at Steiner's lecture series. This group was disbanded in 1941; upon reforming in 1946, the group organized an introductory course to biodynamic agriculture.

The concept of community supported agriculture (CSA) was inspired by the economic teachings of Steiner, and was a natural extension for biodynamic farms. The first CSAs were organized in Europe in the 1960s. CSAs were introduced in the United States in the mid-1980s, influenced by the European model, and were based at two biodynamic farms, Indian Line Farm in South Egremont, Massachusetts, and Temple-Wilton Community Farm in Wilton, New Hampshire. Both of these farms still operate biodynamic CSAs.

THE FARM AS AN ORGANISM

Biodynamic farmers consider farms to be organisms that consist of soil, livestock, crops, people who work on the farm, ponds and streams, wild birds and insects, animals, and the local climate and seasons; all of these aspects of the farm work together. Animals are a critical component of biodynamic farms, for food, manure, and for the creation of biodynamic preparations.

Each farm is unique, with three factors contributing to a farm's individuality: cycle of substances, farm site, and farm organization. An example of the cycle of substances is the carbon cycle, where carbon dioxide from the atmosphere is absorbed by the plant, which then creates nutrients via photosynthesis. Crop residues and manure return to the soil, and are metabolized into carbon dioxide; humans and animals breathe carbon dioxide into the atmosphere, beginning the cycle again. The other nutrients important to plant health (for example, calcium, phosphorous, nitrogen) have different cycles, which occur between the plant and the soil, in contrast to the carbon cycle, which takes place between the atmosphere and the soil. The quantity of nutrients available in the soil for each cycle depends on the quality of humus in the soil. Next, characteristics of the farm site, such as climatic zone, topography, local soil conditions, and micro climate, contribute to the farm's productivity. Farmers choose crops that will thrive on their farms, recognizing that some aspects of each farm site (for example, soil nutrient content) can be improved whereas others cannot be changed without incurring significant costs (for example, soil texture). Finally, the farm organization is based on the needs of plants and animals; only those familiar with farming have the knowledge to speak about the ideal form of farm ownership as well as discuss who profits from the fruits of the land.

Regardless of production technology, yields are influenced by light, warmth, air, water, and soil. Biodynamic farmers adopt special methods for building up soil and working with plants and livestock; the farming practices foster agricultural production by working within the framework of a specific farm's light, warmth, air, water, and soil. Farming methods specific to biodynamic agriculture include special preparations, composting techniques and use of companion plants. The preparations are intended to stimulate nutrient and energy cycling, to raise soil quality, and to increase the quality of the food produced.

There are nine special biodynamic preparations, numbered 500 to 508; each is made from a different substance and has specific, individual directions for preparation, application, and storage. Similar to homeopathy, farmers use extremely diluted forms of these nine preparations. Preparations 500, 501, and 508 differ from the other six preparations, and are used to regulate and stimulate plant growth. Farmers typically make preparation 500 from cow manure collected while animals are fully pastured or when the animals are pasturing with hay supplementation. Preparation 500 is created during the autumn and winter. The farmer fills a cow horn with cow manure and buries the horn in the soil so that the manure ferments. After taking the full season to develop, preparation 500 is sprayed on the soil before sowing seeds or just after the plants emerge. Preparation 500 cannot be applied when it is raining. Preparation 501 is made from ground quartz, which is packed into a cow horn and buried in the soil for 6 months over the spring and a full summer. After the preparation is ripe, it is then sprayed on plant leaves during different stages of growth; this preparation is applied on a sunny or partly sunny morning. Preparation 508 is an antifungal spray for plant leaves and is created from the horsetail plant.

The remaining preparations (502 to 507) are made from yarrow blossoms, chamomile blossoms, stinging nettle, oak bark, dandelion flowers, and valerian flowers, respectively, and are added to the compost. Some of the preparations are created in animal organs: yarrow (bladder of a male deer), bark (skull of a mature cow, sheep, pig, or horse), chamomile (bovine intestine), and dandelion (bovine mesentery). The first step in making the compost is forming the materials for the compost, animal waste, and plant wastes into a windrow. Then, the six preparations are poured into separate holes that are 20 inches deep and 5 to 7 feet apart, and the outside of the pile is then sprayed with valerian. The compost pile is traditionally left undisturbed for 6 months to 1 year, although some farmers do turn the compost pile to shorten the time needed for the compost to be completed.

In addition to using compost, soil fertility is enhanced by rotating crops and planting cover crops and green manures. When rotating crops, farmers alternate between planting soil-exhausting crops (e.g., corn and potatoes) and soil-improving crops (e.g., peas and beans). The planting of deep-rooted crops is followed by the planting shallow-rooted crops, and crops requiring manure are alternated with those not needing manure. Cover crops include rye, vetch, mustard, rapeseed, and oilseed radish; biodynamic farmers plow these crops into the soil when they are green or just after they have flowered. Companion plants are grouped together to either prevent pest damage or to foster growth. Some examples of common plantings are nasturtiums, planted near apple trees to deter woolly aphids, and beans, planted near carrots and cauliflower to promote growth. Other common companions are carrots and lettuce, garlic and potatoes, and strawberries and borage.

WINE

Wine is one of the more popular biodynamic products. As of August 2008, approximately five hundred vineyards across the world were either certified as biodynamic farms, in the process of transitioning to biodynamic practices, using biodynamically raised grapes in their wine, or using biodynamic farming practices. Nearly half of these vineyards are located in France.

Blind taste tests indicate that biodynamic wines are superior to conventional wines; this conclusion results from wine tasters finding that the expression of terroir in biodynamic wines is superior to that of conventionally produced wines. Terroir is the wine aficionado's notion that location, through factors such as climate, geography, and topography of the land, is uniquely manifested in the wine's flavor; quality winemakers seek to bring out terroir in the wine. The concept of terroir is the basis the French rely on when classifying wine according to the region it is grown. One explanation given for the superiority of biodynamic wine is that biodynamic winemakers are the most skilled craftsman, and thus are better able to produce high-quality wine.

CERTIFICATION

Biodynamic products were first marketed in Germany in 1928, when the Demeter Association created the "Demeter" label; this was the first ecological label to be used. The biodynamic label is named in honor of Demeter, the Greek goddess of agriculture and fertility, who was worshipped for her gifts of food and bread. The Demeter Association was also the first to develop an ecological standard for processed foods, which occurred in 1994.

Today, biodynamic farms are still certified by the Demeter Association. Under the umbrella organization, Demeter International, are independent country certification agencies; in 2008, twenty-five countries from around the world were either members or guest members of Demeter. In the United States, Demeter-USA grants biodynamic certification. Farms, processors, and distributors seeking biodynamic certification also receive organic certification, as regulated by the U.S. Department of Agriculture's National Organic Program.

In the United States, biodynamic methods are used for two years prior to receiving Demeter certification. Some prohibited substances are synthetic fertilizers, nitrogen compounds, soluble phosphates, sewage sludge, pesticides, and weed killers. When purchasing organic fertilizers and soil, the farmer consults the Demeter Association. There are limits on the amount of organic fertilizer that can be imported into the farm; a general rule of thumb is that the amount imported cannot exceed the amount of fertilizer created on the farm, which includes compost, green manure (created by cover crops), and animal manure.

A wide range of products are raised by farmers using biodynamic production systems, including fruits, vegetables, dairy products, eggs, wine, spirits, cosmetics, and fibers. There is a logo identifying products produced and processed according the Demeter standards.

According to Demeter International, in July 2006 there were 3,543 farms and 260,506 acres of farmland certified as biodynamic worldwide.

Approximately 429 companies were certified biodynamic processors, and another 154 companies were certified biodynamic distributors. Slightly more than half the biodynamic farmland and 37 percent of biodynamic farms were located in Germany. Small scale, uncertified biodynamic farms in India numbered approximately 2,000 in 2006; these farms raise cotton. In the United States, approximately 1,000 community supported agriculture farms were uncertified but using biodynamic methods.

RESEARCH ON BIODYNAMIC FARMING TECHNIQUES

Research on biodynamic farms includes examining soil quality and long-term cropping systems trials to assess yields. Most of these studies compare biodynamic farms to those managed with conventional methods and to those managed organically.

Soil quality studies show that biodynamic farms have more organic matter, microbial biomass, biological activity, and respiration than conventionally managed soils. However, when comparing soil quality of biodynamic farms with that of organically managed farms, no significant differences are found, suggesting that the soil quality benefits of biodynamic farming result from using compost. By treating compost with biodynamic preparations, the compost had higher internal temperatures and ripened faster than control compost piles. Over the long term, biodynamic compost had much larger earthworm populations than did compost generated under organic and conventionally managed systems.

A 2000 Washington State University study shows that, despite the lack of physical, chemical, and biological differences in soil quality, there was positive plant growth and earthworm behavior in soil amended by biodynamical compost. A 2000 Swiss study of biodynamic, organic, and conventional production through farm trials, conducted over twenty-one years, found that biodynamic and organic plots had slightly lower yields than conventional plots; the biodynamic plots had the greatest biodiversity and microbial diversity.

The current state of research suggests that the relationship among the biodynamic preparations, compost, soil quality, and yields is not yet fully understood.

VIEWPOINTS

There are divergent points of view regarding biodynamic agriculture. Some have called biodynamic agriculture "organic plus." From one perspective, this is an accurate description: the basic requirements of the National Organic Standards in the United States are satisfied for all biodynamic farms. The other requirements regarding composting (both the creation and use of) and use of preparations are unique to biodynamic agriculture, and are in addition to organic farming methods. Others have described biodynamic farming as agricultural voodoo, arguing that farmers are wasting time and money by following this approach. These differences of opinion will likely never be reconciled. Given that there is a segment of the market (both consumers and producers) interested in biodynamic food and wine, reconciliation of these views may be unnecessary.

Further Reading

Carpenter-Boggs, L., A. C. Kennedy, and J. P. Reganold. "Organic and Biodynamic Management: Effects on Soil Biology." *Soil Science Society of America Journal* 64: 1651–59.

Conford, Phillip. *The Origins of the Organic Movement.* Edinburgh: Floris Books, 2001, 287.

Demeter International. www.demeter.net.

Demeter USA. www.demeter-usa.org.

Diver, Steve. *Biodynamic Farming and Compost Preparation.* Alternative Farming Systems Guide. 1999. http://attra.ncat.org/attra-pub/biodynamic.html.

Koepf, Herbert H., Bo D. Pettersson, and Wolfgang Shaumann. *Bio-dynamic Agriculture: An Introduction.* Spring Valley, NY: Anthroposophic Press, 1976.

Mäder, Paul, Andreas Fließbach, David Dubois, Lucie Gunst, Padruot Fried, and Urs Niggli. "Soil Fertility and Biodiversity in Organic Farming." *Science* 296(5573) May 31, 2002: 1694–97.

Reeve, Jennifer Rose. "Effects of Biodynamic Preparations on Soil, Winegrape, and Compost Quality on a California Vineyard." Master's thesis, Washington State University, 2003.

Reilly, Jean K. "Moonshine, Part 2: A Blind Sampling of 20 Wines Shows That Biodynamics Works. But How? (This, by the way, is why we went into journalism.)" *Fortune*, August 23, 2004.

Steiner, Rudolf. *Agriculture.* 3rd ed. London: Rudolf Steiner Press, 1974.

See Also: Agriculture, Alternative; Agriculture, Organic; Agriculture, Sustainable

Carolyn Dimitri

Agriculture, Conventional

Conventional agriculture is an approach to food production, processing, and distribution that has come about through major changes in the structure of the food system starting after World War II. Today, conventional agriculture is distinguished by food production that has a high proportion of food products sold as commodities to food processors and manufacturers rather than food sold directly off the farm. Inputs (seeds, fertilizer, pesticides, etc.) once produced on the farm are now purchased in large quantities from agribusinesses. The primary economic aim is maximizing profits (and output) and minimizing input costs, mainly by reducing labor costs with technology (large machinery, mono-crops, chemical herbicides and pesticides). Conventional farmers have large capital investments in land, machinery, and inputs. Farms are large monocultures of either crops or livestock that rely on off-farm inputs and heavy machinery, rather than nutrient cycling and human farm labor. The outcome is fewer but more specialized farms, most of which do not have ownership of the product they are growing or raising, but are contracted to produce a specific product for a large agribusiness.

Three major developments occurred shortly after the World War II that allowed conventional agriculture to become the dominant form of production. The first was the availability of cheap nitrogen fertilizer, which decreased the need for growing legume crops. Once there was no longer a soil requirement for legume crops, which are typically fed to livestock, farmers were able to specialize production to focus on either crops or livestock. Specialization led to the second development, consolidation in the food supply chain, which came about as a result of the concentration of livestock production. Corporations took advantage of concentrated production and started to locate large processing facilities near livestock facilities in locations with favorable conditions, such as availability of cheap wage laborers and loose environmental regulations. The third development was the creation of machinery designed to harvest one specific crop, which lead the way to standardization and intensification. Improvements in crop varieties and drug enhancements for livestock further expanded large monoculture farms. Major agricultural policies further contributed to the development of an industrial model of agriculture. The agricultural policies encouraged increases to commodity production and created large investments for research on conventional agriculture. As agricultural production levels increased agribusinesses started to consolidate and become vertically integrated. Agriculture evolved to an industrial model of production, processing, distributing, and retailing.

AGRICULTURAL POLICY AND AGRIBUSINESS

The combination of technological change and agriculture policy, specifically price supports or commodity-based subsidies, had a significant effect on the structure of conventional agriculture. Agriculture policy began in North America and Western Europe to provide support to farming incomes, develop export markets, reduce dependency on imported food, and keep food prices low for domestic consumers. The structure of the support meant that the larger farms receive the most benefits. Subsidy payments essentially rewarded those who had the highest output overall. The result was overproduction of crops that were part of the subsidy programs, such as corn, wheat, soybeans, and cotton. Crop prices for these commodities subsequently dropped as the subsidy payments encouraged more production. These prices have never significantly recovered. Improved technology and current agricultural policy have resulted in increasing farm sizes coupled with decreasing numbers of total farm operations and declining profit margins. As capital costs of production increased, the number of farmers who could afford to farm decreased. Although many problems exist with post–World War II government intervention, it remains strong because of economic and political considerations. Factors such as specialization and production efficiency have further supported an agriculture sector that comprises large corporations supported by domestic subsidies and international trade policies.

Specialization and consolidation, supported by agricultural policy and research, were instrumental in creating an agriculture sector that is dominated by agribusiness corporations rather than smaller family farmers. Standardization and specialization allowed corporations to process and distribute mass amounts of food on a global market. Vertical integration created a way for corporations to control a product from seed to store, maximizing profits and minimizing production costs and risk for the agribusiness. Farmers generally did not receive the benefits of increased profits or decreased production costs and risk. The decision-making in a vertically integrated agriculture sector becomes centralized, which influences the market and the decisions made by individual family farmers. Contract farming is a common result of vertical integration and one way corporations maintain control of decision-making at the farm level. Contract farming allows corporations to process and distribute large amounts of standardized crops without having to take on the risks of production. Farmers buy seeds, fertilizer, pesticides, and herbicides from one company. Farmers grow a product and sell it back to the same company, taking on the risks of production while missing out on the profit.

ENVIRONMENTAL CONSEQUENCES

Conventional agriculture has not progressed without consequences. The external costs of conventional agriculture are often the result of short-term gains being considered more important than long-term sustainability. Environmental issues remain a major problem in conventional agriculture. There are three major areas of concern: pollution of soil and water from agrichemical usage, loss of biodiversity and genetic variation, and risk to food safety. These environmental concerns have not been addressed through management practices, national policies, or market structures and remain a problem throughout North America.

The first major concern derives from the fact that the industrial model of agriculture uses large amounts of water, energy, synthetic chemicals, and petroleum. Nonrenewable energy is used to create and transport synthetic chemicals in large quantities, fuel large machinery, and transport food products. This is expensive to both farmers and society in general. It also creates large amounts of wastes and pollution that are difficult to manage. Chemicals used for fertilizers and pesticides contaminate the soil and make their way into the watershed. The contamination of surface and groundwater is a serious concern because much of these sources supply our drinking water. The nutrient runoff also contributes to the degradation of many lakes, estuaries, and ocean environments. Soil degradation and erosion

"Agriculture has become unbalanced: the land is in revolt: diseases of all kinds are on the increase: in many parts of the world Nature is removing the worn-out soil by means of erosion."—Sir Albert Howard, a founder of the organic agricultural movement. From *An Agricultural Testament.* New York: Oxford University Press, 1943.

creates a cycle of need for synthetic chemicals, which in turn further degrade the health of the soil. Animal production is another component of agriculture that can be harmful to the environment when done on a large scale. These operations, called concentrated animal feeding operations (CAFOs) are highly problematic. Concentrations of livestock create waste nutrients in amounts that are concentrated enough to be toxic; these make their way into the watershed. The drugs, used for enhancing growth and preventing disease, are released into watershed as well.

The second major concern is related to conventional crop production practices. Most conventional crop farmers rely on genetically modified organisms produced by the same agribusinesses that provide the associated herbicides and pesticides. Genetically engineered crops reduce genetic diversity of plants through unintended cross-pollination. The use of genetically modified organisms, coupled with monocropping, reduces the variety of beneficial pests available to naturally maintain pest levels. Over time, the beneficial pests are no longer present and there is an increased dependency on synthetic chemicals. This results in a loss of biodiversity and genetic variation. Biodiversity is also affected through loss of viable habitat from the fragmentation of semi-natural areas, overgrazing, mismanagement of interstitial habitats (places like field margins, ditches, and hedgerows) and draining of wetland areas.

Finally, food safety is another major concern within the conventional food system. Some of the major health concerns relating to conventional production practices include salmonella in chicken, bovine spongiform encephalopathy and foot and mouth disease in cattle, and E. coli outbreaks on vegetables and fruit. The problems in livestock and poultry result mostly from the practice of using ground up animal remains to feed chickens, turkey, cattle, hogs, and other animals raised in CAFOs.

Further Reading

Atkins, Peter, and Ian Bowler. *Food in Society: Economy, Culture, Geography*. London: Hodder Arnold, 2001.

Beus, Curtis E., and Riley E. Dunlop. "Conventional Versus Alternative Agriculture: The Paradigmatic Roots of the Debate." *Rural Sociology* 55(4) 1990: 590–616.

Bird, Elizabeth Ann R., Gordon L. Bultena, and John C. Gardner. *Planting the Future: Developing an Agriculture That Sustains Land and Community*. Iowa State University Press, 1995.

Lyson, Thomas A., and Amy Guptill. "Commodity Agriculture, Civic Agriculture and the Future of U.S. Farming." *Rural Sociology* 69(3) 2004: 370–85.

Magdoff, Fred, John Bellamy Foster, and Frederick Buttel, eds. *Hungry for Profit: The Agribusiness Threat to Farmers, Food and the Environment*. New York: Monthly Review Press, 2000.

Robinson, Guy. *Geographies of Agriculture: Globalisation, Restructuring, and Sustainability*. Essex: Pearson Education, 2004.

See Also: Agribusiness; Agricultural Subsidies; Conventionalization; Factory Farming; Farm Workers' Rights; Fossil Fuel Use; Production, Treadmill of

Shauna M. Bloom

Agriculture, Organic

ORIGINS

Farming originated about ten thousand years ago and originally employed only organic agricultural technology. There were no synthetic fertilizers or pesticides at that time. The technology employed was slash and burn agriculture. Trees and shrubs were cut and then burned in place. This added to the nutrients in the soil that had accumulated over many years. At the same time, the burning exterminated the weeds, pest insects, and plant pathogens.

The major handicap to the use of slash and burn techniques today is insufficient land availability. The land would be left fallow for 10 to 20 years in order to accumulate nutrients before it was replanted. Currently, there are too many people on earth to leave cropland fallow for that length of time. Modern organic agriculture has devised new technologies to add nitrogen and other nutrients, without depending on the accumulation of natural nutrients over many years.

GROWTH

Organic agriculture is the fastest growing sector of agriculture in the United States. There has been a doubling of the area used in organic production since around 2003. Currently more than 600,000 hectares (ha) are in organic production. Organic food sales total more than $10 billion per year and are growing at double-digit rates. This growth is stimulated by public concerns about the environment and consumer concerns about chemicals used in food production. As certified organic production increases, there will likely be continued growth in organic food production.

CONVENTIONAL VERSUS ORGANIC CROPPING SYSTEMS

Numerous studies compare conventional and organic production. In general, organic farming is comparable to conventional farming in the areas of economic return and crop yields. Organic crops do even better than conventional crops when there is variation in the ecological conditions, such as drought or other stresses. Organic agriculture aims to augment ecological processes that support plant nutrition and at the same time conserve water and soil resources. Ongoing field experiments comparing conventional and organic production systems are underway at the Rodale Institute in Kutztown, Pennsylvania, on moderately well-drained soils of Comly silt loam. The growing climate there averages 12.4 degrees Celsius with an average rainfall of 1,105 mm per year.

Corn and soybeans were selected for this assessment because these are two of the most important crops produced in the United States. Together they occupy 145 million hectares and have a value totaling more than $120 billion. It is significant that corn production uses more insecticides, herbicides, and nitrogen fertilizer than any other crop grown in U.S. agriculture.

One twenty-two-year study included three cropping systems. The conventional system followed the recommendations of the Cooperative Economic Service of Pennsylvania State University for corn and soybean production. The recommendations included synthetic fertilizers and herbicides. This conventional system is typical of commercial corn and soybean production systems in the United States. No cover crops were grown in the conventional system during the nongrowing season.

The animal manure and legume-based organic system (hereafter referred to as the "organic-animal" system) represented a typical livestock operation in which grain crops are grown for animal feed. This system included a rotation of crops: wheat, corn, soybeans, corn silage, a rye cover crop, and red-clover-alfalfa hay—before the corn silage and soybeans were planted again.

The nitrogen source was aged cattle manure, applied two years out of every five, at a rate of 5.6 tons per ha (dry). It was applied immediately before the plowing and planting of corn. The plow-down of legume-hay crops supplied additional nitrogen. No herbicides were used for weed control. Weed control was carried out by mechanical cultivation plus the weed-suppressing crop rotations.

The "legume-based organic system" represented a conventional cash grain operation. However, this system used no commercial synthetic nitrogen. Instead, this system relied on hairy vetch and rye as fall, winter, and spring cover crops. The hairy vetch cover crop was planted in early spring before the corn was planted. The total nitrogen added from the hairy vetch was 49 kg (or 140 kg per ha for a given year and corn crop). Weed control was similar to the animal-legume system.

INPUTS IN CONVENTIONAL CORN AND SOYBEAN PRODUCTION SYSTEMS

The quantity and energy inputs typical for conventional corn and soybean production systems for the United States are shown in Tables 1 and 2. These data are average values of U.S. conventional corn and soybean production.

In the twenty-two-year Rodale experiment, the conventional technologies followed were similar to those recommended by the Cooperative Economic Service of the Pennsylvania State University. The yields of corn and soybeans in Pennsylvania average 140 bushels/acre, whereas the average yields for the entire United States average 148 bushels/acre. Clearly the average in Pennsylvania is lower than that of the nation. This is probably due to the poorer soils in Pennsylvania. To conserve soil and water resources, crop residues were left on the surface of the land. Thus, there was little exposed soil during the growing season.

Measurements Recorded

The measurements recorded for the three systems included crop biomass, cover crop biomass, weed biomass, grain yields, nitrate leaching, herbicide leaching, water percolation, soil carbon, soil nitrogen, and soil water content.

Table 1. Energy Inputs of Conventional Corn Production per Hectare in the United States

Inputs	Quantity	kcal × 1,000
Labor	11.4 hrs	462
Machinery	55 kg	1,018
Diesel	88 L	1,003
Nitrogen	155 kg	2,480
Phosphorus	79 kg	328
Potassium	84 kg	274
Lime	1,120 kg	315
Seeds	21 kg	520
Irrigation	8.1 cm	320
Herbicides	6.2 kg	620
Insecticides	2.8 kg	280
Electricity	13.2 kWh	34
Transport	204 kg	169
TOTAL		8,228
Corn yield	9,400 kg/ha	33,840
kcal input:output	1:4.11	

Source: Pimentel, David, Alison Marklein, Megan A. Toth, Marissa N. Karpoff, Gillian S. Paul, Robert McCormack, Joanna Kyriazis, Tim Krueger. Biofuel Impacts on World Food Supply: Use of Fossil Fuel, Land and Water Resources. *Energies* 2008.

Corn and Soybean Yields Under Normal Rainfall

For the first five years of the experiment, corn grain yields averaged 4,222, 4,743, and 5,903 kg per ha for the organic-animal, organic-legume, and conventional systems, respectively. The corn yield for the conventional

Table 2. Energy Inputs in Conventional Soybean Production per Hectare in the United States

Inputs	Quantity	kcal × 1,000
Labor	7.1 hrs	284
Machinery	20 kg	360
Diesel	38.8 L	442
Gasoline	35.7 L	270
LP gas	3.3 L	25
Nitrogen	3.7 kg	59
Phosphorus	37.8 kg	156
Potassium	14.8 kg	48
Limestone	2,000 kg	562
Seeds	69.3 kg	554
Herbicides	1.3 kg	130
Electricity	10 kWh	29
Transport	154 kg	40
TOTAL		2,959
Soybean yield	2,890 kg/ha	10,404
kcal input:output	1:3.52	

Source: Pimentel, David, Alison Marklein, Megan A. Toth, Marissa N. Karpoff, Gillian S. Paul, Robert McCormack, Joanna Kyriazis, Tim Krueger. Biofuel Impacts on World Food Supply: Use of Fossil Fuel, Land and Water Resources. *Energies* 2008.

system was significantly higher than the two organic systems. However, after the five-year transition period, corn grain yields were similar for all systems, averaging 6,431, 6,368, and 6,553 kg per ha for the organic-animal, organic-legume, and conventional systems, respectively.

Overall soybean yields for the last twenty years of the experimental study were 2,461, 2,235, and 2,546 kg per ha for the organic-animal, organic-legume, and conventional systems, respectively. The lower yield for the organic-legume system is attributed to the failure of the soybean crop in 1988. If the 1988 year result is removed from the analysis, the soybean yields are not significantly different.

Crop Yields Under Drought Conditions

During 1999 there was an extreme drought, and rainfall was less than half of normal rainfall, 500 mm, for the growing season. The organic-animal system had significantly higher corn yield (1,511 kg per ha) than the conventional system, which had only a yield of 1,100 kg per ha.

Soybean yields responded differently than corn during the 1999 drought. Soybean yields were 1,800, 1,400, and 900 kg per ha for the organic-legume, organic-animal, and the conventional systems, respectively. Significantly more water was held in the soil in the organic systems than in the conventional system, because of the high soil organic matter in the organic systems. The soil organic matter acted as a sponge and held larger quantities of water for use by the crops during the growing season.

Energy Inputs

The energy inputs for the different experimental systems included fossil fuels for the farm machinery, fertilizers, seeds, and herbicides. About 5.2 million kcal of energy per ha were invested in the production of corn in the conventional system. This is lower than the 8.2 million kcal of energy reported for the U.S. average of corn production. The energy inputs for the organic-animal and organic-legume systems were 28 percent and 32 percent less than those of the conventional corn system, respectively. Commercial fertilizers for the conventional corn production system were produced employing primarily natural gas for the nitrogen, whereas the nitrogen for the organic systems was obtained from legumes or livestock manure. Of course, some fossil energy was required to transport the manure to the organic-animal experimental systems. Table 3 shows the potential energy savings for inputs in corn production using organic and conservation practices.

In the Rodale experiments, the energy inputs for all the soybean production systems were similar, about 2.2 million kcal per ha. This is lower than the 3.0 million kcal energy input reported for the U.S. average of conventional soybean production.

Economics of the Organic Systems

The organic farming systems required about 35 percent more labor than the conventional farming system. However, the organic corn systems (without

Table 3. Potential Reduced Energy Inputs of Corn Production per Hectare in the United States Using Organic and Conservation Practices

Inputs	Quantity	kcal × 1,000
Labor	15 hrs	608
Machinery	10 kg	185
Diesel	60 L	684
Gasoline	0	0
Nitrogen	Legumes	1,000
Phosphorus	45 kg	187
Potassium	40 kg	130
Lime	600 kg	169
Seeds	21 kg	520
Irrigation	0	0
Herbicides	0	0
Insecticides	0	0
Electricity	13.2 kWh	34
Transport	75 kg	25
TOTAL		3,542
Corn yield	9,000 kg/ha	31,612

Source: Pimentel, D., Williamson, S., Alexander, C. E., Gonzalez-Pagan, O., Kontak, C., and Mulkey, S. E. Reducing energy inputs in the U.S. food system. *Human Ecology* 2008. DOI: 1007/s10745-008-9184-3.

price premiums) were 25 percent more profitable than conventional corn systems. This was possible because the costs of production of the two organic systems were 15 percent lower than the conventional corn system.

Soil Carbon

Soil carbon, which correlates with soil organic matter, was measured in 1981 and 2002. In 1981, the soil carbon levels in the three systems were not significantly different. However, in 2002 the soil carbon levels in the organic-animal and organic-legume systems were significantly higher than in the conventional system: 2.5 percent and 2.4 percent versus 2.0 percent, respectively. The annual net aboveground carbon input (based on plant biomass and manure) was the same in the organic-legume and the conventional systems (about 9,000 kg per ha), but close to 12 percent higher in the organic-animal system (about 10,000 kg per ha). However, the two organic systems retained more of the carbon in the soil, resulting in an annual soil carbon increase of 981 and 574 kg per ha in the organic-animal and the organic-legume systems, compared with only 293 kg per ha in the conventional system.

Soil Nitrogen

Soil nitrogen levels were measured in 1981 and 2002 in the three experimental systems. Initially in 1981 the three systems had similar percentages of soil nitrogen, or approximately 0.31 percent. By 2002, the conventional system remained at about 0.31 percent, whereas the organic-animal and organic-legume systems significantly increased to 0.35 percent and 0.33 percent, respectively.

Soil Organic Matter and Biodiversity

Soil organic matter provides the basic resources essential for productive organic agriculture and sustainable agriculture. After twenty-two years of organic agricultural management the soil organic matter increased significantly in the organic-animal and organic-legume systems. The increased organic matter ranged from 15 to 28 percent.

The amount of organic matter in the upper 15 cm of soil in the organic farming systems was approximately 110,000 kg per ha. The soil in the upper 15 cm weighed about 2.2 million kg per ha. Approximately 41 percent of the volume of the organic matter in the organic systems consisted of water. The amount of water held in both organic systems is estimated to be 816,000 liters per ha. This large amount of water held in the two organic systems' soils' organic matter helped make both organic systems more tolerant to droughts than the conventional system.

Large amounts of biomass (soil organic matter) significantly increased the soil biodiversity. In good quality soil, with abundant biomass, bacteria and fungi can weigh 3,000 kg/ha each and earthworms can also weigh as much. There can be as many as 1,000 earthworm and insect holes per m² of soil. These holes aid in the percolation of water into the soil and decrease the rate of water runoff.

Overall, environmental damage from agricultural synthetic fertilizers and pesticides was reduced in the organic systems. As a result, overall public health and ecological integrity was improved because of the organic practices followed. Various organic agricultural technologies have been used for more than 10,000 years, making agriculture more sustainable while conserving water, soil, energy, and biological resources.

Further Reading

Colla, Giuseppe, Jeffrey P. Mitchell, Durga D. Poudel, and Steve R. Temple. "Changes of Tomato Yield and Fruit Elemental Composition in Conventional, Low Input, and Organic Systems." *Journal of Sustainable Agriculture* 20(2): 53–67.

Greene, Catherine. "Land in U.S. Organic Production." USDA Datasets Web site. www.ers.usda.gov/Data/organic/.

Lavelle, Patrick, and Alister V. Spain. *Soil Ecology*. Dordrecht: Kluwer Academic Publishers, 2001.

Pimentel, David. *Impacts of organic farming on the efficiency of energy use in agriculture.* Organic Center State of Science Review. August 2006. www.organic-center .org/science.pest.php?action=view&report_id=59.

Pimentel, David, and Marcia Pimentel. *Food, Energy and Society.* 3rd ed. Boca Raton, FL: CRC Press, 2008.

Pimentel, David, P. Hepperly, J. Hanson, D. Douds, and R. Seidel. "Environmental, Energetic, and Economic Comparisons of Organic and Conventional Farming Systems." *BioScience* 55(7) 2005: 573–82.

Pimentel, David, Tsveta Petrova, Marybeth Riley, Jennifer Jacquet, Vanessa Ng, Jake Honigman, and Edwardo Valero. "Conservation of Biological Diversity in Agricultural, Forestry, and Marine Systems." In *Focus on Ecology Research*, A. R. Burk, ed., 151–73. New York: Nova Science Publishers, 2006.

Reganold, John P., Jerry D. Glover, Preston K. Andrews, and Herbert R. Hinman. "Sustainability of Three Apple Production Systems." *Nature* 410(6831): 926–30.

Vasilikiotis, Christos. "Can Organic Farming Feed the World?" 2000. http://nature .berkeley.edu/~christos/articles/cv_organic_farming.html.

See Also: Agriculture, Alternative; Agriculture, Biodynamic; Agriculture, Sustainable; Agroecology; Compost; Conventionalization; Crop Rotation; Cuba; Farmers Markets; Grassroots Organic Movement; Intercropping; Prices; Livestock Production, Organic; ; Rodale Institute; Soaps, Insecticidal; Soil Health; Transition to Organic; Vermiculture; APPENDIX 2: Organic Foods Production Act of 1990

David Pimentel

Agriculture, Sustainable

Sustainable agriculture is an agriculture that is capable of indefinitely maintaining its productivity and usefulness to society. Sustainable agriculture first came to widespread public attention in the late 1980s when the Low Input Sustainable Agriculture program, dubbed LISA, was initiated by the United States Department of Agriculture (USDA). The term was adapted from "sustainable development," which was defined in a 1987 report of the United Nations World Commission on Environment and Development as development that meets the needs of the present without compromising opportunities for the future. The United Nations Commission was a response to growing concerns about the negative impacts of industrial development on the natural environment and its consequent effects on the physical, social, and economic quality of life of people. Questions of agricultural sustainability arose from concerns about soil erosion and depletion of soil productivity, pollution of air and water with agricultural chemicals and wastes, and the consequent effects on the economic viability of farming and the quality of life for farmers, rural residents, and society as a whole.

The productivity of agriculture, and thus its usefulness and economic value to society, must be derived from either natural or human resources, from land or people. To sustain its economic value to society agriculture must maintain the productivity of the natural ecosystems and social systems from which it derives its productivity. Thus, a sustainable agriculture must be ecologically sound, socially responsible, and economically viable; none of these conditions is more important than any other. The ecological, social, and economic dimensions of sustainability are hierarchal only in the sense that societies function within nature and economies function within societies. However, societies are capable of inflicting significant and irreparable damage to nature, and economies are capable of inflicting significant and irreparable damage to society.

Sustainable economies must conform to the principles of healthy ecosystems and healthy societies. Economies that rely on extraction and exploitation may be profitable in the short run but they ultimately destroy the natural and human resources they must rely on for their productivity. However, economies that fail to meet the expectations of societies will not be sustained by those societies. Sustainable societies must conform to the principles of healthy natural ecosystems from which they derive their physical well-being. Societies that pollute and degrade the productivity of the earth may appear to be successful in the short run but they ultimately create an uninhabitable natural environment. However, societies that are unable to meet their basic physical needs inevitably degrade their natural environment. Ecological, social, and economic integrity are but three dimensions of the same whole of sustainability.

SUSTAINABLE APPROACHES TO FARMING

Sustainable agriculture is an overarching approach to farming that includes a variety of different farming methodologies that embrace the ecological, social, and economic principles of sustainability. For example, organic, biodynamic, holistic, alternative, ecological, practical, and innovative farming are all sustainable approaches to farming. Among these, organic farming is perhaps the most widely recognized. People tend to think of organic agriculture as farming without synthetic fertilizers or pesticides, which is accurate. However, in order to maintain their productivity without using synthetic fertilizers and pesticides, organic farms must maintain healthy soils and agroecosystems. The historic purpose of organic farming was to create a permanent agriculture to sustain a permanent society. Furthermore, if organic farming is to sustain a society, that society must find ways to make farming economically viable for its organic farmers. True organic farming is ecologically sound, socially responsible, and economically viable, sustainable farming.

Biodynamic farming is a sustainable farming methodology that emphasizes the self-renewing and regenerative capacity of living agroecosystems. Living systems are essential to sustainability because sustainability ultimately depends on the regeneration of energy. Everything that is of use to humanity ultimately depends on energy. Our houses, clothes, automobiles, and food all require energy to make and energy to use. All physical materials are essentially highly concentrated forms of energy. According to the basic laws of thermodynamics, energy can neither be created nor destroyed, but each time energy is used, some of its usefulness is lost. Whenever energy is used, it always changes from a more concentrated, organized form to a more dispersed, disorganized form, as when gasoline is ignited in the cylinder of an automobile. Energy can always be reused but with each use it becomes less concentrated, less organized, and thus less useful. Physicists refer to this loss of energy usefulness as entropy. Solar energy is the only source of new energy available to offset entropy. Living organisms, biological systems, are the only means of capturing, organizing, and concentrating solar energy; thus validating the importance of biodynamic farmers managing their farm as living biological organisms. Plants, animals, insects, soil organisms, and even farmers are all interrelated aspects of the same biological, organic whole. Biodynamic farms seek to achieve economic and social sustainability through biological renewal and regeneration of energy.

Holistic farmers also manage their farms as ecological, social, and economic wholes. They are guided by a three-part holistic goal which envisions a desired future landscape or agroecosystem, a desired social quality of life, and the economic means of production needed to achieve the desired future. Holistic managers recognize and respect the inviolable principles by which natural ecosystems, societies, and economies function as they develop strategies for achieving their three-part, holistic goal of sustainability. Other sustainable farming methodologies emphasize different aspects of sustainability and suggest different strategies. However, they all seek to operate farms through balance and harmony among the ecological, social, and economic dimensions of farming.

SUSTAINABLE FARMERS

Sustainable farms must function in harmony with the specific natural ecosystems and communities in which they are located and thus are inherently site-specific and individualistic. Sustainable farmers also have unique abilities, aspirations, and social and ethical values. However, most sustainable farms and farmers share common sets of characteristics.

First, sustainable farmers see themselves as stewards of the land. They are committed to passing their land to the next generation at least as healthy and productive as when it was passed to them. They share a deep sense of respect and commitment to caring for the land and the things of nature. They work with nature rather than try to control or conquer nature, by fitting their crops to their ecological landscape rather than trying to manipulate nature. Their farms are more diversified than are conventional farms, because nature is diverse. Diversity may apply to the methods of crop rotations and cover crops or the types of crop and animal produced. By managing diversity, sustainable farmers are able to reduce their dependence on synthetic pesticides, commercial fertilizers, and other costly inputs that squeeze farm profits and threaten the environment on many conventional farms. Sustainable farms are more economically viable, as well as more ecologically sound because they function in harmony with nature.

Second, sustainable farmers build relationships. They tend to have more direct contact with their customers than do conventional farmers. Most either market their products directly to customers at venues such as farmers markets or market through cooperative organizations that connect them with their customers. They realize that their customers value food products differently because they have different needs and different tastes and preferences. They produce the things that their customers value most.

Sustainable farmers have a strong sense of respect for people and an appreciation for the value of human relationships. They are not trying to take advantage of their customers to make quick profits; they are trying to create long-term relationships. They market to people who care where their food comes from and how it is produced—such as food that is locally grown, organic, natural, humanely raised, hormone and antibiotic free, without genetic engineering—and they receive premium prices by producing foods that are highly valued by their customers. Sustainable farms are profitable because they are responsive to their customers as well as to society and nature.

Sustainable farmers challenge the stereotypical image of farmers as being fiercely independent individualists. They freely share information and encouragement with others. They form partnerships and cooperatives to buy equipment, to process and market their products, to do things together that they cannot do as well alone. They are not trying to drive each other out of business; they are trying to help each other succeed. They refuse to exploit each other for short-term gain. They buy locally and market locally in the communities where they live. They help bring people together in positive, productive relationships that contribute to their economic, ecological, and social well-being. They value people, for personal as well as economic reasons, and receive economic and social value in return.

For sustainable farmers, farming is as much a way of life as a way to make a living. They are "quality of life" farmers because they believe that a farm is a good place to live, with a healthy environment and a strong community. For sustainable farmers, these quality of life goals might be more important than the economic bottom line when making farming decisions. Their farms represent the things they believe in as much as the things that yield profits. Many sustainable farmers also find a spiritual sense of purpose and meaning for their lives through farming; they feel they were meant to be farmers. They respect their neighbors, their customers, the land, and animals because it's the ethical and moral thing to do. However, for many, their products are better and their costs are less because by following their passion they end up doing what they do best. Sustainable farmers earn an acceptable income, but more important, they have a higher quality of life because they are living a life they love.

Finally, sustainable farmers are thinking farmers. They must understand nature to work in harmony with nature and must understand people to build relationships with their customers, neighbors, and other farmers. Sustainable farming requires an ability to translate observation into information, information into knowledge, knowledge into understanding, and understanding into wisdom. Agriculture has been characterized by some as the first step beyond hunting and gathering, because farming was often considered a low-skill minimum-thinking occupation that almost anyone could do. However, sustainable farming is the mind work of the future, not the hard work of agrarian agriculture of the past. Certainly, sustainable farming involves some hard work, but its success depends far more on thinking than on working.

ORGANIC/LOCAL FOODS

Sustainable farmers must fit their farming operations to the uniqueness of their "place" within nature and society. They must find ways of farming that fit their abilities, aspirations, and perception of quality in life. More than two decades of experience with sustainable farming provides some insights into the general kinds of enterprises that seem to be working for farmers who employ sustainable approaches to farming.

Organic vegetables, berries, and fruits are perhaps the most widely recognized of all sustainably grown crops. Retail markets for organic foods grew at a rate of more than 20 percent per year during the 1990s and well into the 2000s, with organically grown vegetables leading the way. Most of the smaller organic producers market directly to their customers, through farmers markets, roadside stands, community-supported agriculture (CSA), or other direct marketing methods, and thus, realize the full retail value of their products. However, with the rapid growth in demand, organic foods soon moved into higher volume retail markets.

Organic grain production was the early mainstay of sustainable agriculture in some regions of the United States and Canada. Organic certification of crops was initially carried out by local and regional certification organizations. However, in the mid-1990s, the National Organic Program was established by the USDA, which standardized the requirements for organic production and facilitated organic production in large-scale, specialized operations. Consequently, organic production increased, depressing organic price premiums and limiting opportunities for smaller organic farming operations. Some organic grain producers started growing specific varieties for niche markets that were unattractive to large-scale producers. Others started producing long-neglected specialty grains, such as triticale, spelt, kamut, quinoa, and popcorn. Some began cleaning, processing, and packaging their grains for direct sales to individual customers and to local and regional bakeries, restaurants, and food retailers.

As with organic grains, national certification of vegetables and fruits led to large-scale, specialized production and depressed price premiums. As a result, smaller independently operated organic farms were forced to refocus on direct marketing methods to maintain their economic viability. Eventually, U.S. consumers began to recognize that many "mainstream" organic foods lack the social and ethical authenticity of true organic foods. For many discriminating consumers, "locally grown" has become the "new organic." Locally grown, sustainably produced vegetables, fruits, and berries have the added advantage of being selected, grown, and marketed for freshness, flavor, nutrition, and variety rather than for durability during harvest, transportation, storage, and display. For vegetables and fruits, "locally grown" is becoming the "new sustainable."

For livestock and poultry producers, sustainable farming has been most often associated with grass-based and free-range production, including dairy, beef, poultry, pork, lamb, goats, and others. By utilizing management intensive rotational grazing, farmers are able to reduce the costs of purchased feed, building, equipment, and medication associated with confinement livestock feeding operations. Some grass-based dairy farmers cut costs further by milking only during their pastures' growing season. Pastured and free-range chickens, turkeys, and hogs fit into the same category, although for free-range animals, freedom to roam may be more important than access to grass. It's also easier to realize added economic value from the health benefits of eating grass-fed meats and dairy products by marketing directly to local health conscious customers.

The wide variety of opportunities for sustainable farming is too great to enumerate here. Literally thousands of farmers all across North America, and around the world, have found ways to break away from conventional approaches to farming, finding new and better ways to farm and to live through sustainable agriculture.

Further Reading

Bell, Michael. *Farming for Us All: Practical Agriculture and the Cultivation of Sustainability*. University Park: Pennsylvania State University Press, 2004.

Berry, Wendell. *The Unsettling of America: Culture and Agriculture*. Watertown, MA: Sierra Press, 1982.

Horne, James, and Maura McDermott, *The Next Green Revolution: Essential Steps to a Healthy, Sustainable Agriculture*. New York: Haworth Press, 2001.

Ikerd, John. *Small Farms Are Real Farms: Sustaining People through Agriculture*. Austin, TX: Acres USA, 2007.

Ikerd, John. *Crisis and Opportunity: Sustainability in American Agriculture*. Lincoln: University of Nebraska Press, 2008.

Nickel, Raylene, *A Prayer for the Prairie: Learning Faith on a Small Farm*. Kief, ND: Five Penny Press, 2003.

See Also: Agriculture, Biodynamic; Agriculture, Organic; Agroecology; Compost; Crop Rotation; Family Farms; Soil Health; Vermiculture

John Ikerd

Agritourism

The U.S. agricultural industry is undergoing major structural changes. Almost one in two dairy and beef operations and one in four hog operations have disappeared since 1991, with roughly a 4 to 7 percent decrease in operations per year. Small livestock farm operations have suffered most of the losses. Low product prices, globalization of markets, low efficiency of small farms, and higher feed and operation costs are slowly but substantially eroding small farm incomes.

Kids enjoy a hayride at a pumpkin patch. Courtesy of Leslie Duram.

Rural communities have witnessed a similar decline in the number of farm operations. The decline in farm numbers has led to devastating effects on rural economy and society. To cope with this situation, farmers must recognize the need to diversify their products and revenue sources within current operations. Farmers must also recognize and respond effectively to changing opportunities—including emerging nontraditional markets for farm products and services. As society changes, these new opportunities may be necessary to provide additional farm income.

Tourism is the fastest-growing business in the world, and one of the fastest-growing segments of the industry is rural or agritourism. Agritourism is an alternative form of tourism that allows the visitor to see and experience the primary source of agricultural food production. Although it may encompass numerous activities, agritourism is focused on rural areas and on allowing visitors to see agricultural enterprises, local products, traditional foods, and the daily lives and cultures of rural people. Overall, both the environment and traditional cultures are a focus. Agritourism can include various types of overnight accommodations but may simply involve day trips to farm attractions.

The growth of agritourism across the United States is emerging as an important product and market diversification strategy for farmers. It is estimated that more than sixty-two million Americans, age sixteen or older, visited a farm between 2000 and 2001. An estimated twenty million children under the age of sixteen also visited a farm at some point during this period. Examples of activities that most farm visitors have experienced are farm stays, bed and breakfasts, pick-your-own products, agricultural festivals, farm tours for children, or hayrides. It is projected that between 1997 and 2007, nature- and agriculture-based tourism will be the fastest growing segment of the travel and tourism industry.

An increasing demand for farm or agritourism consumption comes from Europe, Canada, and Australia. Places such as England, France, Germany, and Austria have as many as twenty to thirty thousand agritourism sites in each country. Many of these sites focus on the traditional farm setting, offering high-quality accommodations and farm-related activities that feature the rich and unpolluted environment. In Europe and parts of North America, governments and local authorities view farm tourism as a resource to rejuvenate rural economies and preserve a way of life.

Motivations to participate in agritourism developments differ depending on the size of the farm, the age, and lifestyle of the farm operator. Larger farms may be more concerned with the additional income farm tourism provides, as well as with educating the visitor, whereas smaller farm owners more likely have additional time due to retirement and are subsequently concerned with retirement income. Farm owners are often motivated to participate in agritourism enterprises by the idea of sharing the rural experience with outsiders, as well as by the opportunities to socialize and meet new people. Family farm interest in agritourism include additional income, utilization of resources, and an opportunity to educate consumers. Income and socialization are common motivations for farm tourism operations, and these findings are similar among farmers in different countries.

Agritourism represents an opportunity and a practical tool to assist farmers to generate supplemental income. Farm operators engaged in agritourism are more likely to attain higher income levels than farmers who do not undertake these activities.

Whereas important research has been conducted concerning the supply-side motivation of farmers to participate in agritourism, alternative research has examined the demand-side, the consumer's evaluation of the agriculture product. Farm-stay visitors have been grouped into distinctive markets based on the activities participated in while staying overnight. Agritourists have been identified as having three distinctive interests: passive recreation, farm-related activities, and active recreation. Many agritourists are interested in the cultural, natural, and family appeal that agritourism visits offer. First impressions usually have the greatest effect on the visitor's overall satisfaction, and visiting a farm is no exception.

An underlying goal of agritourism is to achieve a more balanced approach to tourism. By diversifying the seasonality and high demands of one particular tourism product, a higher quality experience can be achieved. Agritourism also spreads the benefits of tourism to stakeholders beyond the beaches and major cities. Thus, a more significant proportion of the tourism income can be channeled to rural communities. In addition, agritourism protects and conserves the existing natural environment and preserves and strengthens traditional cultures and lifestyles. Agritourism is gathering strong support from small communities as rural people realize the benefits of sustainable development brought about by similar forms of nature travel. Agritourism embraces the aims of rural tourism development.

Further Reading
www.unwto.org/index.php.

See Also: Family Farms; Farmers Markets; Marketing, Direct

Sylvia Smith

Agroecology

The term "agroecology" is a compound of the words "agronomy" and "ecology." Agronomy is the science of agriculture. Ecology is the science and study of the interactions and interrelationships of species and their environment. Agroecology is the study of the ecology of agricultural production systems, and provides the basis for an ecological approach to farming.

Historically, agronomy and ecology have been at odds. The science of ecology is focused almost exclusively on the study and understanding of natural systems. Agronomy on the other hand, has been the application of scientific methods to the development of technologies and practices leading to higher crop and livestock yields. In the 1920s the two fields came together briefly in what was termed "crop ecology." By the 1930s, the field of ecology had become more experimentally devoted to natural systems, and the applied field of crop ecology was left to agronomists. Over time agronomists abandoned the ecological perspective as a more industrial agricultural model took hold. During the 1960s and 1970s interest in applying ecological principles to agriculture increased with growing awareness of environmental issues. By the early 1980s, more and more ecologists came to view agroecology as a legitimate field of study, and more agronomists recognized the value of an ecological, whole-systems approach to studying and understanding agriculture. Since that time, the conceptual framework of agroecology has contributed greatly to the emerging issues of agricultural sustainability.

An agroecosystem—a designated area of agricultural production, understood as an ecosystem—is subject to, and dependent on the same peculiarities of all natural systems, such as climate variation, water and nutrient cycles, and pests and soil dynamics. It has a unique, yet recognizable structure of biotic and abiotic components that interact and function together within the system. As with natural ecosystems, agroecosystems are structured in a hierarchal concept starting with individuals (i.e., crop plant), grouped into populations (i.e., field or plot of the same species), which are part of a community (i.e., collection of the crop plants and all other species), which with other communities make up the local ecosystem. Unlike natural systems, agroecosystems include a strong human management factor working to promote production of harvestable biomass useful for human consumption.

A brief survey of some key ecological concepts lends insight to the chronic problems facing industrial agricultural and provides ideas for farming from an ecological perspective.

SUCCESSION

When land is damaged, nature repairs itself with what ecologists term "ecological succession." It is a trajectory of change leading to ecosystem communities of one or more steady states. The specific details of succession differ depending on location and conditions, weather, and the type of damage imposed on the landscape. Disturbance in ecosystems is normal. Fire, floods, and tornados can destroy plant and animal communities and leave soil bare. In some cases, the restored ecosystem differs from the one that was disrupted.

Ecological succession is a sequence of changes triggered by disruption. The process begins when pioneer species establish themselves within a disrupted ecosystem. These "invader" species are particularly well adapted to the disturbed environment. Over time, the presence of these organisms change conditions within the ecosystem and render it more suitable for longer-living transitional species. The process continues in this manner until an equilibrium is reached at very mature stages of ecosystem development.

Agricultural systems are disturbed ecosystems. Tillage, burning, mowing, grazing, or harvesting removes most, if not all, vegetative cover and even disrupts soil structure on a regular basis. When this occurs, succession must begin again. This intentional disturbance maintains the ecosystem in a perpetual state of immaturity. Systems in early stage succession are highly productive, able to capture the maximum amount of solar energy per unit of land area. Available moisture and nutrients are plentiful. Annual species grow and produce a harvestable portion quickly. Once the annual crop is harvested, the farmer tills the soil and sets the succession process back to square one. Weeds and insect pests are good invader species, too. The battle against weeds and other pests will never be completely won as long as we insist on farming in early-stage, succession systems.

BIODIVERSITY

An ecosystem is said to have greater biodiversity when it contains more species. The richness of any natural system is measured in terms of its biodiversity. When the number of species within a system diminishes through natural or human-induced disruption, it is said to be "impoverished." As ecological succession advances, biodiversity within the system generally increases, and the system becomes more stable. Monocultures (single species grown across large areas) are extremely rare in nature.

Modern intensive agriculture's effect on an ecosystem is primarily to impoverish it in terms of biodiversity. By design, agriculture purposely eliminates the diversity of species within the system. Consequently, nutrient cycles and energy flows are affected, water and soil quality diminished, and the long-term stability of the system is undermined.

Diversity increases as succession progresses. Over time, species find niches within ecosystems wherein they can become established and function at a particular point within the system's food web. As the ecosystem moves towards equilibrium, it is characterized by the lack of any single species able to gain dominance that threatens the health and balance of the whole. Every species and organism is kept in check by one or more other species. At an optimum level of diversity, an ecosystem is kept stable (resistant to disruption) and sustainable.

ENERGY FLOW

In natural systems, energy flows from the sun in the form of light and is captured by primary producers, photosynthesizing plants. Stored energy in plants then flows to herbivore species. Stored energy in herbivores flows to carnivores. A portion of the energy captured by the system is passed to decomposers at each stage of the food chain. Another portion of that energy is stored in the tissue of species persisting in the system.

In agroecosystems, crops are arranged in the field to capture solar energy and convert it to the harvestable portion of the crop. To accomplish this, additional energy is needed to till, plant, weed, spray, and harvest the crop. This additional energy comes from outside the system in the form of human labor and carbon-based fuels such as gasoline and oil. These outside sources of energy are necessary, but the system's dependence on external inputs renders it less sustainable than a natural system operating solely on energy captured from the sun during the growing season.

An ecologically based farming system is less dependent on nonrenewable energy and is designed to maintain a balance between energy required to maintain the processes of the system and that removed from the system by harvest.

NUTRIENT CYCLING

Nutrients within a natural system cycle in a manner similar to energy. Plants take up nutrients from the soil. From there, nutrients flow to herbivores, carnivores, and decomposers. Decomposers return nutrients to soil in forms that can be taken up and used again by plants. In a mature ecosystem, nutrients flow in a closed loop, with a relatively small portion lost.

To increase yields in agroecosystems, additional nutrients from outside sources are added as inputs. Enhanced nutrient levels allow for higher yields, but because of the simplicity of the system and the lack of filled niches, agroecosystems are typically "leaky," losing nutrients via leaching. The leakiness of agroecosystems requires additional outside inputs, continuing the cycle. In addition, the lost nutrients can accumulate and cause imbalances in downstream ecosystems.

Sustainable farming systems based on sound ecological principles are designed to close nutrient loops, minimizing outside inputs and losses. Building soil health increases its capacity for water and nutrient storage and reduces its loss of nutrients by leaching. Creating more ecologically complex farming systems (i.e., species diversity) also reduces leakiness and increases storage of nutrients and energy in the system within living biomass present in systems year round.

POPULATION REGULATION

Due to the structural and trophic complexity found in natural ecosystems, species populations are typically self-promoting and self-regulating. Populations are kept in balance by the carrying capacity of the habitat and by interactions with other species. High levels of biodiversity provide stronger competition for resources and narrower trophic niches within the system. Each species plays a role in the ecological functioning of the system, one of which may be to control the population of a target food species.

Agroecosystems, on the other hand, are maintained at a far lower level of biodiversity, many times as monocultures (single-crop species across a large area). As a result, many ecological functions, including population regulation, must be maintained by labor, equipment, and chemical inputs. Farmers maximize the population of the desired crop species and minimize or eliminate the populations of species that compete with the crop for resources or threaten it with herbivory or disease. The annual disruption of the agroecosystems creates an early-succession environment in which opportunistic species naturally thrive. It is no coincidence that these species types are agriculture's most pervasive pests.

Practices that increase agroecosystem biodiversity such as strip-cropping, intercropping, and cover crops, can, if managed properly, restore a measure of ecological function to the system and reduce the need for external inputs.

Further Reading

Altieri, Miguel. *Agroecology: The Science Of Sustainable Agriculture*, 2nd ed. Boulder, CO: Westview Press, 1995.

Altieri, Miguel. Agroecology in Action. 2004. www.agroeco.org.

Gliessman, Stephen R. Agroecology. 2005. www.agroecology.org.

Gliessman, Stephen R. *Agroecology: The Ecology of Sustainable Food Systems*, 2nd ed. Boca Raton: CRC Press, 2006.

Jackson, D. L., and L. L. Jackson, eds. *The Farm as Natural Habitat*. Washington, D.C.: Island Press, 2002.

Rickerl, D., and C. Francis, eds. "Agroecosystem Analysis." Agronomy No. 43. American Society of Agronomy. 2004.

Ryszkowski, Lech, ed. *Landscape Ecology in Agroecosystems Management.* Boca Raton FL: CRC Press, 2002.

Shiyomi, M., and H. Koizumi, eds. *Structure and Function in Agroecosystem Design and Management.* Boca Raton, FL: CRC Press, 2001.

See Also: Agriculture, Organic; Agriculture, Sustainable; Biodiversity; Environmental Issues; Compost; Insects, Beneficial; Integrated Pest Management; Intercropping; Pest Control, Biological; Pesticides; Soil Health; Vermiculture

Dan Anderson

Animal Welfare

In the United States, there has been an increasing public concern for the humane treatment of animals in agriculture, including the way they are raised, transported, and slaughtered. Corresponding to this has been an increase in the market for alternative meat products (such as "free-range" and "grass-fed"), providing growing opportunities for farmers using alternative livestock production practices.

Much of the public concern over animal welfare for farm animals is in response to the intensification of the livestock sector over the last fifty years and the growing number and concentration of livestock confinement operations and animals being processed at slaughtering plants. Particular farm animal production practices have especially created public protest—for example, the crating of veal calves, the caging of layer hens, the debeaking of poultry, and the slaughter of "downer" animals or animals that cannot walk due to injury or sickness.

In the United States, there are no federal laws regulating the treatment of animals on farms. Although farm animals are regulated under the Animal Welfare Act (AWA), they are only regulated when used in biomedical research, testing, teaching, and exhibition; those used for food and fiber or for food and fiber research are not included. At the state level, few states have on-farm standards for farm animals, although in some states voters have approved ballot measures to outlaw some practices (e.g., Arizona's ban of veal stalls). At the processing level, the Humane Methods of Slaughter Act regulates the humane slaughter and handling of livestock (excluding poultry) at processing plants.

Public pressure regarding animal welfare for farm animals has led to some changes at the market level. Many retailers, such as Whole Foods, Safeway, and even fast food outlets like Burger King, have established animal welfare standards for their suppliers, although these standards vary significantly among different retailers. Retail and industry groups have also been active in developing standards. In addition, third-party certification programs have been developed by groups such as the American Humane

Association (American Humane Certified) and the Animal Welfare Institute (Animal Welfare Approved) in an effort to convey animal welfare attributes to consumers.

Smaller-scale and sustainable or organic farmers are well placed to take advantage of these marketplace changes, and the market for niche meat products with humane treatment attributes seems to be growing rapidly. For example, organic meat and poultry products (often regarded by consumers to have these attributes) are currently the fastest-growing organic sector, with annual average growth rates over 40 percent from 2004 through 2007. Another indication of the growing market for alternative livestock products is that the USDA's Agricultural Marketing Service instituted a voluntary standard for a "grass-fed" livestock marketing claims for ruminants in 2007. In general, consumers are looking for farming systems that allow outdoor access for livestock (e.g., free-range, nonconfinement) and environments that allow for animals' natural behaviors. Farmers are generally able to obtain a price premium for these products, and start-up costs for these systems are generally lower than for conventional practices. In addition, many sustainable livestock systems can reduce some costs such as feed, although management and other costs can be higher.

Further Reading

Animal Welfare Information Center of the National Agriculture Library. http://awic.nal.usda.gov/nal_display/index.php?info_center=3&tax_level=1.

"Appropriate Technology Transfer for Rural Areas." National Center for Appropriate Technology. 2008. http://attra.ncat.org/livestock.html.

Becker, Geoffrey. *Humane Treatment of Farm Animals: Overview and Issues.* CRS Report for Congress. Washington, D.C.: Congressional Research Service, 2008.

See Also: Factory Farming; Free-Range Poultry; Vegetarian Diet

Lydia Oberholtzer

Antibiotics and Livestock Production

The discovery and development of antibiotics has been considered by many to be the greatest advance in the history of medicine. Antibiotics are used to therapeutically treat both humans and animals. In addition, antibiotics are used subtherapeutically in livestock production systems for disease prevention and growth promotion. Subtherapeutic livestock antibiotics are generally administered as a dry mix with the animal feed. A typical dose ranges from 1 to 200 grams of antibiotic per ton of feed, depending on the antibiotic and animal.

Antibiotics are generally poorly adsorbed in the stomach of animals, resulting in substantial excretion. Although highly variable, as much as 70 to 90 percent of the antibiotics administered to animals are excreted in manure. Antibiotic concentration in manure can vary from parts-per-billion to parts-per-million levels. The vast majority of manure is applied to agricultural lands as a nutrient source for crop production.

Although the use of antibiotics has decreased animal mortality rates, it has raised significant health and environmental risks. These include (1) antibiotic residues in animal products, (2) antibiotic residues in plant products, (3) aquatic contamination from land application of manure, and (4) antibiotic resistance in the environment.

ANTIBIOTIC RESIDUES IN ANIMAL PRODUCTS

Antibiotics are known to accumulate in animal tissues. To ensure food safety, the United States Food and Drug Administration set Maximum Residue Limits (MRLs) for antibiotic compounds in animal tissues. These limits vary according to the specific antibiotic compound, animal species, and type of tissue. However, antibiotics account for a significant portion of drug and chemical residue violations in the United States, which are mainly a result of failure to observe withdrawal times, drug administration errors, and dosage errors. Potential adverse effects of antibiotic residues in animal products include allergic/toxic reactions (rare), chronic effects as a result of prolonged low-level exposure, development and spread of antibiotic resistant bacteria, and disruption of digestive system functioning.

ANTIBIOTIC RESIDUES IN PLANT PRODUCTS

Antibiotics that are land-applied with manure can persist in soil from a few days to several hundred days depending on the chemical characteristics of the antibiotic and the environmental conditions. These antibiotics can be toxic to plants resulting in stunted growth. Antibiotics have also been shown to bioaccumulate in plant tissues when food or vegetable crops are produced on soil amended with antibiotic-containing manure. This raises the potential of veterinary antibiotics contaminating the food supply. Health risks associated with consumption of contaminated plant products are similar to those associated with consumption of contaminated animal products.

AQUATIC CONTAMINATION FROM
LAND APPLICATION OF MANURE

Antibiotics in manure can contaminate surface and ground water, by runoff and leaching from manure-applied land. Although antibiotics are generally tightly adsorbed to soil particles, concentrations in the part-per-billion range have been detected across the United States in rivers, lakes, groundwater, and drinking water supplies. However, management practices such as timing of manure application and type of tillage can help mitigate these risks. Although antibiotic transport from agricultural land is generally less than 10 percent of the amount in manure, low-level contamination of aquatic environments can promote the development and spread of antibiotic resistant bacteria in the environment.

ANTIBIOTIC RESISTANCE IN THE ENVIRONMENT

The main concern in the usage of antibiotics in livestock production is the growing global problem of antibiotic resistance in the environment. Low-level antibiotic use results in the natural selection of antibiotic resistant bacteria. These antibiotic resistant bacteria in turn are a substantial human health threat because they can impart their resistance to human pathogens, thus rendering many of the human antibiotics ineffective. This results in significant increases in health care costs either in finding alternative treatments or development of new antibiotics. A large number of the antibiotics used for livestock production are either identical or similar to human antibiotics. Antibiotic resistant bacteria that have developed in animals may also contaminate food products and thus transmit the resistance through the human food chain. Antibiotics applied to soil with manure can also increase the antibiotic resistance of bacteria in soils, which in turn can spread to surface and ground water through leaching and runoff from agricultural land.

Further Reading

Gustafson, R. H., and R. E. Bowen. "Antibiotic Use in Animal Agriculture." *Journal of Applied Microbiology* 83: 531–41.

Kumar, K., S. C. Gupta, Y. Chander, and A. K. Singh. "Antibiotic Use in Agriculture and Its Impact on the Terrestrial Environment." *Advances in Agronomy* 87: 1–54.

Kümmerer, K. "Significance of Antibiotics in the Environment. *Journal of Antimicrobial Chemotherapy* 52: 5–7.

National Research Council. *The Use of Drugs in Food Animals: Benefits and Risks.* Washington, D.C.: National Academy Press, 1999.

See Also: Agrichemicals; Factory Farming; Growth Hormones and Cattle

Holly A. S. Dolliver

Antioxidants

Several studies show that organic foods have higher levels of key antioxidants than foods grown with chemicals. Although oxidation is an important process for sustaining life, it produces free radicals ("hydroxyl radical" or "superoxide anion" or "hydrogen peroxide"), which cause chain reactions in the body that eventually damage the cells. Compounds known as "antioxidants" can prevent this destructive oxidation of other molecules in the body. Antioxidants stop the chain reaction by removing these free radicals and by stopping other oxidation reactions.

Oxidative damage is associated with chronic conditions such as rheumatoid arthritis, Alzheimer's disease (AD), Parkinson's disease (PD), and the pathologies caused by diabetes. Many studies report benefits of antioxidants in preventing heart disease, neurological diseases, macular degeneration, and even some cancers. Oxidation of low density lipoprotein (LDL) can cause atherosclerosis, and lipid peroxidation damage to DNA can cause cancer.

Eating fruits and vegetables high in antioxidants can protect us from exposure to these chronic diseases. Different types of antioxidants distributed throughout foods work together as an "antioxidant system" in the human body. Major antioxidants are vitamin C (found in citrus fruits and other fruits and vegetables), vitamin E (found in vegetable oil, salad dressings, nuts, and avocados), vitamin A as carotenoids (found in green and yellow vegetables), and phytochemicals such as flavonoids (found in citrus and other fruits), lycopene (found in tomatoes, grapefruit, and watermelon), quercetin (found in tea, onion, apple, red wine), and glucosinolates (found in broccoli and kale). Antioxidants such as vitamin C react with oxidants present in the cell cytoplasm and blood plasma, whereas the lipid-soluble antioxidants like vitamin E and carotenoids protect the cell membranes against lipid peroxidation. Together, they prevent oxidative damage to DNA, proteins, and lipids.

Fruits and vegetables grown organically typically have higher levels of antioxidants compared with those grown conventionally. Many studies have found organically produced corn, strawberries, blackberries, and tomatoes have significantly higher levels of antioxidants than nonorganic varieties. For example, the antioxidants quercetin and kaempferol in organic tomatoes are reported to be 79 and 97 percent higher than conventional tomatoes. It is thought that herbicides and pesticides may prevent plants from producing certain antioxidants such as flavonoids, which the plant produces in response to environmental stressors like insects or competing plants. Another reason for higher levels of antioxidants in organic produce may be that organically produced fruits are picked when they are ripe and at the stage when they can absorb trace vitamins and minerals. On the other hand, conventionally produced fruits are picked when they are still green and lack these trace vitamins and minerals. It is recommended to eat fresh organic fruits and vegetables whenever possible because some antioxidants may be lost during processing.

People who do strenuous exercise and take antioxidant supplements to prevent oxidative damage could benefit from eating organically produced fruits and vegetables instead. For example, one organic tomato has the nutritional equivalent of four conventionally produced tomatoes. Thus, a person could get high-quality nutrition from and maybe even save money by eating one large organic tomato instead of taking antioxidant supplements, which can cost over $1.00 per capsule.

Further Reading

Grosvendor, Mary, and Lori Smolin. *Nutrition: Everyday Choices*. New York: Wiley and Sons, 2006.

ShapeFit.com. "Organic Foods—Natural Foods with No Preservatives or Fertilizers." 2009. http://shapefit.com/organic-foods.html.

See Also: Health Concerns; Organic Foods, Benefits of

Sharon Peterson

B

Bacillus thuringiensis (Bt)

Bacillus thuringiensis (Bt) is a spore-forming, Gram-positive bacterium that is found naturally in soils. Since its discovery in 1901, strains of Bt have been found to be effective in the control of some insect larvae pests in crops. Thus, Bt is an important natural method of insect control that is used by organic farmers.

When sprayed on leaf surfaces, Bt spores are ingested by larvae that feed on the leaves. When enough Bt has been consumed by the larvae, crystalline protein toxins produced by the Bt works to paralyze the mouth and gut, causing the larvae to cease feeding. The toxins continue to break down the gut, eventually allowing Bt spores and the normal gut flora to invade the body cavity, causing death within twenty-four to forty-eight hours. Bt products are nontoxic to humans, fish, and animals and do not persist in the environment; they are also very specific in targeting certain types of insects, with few reported effects on beneficial insects. Several Bt products, such as DiPel, are approved for use in certified organic systems. Different subspecies are available for the control of different pests. Bt *kurstaki* and Bt *aizawai* are commonly used for control of butterfly and moth caterpillars in cabbage and other susceptible vegetables; Bt *tenebrionis* is used for control of Colorado potato beetle. In sweet corn, a method for control of corn ear worm has been developed in which Bt *kurstaki* is mixed with vegetable oil and injected into the silks of developing sweet corn ears using a device called the Zea-later™. Resistance to Bt has been documented in some pests—which is the outcome of all chemical- or toxin-based control methods. For this reason, continuous applications of Bt should be avoided, early detection of insect problems should be pursued, and Bt should be used in conjunction with complementary control methods. Best results are achieved when insect larvae are small and economic thresholds have not yet been reached. Products may break down in direct sunlight, and rain can wash Bt from leaves; therefore, sprays should reach both sides of the foliage and should be reapplied after heavy rain.

In addition to the use of Bt as a direct form of insect control in crops, gene sequences from Bt have been used to create genetically modified crops, such as corn, potato, and cotton, which express the bacteria's toxic proteins in plant tissue, killing insect larvae without the need for additional pesticides. Although reducing the need for chemical pesticides in large-scale production of these crops, the use of Bt crops is controversial because of concerns about cross-pollination, resistance, and safety. Genetically modified crops cannot be used in certified organic systems, and some regions of Europe have banned their use.

Further Reading

Rowell, Brent, and Ricardo Bessin. "Bt Basics for Vegetable Integrated Pest Management." University of Kentucky Cooperative Extension Service, 2008.

See Also: Integrated Pest Management; Pesticides

Mark Williams and Audrey Law

Bacillus thuringiensis (Bt) Corn

Although genetically engineered (GE) crops are not allowed in certified organic production, GE crops now make up most of the conventional production of some crops, most notably corn and soybeans, in the United States. *Bacillus thuringiensis* (Bt) corn is a genetically engineered variety designed to protect plants from the European corn borer, a caterpillar that causes millions of dollars in crop loss annually in the United States. Through recombinant DNA technology, the DNA from naturally occurring Bt bacteria is inserted into the corn's genetic structure, allowing the plant to express the bacterium's pesticide in every cell. Pests such as the corn borer perish within minutes of consumption, thereby protecting crops from damage and preserving yields.

The first Bt plant, tobacco, was developed by the Belgian company, Plant Life Science, in 1985. Since then, the technology has been used in cotton and potatoes. Bt corn was introduced jointly by Ciba Seeds (now

Novartis) and Mycogen Seeds in 1996. Several companies, including Monsanto Corporation have introduced their own Bt corn hybrid lines.

Along with Roundup ready soybeans, Bt corn and cotton account for the largest acreage of genetically engineered crops worldwide. In 2008, 57 percent of the U.S. corn harvest was planted with Bt varieties.

The environmental and economic advantages of Bt corn are highly disputed. Supporters of the technology argue that endogenous pesticide production reduces the frequency of broadcast pesticide spraying, thereby protecting human and ecological health. Protecting yields and reducing chemical inputs is also claimed to increase farmer income. There is some evidence suggesting that synthetic pesticide use has decreased with the introduction of Bt varieties. However, there is little evidence supporting the claims that genetically engineered crops improve financial returns to farmers. On the contrary, patent rights on Bt seeds strip users of the right to save seeds and render them vulnerable to legal challenges by seed firms. Initial improvements in yields have dropped off as insect populations develop resistance to Bt toxins—a fact that is particularly problematic to organic farmers who rely on Bt sprays.

While Bt corn is specifically engineered to protect against the European corn borer, the pesticide is also toxic to many other Lepidoptera species (moths and butterflies), as well as Diptera (flies and mosquitoes) and Coleoptera (beetles). Critics of the technology fear that widespread use threatens nontarget insect populations. In 1999, a group of Cornell scientists led by John Losey published findings of a preliminary study on the effects of Bt toxin on Monarch butterflies. The short article, printed in *Nature*, indicated that pollen from Bt corn could kill larvae if it landed on milkweed plants, the caterpillar's primary food source. The study sparked an international scientific controversy that has spurred thousands of published articles and has linked the Monarch butterfly to the antibiotechnology movement. Such opposition to biotech crops has fueled increased interest in organic farming, which prohibits the use of genetically modified seeds and inputs.

Further Reading

Kloppenburg, Jack. *First the Seed: The Political Economy of Plant Biotechnology, 1492–2000.* 2nd ed. Madison: University of Wisconsin Press, 2004.

Losey, John, Laura Rayor, and Maureen E. Carter. "Transgenic Pollen Harms Monarch Larvae." *Nature* 399 (1999): 214–15.

National Agricultural Statistics Service. 2008 Acreage Report. U.S. Department of Agriculture Web Site. 2008. http://usda.mannlib.cornell.edu/usda/current/Acre/Acre-06-30-2008.pdf.

Tokar, Brian, ed. *Redesigning Life: The Worldwide Challenge to Genetic Engineering.* London: Zed Books, 2001.

See Also: *Diamond v. Chakrabarty*; Genetically Engineered Crops; Roundup Ready Soybeans; Substantial Equivalence; Terminator Gene

Robin Jane Roff

Balfour, Lady Eve (1898–1989)

Lady Eve Balfour was one of the early leaders of the organic movement in the United Kingdom. She is best known for the Haughley Experiment, the classic book *The Living Soil*, and for cofounding the Soil Association. She was the fourth child of the second Earl of Balfour, Gerald William Balfour, and Lady Elizabeth Edith Balfour. Balfour graduated from the University of Reading as one of the first women to study agriculture. Balfour began farming in 1919 on New Bells Farm, Haughley Green, Suffolk, England.

In the 1930s, Balfour became critical of conventional farming methods. Leading critics of agriculture and its effect on human health, including Gerard Vernon Wallop, ninth Earl of Portsmouth, Lionel Picton, Sir Robert McCarrison, and Sir Albert Howard, influenced her ideas and concerns about agriculture. *Famine in England* (Wallop, 1938), *Medical Testament* (Picton, 1939) and *Agricultural Testament* (Howard, 1942) questioned the sustainability of agriculture and linked soil conditions to human nutrition and health.

In 1939, Balfour established a long-term study at Haughley Green comparing nutrition and productivity of organically grown and conventionally grown crops. It was the first study to compare organic and "chemical-based" farming systems, side-by-side. The study compared: arable farming with no animals; a "Mixed Section" consisting of pasture alternated with crops treated with synthetic fertilizers and pesticides; and an "Organic Section" that did not use synthetic fertilizers or pesticides. Both the "Mixed Section" and "Organic Section" had dairy cows, poultry, and sheep feeding on the pastures, and all crop and animal wastes were returned to the land. Balfour found similar productivity in the cropping systems and very little of the synthetic fertilizer was used by the plants. The soils, crops, and animals were best considered as one organic whole. In 1943, Balfour published *The Living Soil* based on the first four years of the Haughley Experiment and presenting the case for organic agriculture.

"Contrary to the views held by some, I am sure that the techniques of organic farming cannot be imprisoned in a rigid set of rules. They depend essentially on the outlook of the farmer. Without a positive and ecological approach it is not possible to farm organically. The approach of the modern conventional farmer is negative, narrow, and fragmentary, and consequently produces imbalance. His attitude to 'pests' and 'weeds,' for example, is to regard them as enemies to be killed—if possible, exterminated. When he attacks them with lethal chemicals he seldom gives a thought to the effect this may have on the food supply or habitat of other forms of wildlife among whom he has many more friends than foes. The predatory insects and the insectivorous birds are obvious examples. The attitude of the organic farmer, who has trained himself to think ecologically, is different. He tries to see the living world as a whole. He regards so-called pests and weeds as part of the natural pattern of the Biota, probably necessary to its stability and permanence, to be utilized rather than attacked. Throughout his operations he endeavors to achieve his objective by cooperating with natural agencies in place of relying on man-made substitutes. He studies what appear to be nature's rules—as manifested in a healthy wilderness—and attempts to adapt them to his own farm needs, instead of flouting them.

Besides biological balance, the ecologically minded organic farmer takes note of, and tries to apply, other

In 1946, Balfour cofounded and was the first president of the Soils Association, which promoted and developed information about organic agriculture. The Soils Association conducted the Haughley Experiment from 1947 until 1969. As she got older, Balfour faded into the background of the Soils Association, retiring in 1984. Balfour's lifetime contributions to organic agriculture were recognized with an Order of the British Empire shortly before her death in 1989. Her legacy, the Soils Association, is at the forefront of organic agriculture in the United Kingdom. It works to promote sustainable and organic farming through development of organic standards and by research and sharing knowledge about soil fertility. In a fitting tribute, the day after Balfour's death, the British government announced the first grant available for farmers to convert from conventional to organic farming methods.

apparent biological roles. For example, nature's diversity of species he adapts through rotations, under-sowing, and avoiding monoculture of crops or animals. Nature's habit of filtering sunlight and rain through some form of protective soil cover, he adapts by such practices as cover-cropping and mulching. Top soil on the top appears to be nature's plan. Organic matter is always deposited on the surface. It is left to the earthworms and some insects to take it below. The organic farmer also puts his compost and farmyard manure on, or very near, the surface and in carrying out mechanical cultivations keeps soil-inversion to a minimum, the tine cultivator being preferred to the plough."

—Lady Eve Balfour, researcher and proponent of organic farming methods. From her address to the 1977 International Federation of Organic Agriculture Movements (IFOAM) Conference, Sissach, Switzerland.

Further Reading

Balfour, Lady Eve. *The Living Soil*. London: Faber and Faber, 1943.

Conford, Philip. "Eve Balfour: The Founder of the Soil Association." www.soil association.org.

Conford, Philip. *Origins of the Organic Movement*. Edinburgh, U.K.: Floris Books, 2001.

Lauren Flowers and John Masiunas

Bees

The movement of pollen from anther to stigma—pollination—is central to the lifecycle of flowering plants. Although some plants can produce seed (and thus fruit) through self-pollination, the majority of flowering plant species—more than 80 percent—rely on animal pollinators to carry pollen grains from flower to flower.

Bees are the most important group of pollinators. With the exception of a few wasp species, only bees deliberately gather pollen to bring back to their nests as food for their offspring. Because of this habit, bees are covered with thick, pollen-collecting hairs, giving them a characteristically fuzzy appearance. In the process of collecting this food source, some pollen

Bee on echinacea flower. Courtesy of Jon Bathgate.

is accidentally rubbed off of the bee's body, and pollen grains are thus transferred between flowers. During a single foraging trip, a female bee may visit hundreds of flowers, transferring large amounts of pollen along the way.

Roughly 35 percent of all crop species grown in the world are insect pollinated. In North America, bees were responsible for roughly $20 billion in agricultural production in 2000. Most of these crops are pollinated by managed hives of the European honey bee *(Apis mellifera)*. However, the number of managed honey bee hives is declining because of diseases, pests, aggressive strains of honey bees, and a new, little-understood syndrome called Colony Collapse Disorder.

In the past, native bees and feral honey bees could meet all of a farmer's pollination needs for orchards, berry patches, squash, melons, vegetable seed, sunflowers, and other insect-pollinated crops. These farms were relatively small and close to areas of natural habitat that harbored adequate numbers of wild pollinators to accomplish this task.

Today, however, many agricultural landscapes are much more extensive and lack sufficient habitat to support pollinators. Subtle changes in farming practices, such as identifying and protecting nest sites, and providing alternative forage sources, can increase the number of native bees in farm landscapes. The reduction of pesticides, and the implementation of less intensive, more sustainable agricultural practices, such as the adoption of organic farming practices, can also be beneficial.

The value of these efforts is supported by ongoing research conducted across North America, which has demonstrated that native bees can still play an important role in crop pollination and can help fill the niche left by the decline of honey bees. The value of these native bees to U.S. agriculture is currently estimated to be about $3 billion per year.

North America is home to about 4,000 species of native bees, ranging from tiny miner bees less than a eighth of an inch long to bumble bees bigger than an inch. The majority of these insects are overlooked because they are docile, nonaggressive, and may not look like stereotypical bees.

About 70 percent of these native bees nest underground as solitary individuals, with an individual female bee excavating and provisioning her brood. Almost 30 percent of native bees nest as solitary individuals in pre-existing wood cavities, such as those made by wood boring insects, or in the hollow stems of pithy twigs. Only bumble bees form social colonies that include a queen and her daughter-workers. Bumble bees often nest in abandoned rodent burrows or below grass tussocks.

Because of their importance in pollination, the health of bees is critical to agricultural production. Farming systems, such as organic methods, that reduce pesticides are beneficial to the long-term survival of bee species.

Further Reading

Buchmann, Stephen, and Gary Nabhan, *The Forgotten Pollinators*. Washington, D.C.: Island Press, 1997.

O'Toole, Chris, and Anthony Raw, *Bees of the World*. Facts on File, 1991.

Shepherd, Matthew, Stephen Buchmann, Mace Vaughan, and Scott Hoffman Black, *Pollinator Conservation Handbook*. Portland, Oregon, Xerces Society, 2003.

The Xerces Society. www.xerces.org.

See Also: Insects, Beneficial

Eric Mader

Biodiversity

In general, the term "biodiversity," which is a shortened form of the term "biological diversity," refers to the variety of plant and animal life found in a given ecosystem or region. Organic and sustainable farms tend to have higher biodiversity because of their complex cropping systems; rather than growing just one or two crops, they grow many different crops at the same time or in a rotation.

The term "biodiversity" is more or less equivalent with "species richness." More specifically, the term encompasses not only the diversity of species, but also genetic variability within species, and the diversity of

Songbird eating seeds on a sunflower. Courtesy of Jon Bathgate.

habitats or ecosystems within a given area. Knowledge of the actual number of species on earth is quite crude, with estimates ranging over an order of magnitude from perhaps ten million to one hundred million species, with insects and bacteria representing the greatest percentage of the estimates.

Biodiversity is generally highest in tropical regions (tropical forests account for more than half the total number of species on the globe) and decreases towards the poles. This is a result of the influence of both decreasing average absolute temperatures and greater temperature variability in the temperate and polar regions. High biodiversity is often found in unusual environments with unique associations of climate, soils, and other variables. Unique habitats also often have high numbers of endemic species: that is, species found only in that unique environment. Areas with high biodiversity and high levels of endemism are often referred to as "hotspots," though this term more specifically refers to such areas that are also undergoing threatening levels of habitat destruction. Many of the world's tropical islands are hotspots because of their unique geography, and the study of island biogeography by the imminent ecologist E. O. Wilson and his colleagues led to much of the current understanding of the relationship between habitat and diversity. Of the many indices used by scientists to express values of biodiversity, two of the most common include Simpson's index and the Shannon index, which account for both species richness (relative abundance of different species) and species evenness (numbers of individuals per species). Rareness is also an important and not completely straightforward concept in the study of diversity.

Leaving aside the intrinsic value of species diversity, humans are also entirely dependent on the earth's natural diversity for basic needs. In addition to obvious benefits such as the diversity of plant- and animal-based food, plant-based medicine, for example, has become a multibillion dollar industry and saves tens of thousands of lives each year in the United States alone. The importance of biodiversity has also become apparent to the travel industry. Ecotourism, for example, is a global, rapidly growing, multi-hundred-billion dollar pastime. In addition to material goods and services, ecosystems provide literally trillions of dollars worth of ecosystem services, which could be threatened by loss of biodiversity at regional and global scales. In total, ecologists have estimated that the dollar value of biodiversity ecosystem services exceeds that of the global economy, perhaps by more than $30 trillion per year.

There is general agreement in the scientific community that biodiversity is greatly threatened. Estimates of species extinction rates vary greatly, but the consensus is that human activities are currently causing species extinctions faster than at any time in history, at a rate hundreds or even thousands of times greater than the background rate. Important threats to biodiversity include overharvesting or overconsumption by humans, competition from introduced invasive species, anthropogenic alterations of various biogeochemical cycles (e.g., carbon dioxide and nitrogen), direct habitat destruction in the form of conversion (e.g., from natural forest to pasture, cropland, or forest plantation), and combinations of all of these.

Examples of overharvesting to the point of extinction of even some of the most abundant species are commonplace throughout the human record. There is still scholarly debate as to whether human harvesting of Pleistocene megafauna, in the Americas for example, directly led to their extinction, but there is clear evidence that this occurred on many islands. More recent popular examples include the demise of the Passenger Pigeon, once the most common bird in North America, hunted to extinction for cheap meat within the span of a century, or the near-extinction of the Black Rhino, killed for its horns by poachers. Many current overharvesting issues are focused on marine resources, particularly fish stocks, which are in decline globally. Even once robust fisheries, such as North Atlantic cod or Northern Bluefin tuna are nearing commercial extinction.

Although true for some species, overharvesting is not the paramount threat to biodiversity for all species. Human land use that results in habitat conversion or destruction poses the greatest overall threat to biodiversity. Historically and currently, agriculture is the primary driving force behind land conversion. Estimates of the total area of land conversion by humans for agriculture vary, but it is clear that agricultural land use has had a profound effect on natural habitats since the beginning of human agriculture and continues today at a rapid pace. About one-third of the earth's land surface is cropland and pasture, and most of this area was formerly forested. The bulk of this conversion took place over the past couple of centuries, particularly in the now-developed countries, but conversion continues unabated today in the developing world, especially in Asia. The effects of changing land use include not only direct habitat destruction, but also degradation and fragmentation. In addition, land cover conversion has a profound effect on global biogeochemical cycles (e.g., carbon and nitrogen), which also contribute to biodiversity loss through habitat alteration via both global climate change and alteration of ecosystem functions.

Although not widely perceived by the public, one of the driving forces behind much of the current land conversion to agricultural, is increased meat consumption, particularly in the developing world and especially in Asia. Global meat demand has roughly doubled over the past half century and will likely continue growing at that rate for decades. This demand in turn has driven increased land conversion for the production of crops such as maize and soya that are the main feed crops for livestock. In addition to negative effects on biodiversity as a result of land conversion, agriculture in general, and industrial livestock production specifically, result in an array of additional environmental problems that reduce biodiversity. Like land conversion itself, livestock production is a major element of anthropogenic production of greenhouse gases, such as carbon dioxide, nitrous oxide, and methane, which are predicted to cause enhanced greenhouse warming that is expected to result in the extinction of perhaps as much as 30 percent of the world's species. The production of biofuels as a possible answer to the greenhouse warming issue will also play a role in land-use change and agricultural production. In addition, agriculture, especially livestock production, is a major contributor to the overload of nitrogen in aquatic systems. This results in habitat loss and degradation both to inland aquatic systems as well as coastal marine ecosystems with attendant loss of biodiversity.

Agricultural biodiversity loss has also become a topic of concern in recent years. Many of the world's heritage breeds of plants and animals have been lost because of increased reliance on standard hybrids. Conserving agricultural genetic resources has become an important priority in the agricultural community. There are strong international efforts aimed at developing and improving seedbanks as well as saving endangered livestock breeds. Other agriculturally relevant biodiversity issues include transgenic organisms and intellectual property rights.

The continued growth of the human population suggests that pressure for increased agricultural output will also grow. This will most likely lead to additional land conversion for crops and livestock as well as continued growth of other negative effects on the environment and biodiversity. Other social and economic pressures, such as energy costs and government subsidy policies, will also play a role in the future development of agriculture and its effect on biodiversity. Organic and sustainable agricultural production are more ecologically varied, thus promote biodiversity to a greater extent than large-scale conventional production.

Further Reading

Novacek, M. J., ed. *The Biodiversity Crisis: Losing What Counts.* New York: New Press, 2001.

Pimm, S. L., G. J. Russell, J. L. Gittleman, and T. M. Brooks. "The Future of Biodiversity." *Science* 269: 347–50.

Wilson, E. O., ed. *Diversity.* Washington, D.C.: W.W. National Academies Press, 1988.

Wilson, E. O., *The Diversity of Life.* New York: Norton, 1992.

See Also: Agroecology; Buffers; Fossil Fuel Use; Seeds, Extinction of; Vegetarian Diet

Matthew D. Therrell

Brands

The organic industry is increasingly dominated by large corporations. By 2007, fourteen of the twenty largest food processors in North America had either acquired organic brands or introduced organic versions of their own brands. Just a decade before, only one, General Mills, sold organic foods. Larger food processors have been attracted to the industry by consumers' willingness to pay higher prices for these foods, and sales growth rates that have increased approximately 20 percent annually since 1990. In contrast, the rest of the food industry's sales grew an average of less than 4 percent annually during this time. Another event that motivated corporate involvement was the introduction of a national organic standard overseen by the U.S. Department of Agriculture; it replaced differing state and regional standards, beginning in 2001. A uniform standard made it easier for large corporations to market and distribute organic foods nationally, as well as internationally.

ACQUISITIONS

One strategy large food processors use to enter the organic industry is to acquire an existing organic brand. Examples include Coca-Cola's buyout of Odwalla, General Mills' takeover of Cascadian Farm, and Dean's purchases of Horizon, Alta Dena, and White Wave/Silk. Hain Celestial Group, which at one time was 20 percent owned by Heinz, has acquired more than twenty organic brands. Some of these include Walnut Acres, Health Valley, Spectrum, Garden of Eatin', ShariAnn's, Millina's Finest, Earth's Best, DeBole's, Bearitos, Westbrae, and Imagine/Rice Dream/Soy Dream. By 2007, Hain Celestial Group had become the eighty-fifth largest food

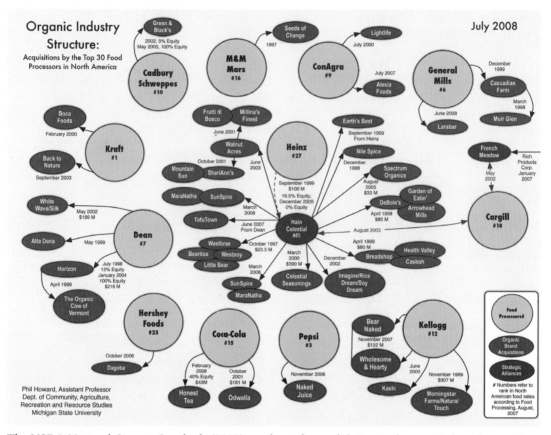

The USDA National Organic Standards (2002) accelerated consolidation in the organic brand sector, but few companies identify these ownership ties on product labels. Courtesy of Phil Howard.

processor in North America by sales, which amounted to nearly $1 billion annually.

Sale prices for these acquisitions are typically much higher than average for the food industry. These inflated prices reflect the strong motivation of some of the largest food processors in North America to become involved in the profitable and rapidly growing organic sector. Buying an established company that already has brand recognition is often easier than introducing a new brand. This is particularly true for organic foods—finding a stable supply of certified organic ingredients for food products can be more difficult than for undifferentiated foods because they are grown on a much smaller number of farms.

Acquisitions of most pioneering organic brands occurred between 1997, when a draft U.S. national organic standard was released, and 2002 when the standard was fully implemented. By the end of this period there were few independent organic brands with national distribution remaining as potential takeover targets. More recent acquisitions, since 2006, tend to involve companies that started in 2001 or 2002, the period when the national organic standard was phased in. Examples include Dagoba (chocolate) by Hershey's, Bear Naked (granola) by Kellogg, Larabar (energy bars) by General Mills, and Alexia Foods (frozen foods) by ConAgra.

Investment firms, or venture capitalists, often play a catalyzing role in these acquisitions. They invest money in an organic brand to help expand distribution and sales, in return for a share of ownership in the business. Investors expect to make a profit by selling the business to a large food processor, typically within seven years. Some investment firms, recognizing corporate interest in dominating sales for a given type of product, are engaging in a strategy of acquiring multiple organic brands in the same category. The Charterhouse Group, for example, acquired Rudi's Bakery, Adams Baking, and Vermont Bread Company in 2005. American Capital Strategies Ltd. has acquired several organic chocolate and candy companies, and helped another investment firm, Booth Creek Management Corporation, acquire several organic meat companies.

Investment firms describe the structure of the organic industry as fragmented, because a large number of companies are currently competing with each other for market share. They see opportunities to make large profits by consolidating the industry, so that it is controlled by a much smaller number of companies. Consolidated markets are more stable, and with a smaller number of competitors, firms are less likely to compete on the basis of price.

Foreign companies have also made investments in U.S. organic brands. Pulmuone, which is based in Korea, acquired the soyfoods company Wildwood/Midwest Harvest from investment firms. French company Danone, makers of Dannon yogurt, acquired 85 percent of pioneering organic yogurt company Stonyfield Farm. An Irish firm, the Kerry Group, bought Oregon Chai. Companies based in the U.S. have acquired foreign organic brands as well, such as Dean's buyout of the United Kingdom's first certified organic dairy, Rachel's.

Most of these acquisitions have remained hidden from consumers. Few acquiring corporations list their ownership ties anywhere on the product label. This has been referred to as "stealth ownership." The practice is motivated by the perception that the most committed organic consumers distrust large corporations. Some retailers, such as Whole Foods, will carry brands with hidden ownership ties, but not organic versions of more mainstream brands produced by these same companies. Representing pastoral or environmental ideals on product labels, despite being owned by gigantic companies, is also common. Dean subsidiary Horizon Organic, for instance, depicts a smiling cartoon cow in front of the planet earth on its milk cartons.

When organic food companies become part of much larger organizations, the minimum USDA guidelines for organic products must still be met. Other values not embodied in these standards may not always be retained, however. Some companies that once bought organic ingredients entirely from U.S. farmers are now importing them to cut costs and ensure stable supplies, such as Dean subsidiary White Wave/Silk. Another possibility is a reduction in commitment to organic in wake of being acquired.

Examples of expansion into nonorganic products after a change in ownership include Dean's Rachel's and Silk brands, Danone's Stonyfield brand, and Coca-Cola's Odwalla brand. Odwalla now uses less than 5 percent organic ingredients in its entire product line.

INTRODUCTIONS

A second strategy large corporations use to enter the organic industry is to develop organic versions of existing products. Processors have introduced dozens of organic versions of very mainstream brands such as Kellogg's Rice Krispies, Nabisco Oreos, and Heinz Ketchup. Although a number of such introductions followed the implementation of a USDA national organic standard in 2002, the majority occurred in 2006 or afterward. Contributing to this trend was Walmart's announced intention in 2006 to greatly expand the number of organic products it sold, and its encouragement of manufacturers to assist in achieving this goal. An organic Dove chocolate bar, for instance, was developed exclusively for Walmart by M&M Mars.

Although most introductions are clearly associated with large food processors, there are exceptions. Anheuser-Busch created a new Green Valley Brewing Company label for its introductions of Wild Hop and Stone Mill beers. Its Web site did not list any ties to its corporate parent when test marketing these beers in 2006. When a story in the *San Francisco Chronicle* pointed out these connections, the Web site was changed to make Anheuser-Busch's ownership much more apparent.

Retailers and distributors have also created organic product lines through private labels or store brands. Private labels may carry the company's name, but it is more common to establish a unique name for organic foods, frequently with pastoral or environmental themes (e.g. Grateful Harvest, Naturally Preferred, Nature's Best, Greenwise). Most of these are co-packed, produced under contract by other companies. An example is Aurora, which was established specifically to supply retailers and distributors with organic dairy products packaged to their specifications. Some of the companies they have contracts with include Walmart (Great Value), Target (Archer Farms), Costco (Kirkland Signature), Safeway (O Organics), Giant (Nature's Promise), and United Natural Foods (Woodstock Farms).

Sales of private label organic food products are currently increasing four times faster than brand name products. Their success is primarily a result of being sold at lower prices than similar, branded foods. Private label products are cheaper because less money is devoted to marketing, and to the fees that some retailers charge to place products on their shelves. Market research indicates that organic consumers tend to be less brand conscious when compared to others, with organic as essentially an umbrella brand. Because organic certifiers cannot require farming practices above the USDA minimum standards, it makes financial sense for consumers to purchase the lowest priced product. Safeway's "O Organics" private label, which was not introduced until December 2005, now includes more than 350 products, and is among the world's best-selling organic brands.

Safeway is one of several distributors and retailers licensing their organic private label products to other companies. O Organics are sold through the distributor Sysco in California, and in stores owned by the French retailer Carrefour in Taiwan. Wild Oats, before being acquired by Whole Foods, distributed its private label organic products through Giant, Stop & Shop, Peapod, Pathmark, and Price Chopper.

INDEPENDENTS

Despite the large number of acquisitions and increasing competition from multinational corporations, some pioneering organic brands have resisted lucrative buyout offers. At least sixteen companies with $15 million or more in annual sales remain independent. Four of these are organized as cooperatives. Equal Exchange and Alvarado Street Bakery are employee cooperatives, Organic Valley is a producer cooperative, and Frontier Natural Products is a wholesaler cooperative. Although other companies are private and could more easily accept an offer to be acquired by another company, their refusal to do so is often based on values that go beyond making money. Arran Stephens, the founder of Nature's Path, for example, has expressed his interest in having his children involved in the business, rather than accepting a buyout. Gary Erickson nearly sold his energy bar company, Clif Bar, for $120 million, or approximately three times annual sales in 2000. He walked away from the deal and risked the possibility of going out of business in order to further noneconomic goals, such as philanthropy, paid time for employees to do volunteer work, and reducing the company's resource use.

Some independent companies go further, and take principled opposition to recent trends in the organic food industry. Eden Foods, for instance, refuses to use the USDA organic seal on its certified organic products, stating that the USDA standard does not approach the company's own high standards. As one example, they point out that the Organic Trade Association successfully lobbied Congress in 2005 to allow synthetic ingredients in food that is labeled with USDA organic seal. Eden Foods is an outspoken opponent of the use of synthetics in organic food. Dr. Bronner's, a soap company, is another critic within the organic industry. They have opposed the use of the word "organic" in nonfood products unless those products follow the USDA organic standards.

Maintaining an independent organic brand can be difficult when competing with much larger corporations. The size of larger companies provide the capacity to spend much more on advertising and retailer fees. Large corporations can also afford to temporarily sell products for less than it costs to make them, particularly if it reduces the sales of their competitors; this practice is referred to as predatory pricing. Although four of the top ten organic brands in 2007 were independent (Organic Valley, Amy's Kitchen, Nature's Path, and Clif Bar), the number of independent organic companies is expected to decline further.

DISTRIBUTION AND RETAILING

Distribution of processed organic brands in the United States is controlled primarily through just two companies, United Natural Foods and Tree of Life. Both companies grew by acquiring competitors, with United Natural Foods buying out more than two dozen distributors beginning in the late 1980s. The natural and organic distribution industry was composed of twenty-eight regional, cooperatively owned businesses in the early 1980s, but by 2004 only three of these remained. Retailing shows a similar pattern, with Whole Foods and Trader Joe's accounting for the largest percentage of organic food sales. Whole Foods is also a product of more than two dozen mergers and acquisitions, most recently acquiring its largest competitor, Wild Oats, in 2007. The dominance of these firms is declining, however, as mainstream supermarkets continue to expand their organic offerings, often bypassing distribution companies and sourcing directly from manufacturers. Establishing new distribution and retail channels is a key challenge for independent brands looking to grow beyond current markets, which is made even more difficult by consolidation in these sectors.

Further Reading

Erickson, Gary. *Raising the Bar: Integrity and Passion in Life and Business.* San Francisco: Jossey-Bass, 2004.

Fromartz, Samuel. *Organic, Inc.: Natural Foods and How They Grew.* New York: Harcourt, 2006.

Howard, Philip H. "Organic Industry Structure." Department of Community, Agriculture, Recreation and Resource Studies, Michigan State University, 2009. www.msu.edu/~howardp/organicindustry.html.

Pollan, Michael. *The Omnivore's Dilemma: A Natural History of Four Meals.* New York: Penguin, 2006.

Sligh, Michael, and Caroline Christman. *Who Owns Organic? The Global Status, Prospects and Challenges of a Changing Organic Market.* Pittsboro, NC: Rural Advancement Foundation International-USA, 2003.

See Also: Agribusiness; APPENDIX 2: Organic Foods Production Act of 1990; Hain Celestial Group; Newman's Own; Trader Joe's; United Natural Foods; Walmart; Whole Foods

Philip H. Howard

Buffers

In the context of agriculture, buffers are vegetated areas of transition between two different land areas. Buffers are established in field edges, along stream banks, or within the crop field. Buffers serve multiple purposes in agricultural systems. They absorb agricultural runoff, stabilize stream banks

Grassy buffer strip near plowed farm field. Courtesy of Leslie Duram.

and sloped terrain, offer windbreaks, add aesthetic beauty to the farm, and provide food and shelter for wildlife. Buffers established along stream and river banks are called riparian buffers and are critical for the protection of freshwater systems from the effects of adjacent crop and pasture land.

Buffers play an especially important role in organic systems by absorbing pesticide drift from neighboring crop fields under conventional management, which if left unchecked would contaminate organic certified crops. Although a minimum area is not defined under the USDA organic rules, section 202.202(c) requires "distinct, defined boundaries and buffer zones to prevent the unintended application of a prohibited substance to land under organic management."

In 1997, the National Conservation Buffer Initiative (NCBI) was created to improve and conserve soil, air, water, and biodiversity in agricultural landscapes. The NCBI, managed by the Natural Resources Conservation Service (NRCS), has been instrumental in protecting streams and rivers from agricultural pollutants through the establishment of grass buffer strips. In 2004, the use of buffers for wildlife conservation became a focus in the new program, Habitat Buffers for Upland Birds, under the Continuing Conservation Reserve Program (CCRP). The goal of the initiative was to restore habitat for bobwhite quail and other bird species establishing 30 to 120-foot-long grass buffers along the edges of crop fields.

Buffers range in plant composition from predominantly grass species to more complex assemblages that include grasses, forbs, shrubs, and small trees, which are commonly referred to as hedgerows. Due to their extensive root systems, grasses are particularly important in erosion control and nutrient capture. Grassland habitat also provides shelter and nesting sites for quail, small mammals, spiders, and ground dwelling insects. The plant diversity and architectural complexity of hedgerows provides a diversity of niches and food resources for beneficial insects, mammals, reptiles, and birds.

Agricultural buffers offer an opportunity to restore some of the native plant diversity that was lost in the transition of grasslands, woodlands, and wetlands into farmland. The incorporation of buffers in farms does not necessarily translate into a loss of production, as they are typically established on marginal land that would not otherwise be cultivated. In fact, there is growing evidence that suggests conservation buffers have the potential to restore ecological services such as biological control and nutrient sequestration or cycling, which add to the long-term sustainability of the farm.

Further Reading

Earnshaw, Sam. *Hedgerows for California Agriculture: A Resource Guide*. Davis, CA: Community Alliance with Family Farmers, 2004.

Natural Resources Conservation Service. "Buffer Strips: Common Sense Conservation." Natural Resources Conservation Service Web site. www.nrcs.usda.gov/feature/buffers/.

See Also: Biodiversity; Endocrine Disruptors

Tara Pisani Gareau

C

Campus Programs

The rise in consumer interest in local and organic foods is a key component of modern student life and campus activism. Generally following the model established by the University of Wisconsin–Madison, campuses around the nation now offer sustainable and local choices, and food processing and distribution companies are developing processed organic foods for campus sale. Although these programs are universally popular with students, administration, and food suppliers, there are ongoing challenges of adequate and timely supply and higher costs per unit. Change in food selection was and is driven by faculty, student, and staff activists, and encompasses a multitude of forms, from campus-based farmers markets to farm-to-school direct initiatives. The emphasis on these foods is being driven by concerns for health and safety, for environmental sustainability, and for fair and equitable trade in the era of globalization. Many universities and colleges are directly linked to larger organizations, such as Slow Food, Fair Trade, Farm to School, as well as to multiple state or region organic and sustainable agricultural organizations. These alliances have created new university community partnerships in many states. In addition to direct involvement in food choices at the university, students are taking advantage of experiential

Students work at the Oregon State University student-run organic farm. (AP Photo/Rick Bowmer)

programs in organic agriculture production, which allow them to learn and experience life and work on organic and sustainable farms in the United States and throughout the world.

Campus programs share an emphasis on local foods and on organic foods. Local foods are those specifically grown within a certain radius of the region, and are often limited in their utility to large campus programs by seasonal availability and quantity of production. Organic foods are produced without, or depending on the certification guidelines, with a minimum of chemical inputs and additives. Organics are broadly available in the United States, in both processed and unprocessed forms. For many involved with campus food programs, local foods are preferred over organics because of the environmental and health effects of food transportation and processing, even in the organic industry.

Student dining services were the first area of noticeable change. Beginning in 1994, the University of Wisconsin–Madison was the first in the nation to offer local and organic foods on campus, albeit mostly through specialty events and activities and initially for small events. The impetus for changing to organic and local foods originally came from faculty and staff, in contrast to later campus efforts, which came about as a result of student action. Like many other universities, UW–Madison has two separate food services, one for student and staff meals and one for catering and special events. In what would become a common model, some of the earliest introduction of organic and local foods was in catering and special events through what was called "Consciousness Catering." This form of food service combined a meal that mostly comprised local and sustainable foods with educational efforts and community links, particularly with local chefs.

The "local foods banquet" debuted at UW–Madison in 1996 and has emerged as the vanguard effort to integrate local foods throughout the country.

The university food service at Madison began annual service of a local foods dinner for students in 1997, another common model adopted by other campuses. Although organic and local foods have now integrated as single items in the food service year round, the continued special service of an entirely local meal serves to both highlight sustainable production and to educate students.

UW–Madison was also a leader in addressing the key challenge of integrating local and organic foods into institutional dining—establishing a system to expedite adequate farm-direct supply of needed items, particularly across seasons. Working with the catering division in 1996, the Center for Integrated Agricultural Food Systems established a sales network with farm-direct links through the assistance of a Sustainable Agriculture Research and Education grant. Key to the ongoing success of UW–Madison was linking with larger organic cooperatives and alternative distributors who could provide the quantities necessary to supplement the 15,000 meals served daily. Homegrown Wisconsin, Organic Valley, Wisconsin Pasturelands, and North Farm Cooperative are among those identified who have successfully partnered to provide meat, dairy, and vegetables.

Since 2000, the number of campuses across the United States engaging in some form of local, organic, or sustainable foods education and offerings has exploded. Colleges and universities in Iowa, Hawaii, New Hampshire, Minnesota, Montana, North Carolina, New York, Wisconsin, Arizona, Connecticut, Maine, California, Vermont, Massachusetts, Alaska, and possibly elsewhere, have highlighted local and sustainable foods. Private universities and colleges, too, are leading efforts to integrate local, organic, and sustainable foods, as can be seen at Grinnell, Oberlin, and Pacific Lutheran University, to name just a few.

Although some of these efforts are idiosyncratic to the institution, the majority follow much of the structural model established by UW–Madison. Local and sustainable foods are generally offered as an annual event, with some individual organic or local products integrated more broadly. Integration of organic and local foods first appears at catered events, then moves into student and staff meal plans. In general meal plans, local and organic foods most often appear as single items, for example, apples in season or organic tortilla chips. Most campus programs begin with a local food presentation or food fair to educate students and administrators, as well as to bring local farmers to campus. This food fair may take the form of a small farmers market or, more commonly, as a day-long conference and forum. These initial efforts serve to set the stage and connect dining services and local producers.

The recent impetus for new efforts at colleges and universities is more often than not student driven, though it sometimes occurs in conjunction with student coursework or other community activity. The University of

Minnesota, Morris, established an annual foods dinner and farmers market in conjunction with a food systems course, for example, which went on to become an annual community and university event.

Not all campuses are moving toward local production. For many, student demand for organic foods is being met through multiple new forms of organic preprocessed and convenience foods. Companies such as Amy's Kitchen sell processed foods directly to campus food services and in vending outlets, and focus their organic products on convenience. Traditional campus food service companies, such as AraMark, are also expanding their organic offerings in response to student demand. In addition, many food service companies are committing to increasing involvement through new systems emphasizing food supply and sourcing, transportation, and waste management.

Campus programs often link to larger national organizations and extend their reach into the community. For example, cooperative programs such as the state-level Farm to School program based at the University of Montana, Missoula, extend local and organic food networks into primary and secondary schools in the state. The program connects local farmers and ranchers to school food service staff and expedites the exchange. In addition, the program offers curriculum for schools to educate young pupils on the economic, environmental, and social benefits of local and organic foods. The state program is part of the larger Farm to School national organization promoting local and organic foods in schools throughout the United States.

Student interest in local and organic food also extends to production, and there has been equally large growth in direct production activity by students. On-campus efforts include the establishment of student-operated farms, such as that at Yale University, although off-campus individual internships are more common. Students interested in organics have a variety of choices worldwide through organizations such as World Wide Opportunities on Organic Farms (WWOOF), which connects student workers with farmers in Australia, New Zealand, France, Japan, and the United States, as well as with independent farmers in other nations.

Further Reading

Amy's. www.amys.com/buy/on_campus.php.

Center for Integrated Agricultural Food Systems. The College Food Project: UW-Madison Case Study. July 2001. www.cias.wisc.edu/farm-to-fork/collegefood/the-college-food-project-uw-madison-case-study/.

Farm to School. www.farmtoschool.org.

Heinrichs, Claire. "The Practice and Politics of Food System Localization." *Journal of Rural Studies* 19(1) January 2003: 33–45.

World Wide Opportunities on Organic Farms. www.wwoofusa.org.

See Also: Farm to School; Gardens, Children's; Gardens, Community

Meredith Redlin

Certification, Organic

Organic certification allows producers (farmers and ranchers) and handlers (processors) to label their products as "organic" and is achieved by using organic methods that meet certain standards. In the United States, organic certification is regulated by the Organic Foods Production Act (OFPA) of 1990, which was implemented in 2002 when the National Organic Program (NOP) Final Rule went into effect.

The NOP covers all agricultural products labeled and sold as "organic" or "organically produced." With NOP regulation, any agricultural product can be produced using organic methods. The regulation governs organic vegetable growers, orchardists, row crop farmers, livestock producers, ranchers, processors, and retailers who conduct processing activities.

All producers and handlers who sell over $5,000/year of organic products must be certified to include "organic" labeling on their products. Producers and handlers who sell under $5,000/year, and retailers who do not conduct processing activities, do not have to be certified, but they still have to follow NOP requirements. Noncertified organic producers are allowed to sell their products directly to customers or to retail stores. However, their products cannot be used as organic feed by other farmers or as organic ingredients in products. Most important, noncertified organic producers cannot use the "USDA Organic" seal.

Although the NOP regulations only took effect in 2002, organic standards and certification have existed in the United States since the mid-1970s. The initial certifying organizations were California Certified Organic Farmers, Oregon Tilth, the Organic Growers and Buyers Association (in Minnesota), and the Northeast Organic Farming Association. As the markets for organic products grew, so did the number of organic certification agencies. Most of the organizations had similar standards, but there were some differences. These variations sometimes led to problems for export markets and arguments over which standards were better.

OFPA was passed by Congress in 1990 to begin the process of resolving the differences of these organizations and establishing one set of national standards. OFPA and the NOP regulations are enforced by the United States Department of Agriculture (USDA). The USDA does not certify organic operations. Instead, the USDA accredits "certifying agents," who carry out certification activities. All certifiers who operate in or certify product sold in the United States must follow the NOP regulations. They must be accredited by the USDA to certify organic farm and handling facilities.

Organic certification is the outcome of a process-based system, under which the production or handling system is certified as following NOP requirements. Once the process is certified, the resulting products can carry the "organic" label. "Organic production" is defined by the regulations as "a production system that is managed . . . to respond to site-specific conditions by integrating cultural, biological, and mechanical practices that foster cycling of resources, promote ecological balance, and conserve biodiversity."

Simply stated, NOP regulations require the following components for crop farms: three years (36 months prior to harvest) with no application of prohibited materials, which include synthetic fertilizers, pesticides, or genetically modified organisms (GMOs) prior to certification; distinct, defined boundaries for the operation; proactive steps to prevent contamination from adjoining land uses; and implementation of an Organic System Plan (including proactive fertility management systems, conservation measures, and environmentally sound manure, weed, disease, and pest management practices). In addition, organic farmers must have systems to monitor their management practices to ensure compliance. They must use only natural inputs or approved synthetic substances on the USDA National List, provided that proactive management practices are implemented prior to use of approved inputs.

Overall, organic certification ensures that none of the following are used in organic farming operations: prohibited substances; GMOs (which the regulations specifically define as "excluded methods"); sewage sludge as fertilizer; and irradiation. Furthermore, organic farms must use organic seeds when commercially available, must not use seeds treated with prohibited synthetic materials (such as fungicides), and must use organic seedlings for annual crops. Certification rules restrict the use of raw manure and compost. They enforce the maintenance of the physical, chemical, and biological condition of the soil; the minimizing of soil erosion; and the implementation of soil-building crop rotations. Fertility management is not allowed to contaminate crops, soil, or water with plant nutrients, pathogens, heavy metals, or prohibited substances. The farm must maintain buffer zones appropriate to the risk of contamination. The entire farm does not have to be organic as long as sufficient measures are in place to segregate organic from nonorganic crops and production inputs (commingling must be prevented on split operations). Residues of prohibited substances cannot exceed 5 percent of the EPA tolerance. The organic certifier may require residue analysis if there is reason to believe that a crop has come in contact with prohibited substances or was produced using GMOs.

For livestock operations, the farm must implement an Organic Livestock Plan and monitor management practices to ensure compliance. The organic management must be followed from the last third of gestation for slaughter stock or from the second day after hatching for poultry. Dairy cows must have one year of organic management prior to the production of organic milk. All species must have outdoor access when weather is suitable, and ruminants must have access to pasture. Feed and approved feed supplements must be 100 percent organic. No antibiotics, growth hormones, or GMOs may be used. The farmer must implement preventative health care practices.

Animals must not be rotated between organic and nonorganic production. Farmers must not withhold treatment to preserve an animal's organic status, but any animal treated with a prohibited substance must not be used or sold as organic. Manure must be managed to prevent contamination of crops, water, and soil, and to optimize the recycling of nutrients.

Certification rules for processing and handling operations include implementing an Organic Handling Plan; using proactive sanitation and facility pest management practices to prevent pest infestations; taking steps to protect organic products and packaging from contamination, if pesticides are used in the processing facility; keeping records of all pesticide applications; and using organic minor agricultural ingredients in products labeled "organic," unless such ingredients appear on section 205.606 of the National List and are not commercially available from organic sources. The regulations allow using mechanical or biological processing methods, but forbid the following: commingling or contamination of organic products during processing or storage; using GMOs or irradiation; and using packaging materials that contain fungicides, preservatives, or fumigants. Processing and handling operations must use approved label claims for "100% organic" (100 percent organic ingredients, including processing aids), "organic" (at least 95 percent organic ingredients), and "made with organic ingredients" (at least 70 percent organic ingredients). In addition, the product's information panel must include the name of the certifier of the final handling operation.

All operations producing or selling organic products must keep records to verify compliance with the regulations. Such records must be adapted to the particular operation; fully disclose all activities and transactions of the certified operation in sufficient detail as to be readily understood and audited; be maintained for at least five years beyond their creation; and be sufficient to demonstrate compliance with the regulations. The operator must make the records available for inspection.

All certified operations must complete and submit Organic System Plans, which are typically provided by certifying agents as part of the application process. The plans must be updated annually, and operators are required to notify their certifying agents of all changes to the operations, which might affect the operation's certification status. Organic operations must follow their Organic System Plans and be inspected at least annually.

Further Reading

USDA Agricultural Marketing Service. "National Organic Program." www.ams .usda.gov/nop/NOP/standards.html.

USDA Economic Research Service. "Organic Agriculture: Organic Certification— Steps and Links." www.ers.usda.gov/Briefing/Organic/certification.htm.

See Also: Organic Labels; Transition to Organic; USDA National Organic Program; USDA National Organic Standards Board; APPENDIX 2: Organic Foods Production Act of 1990; Appendix 3: U.S. Organic Farming Emerges in the 1990s: Adoption of Certified Systems, 2001

Jim Riddle

Climate Change

"Climate change" is a term used by scientists to describe natural variations in the behavior of the climate system across timescales ranging from decades to millennia and beyond. In a contemporary context, the term is commonly used by the general public to refer to the "enhanced greenhouse effect" or "global warming," both of which refer to the warming of the lower atmosphere during the last one to two centuries and the accelerated warming during the last several decades. Large-scale climate change is normally a result of changes in so-called boundary conditions and may be a change in the amount of solar energy received by the planet, a change in the amount of sunlight reflected by the earth (i.e., its albedo), or a change in the distribution of heat-trapping greenhouse gases (GHGs) such as carbon dioxide (CO_2) among climate system reservoirs, which include the atmosphere, hydrosphere, biosphere, and lithosphere.

Current research indicates that changes in solar radiation are not a significant influence on the recent warming trend, and that there have clearly been large, human-driven changes in both the earth's albedo and the distribution of GHGs. Although changes in earth's reflectivity and the associated effects on climate change are quite complex and difficult to model, the effect of increasing GHGs is more straightforward. Extraction and combustion of fossil carbon has clearly resulted in a large net transfer of carbon from the lithosphere to both the oceans and the atmosphere. Atmospheric CO_2 concentrations have increased from preindustrial values of around 280 parts per million to current values approaching 385 parts per million. The additional atmospheric CO_2 burden has so far resulted in a global near-surface temperature increase of around 0.75°C over the last one hundred years, with accelerated warming since the mid-1970s. Both the increased CO_2 concentrations and greenhouse warming are predicted to, or have already led to, changes that are relevant to agriculture. This includes changes in plant phenology, increases in growing season length, and increases in precipitation variability.

Food production and transportation contribute to changes in both albedo and GHG emissions in several ways. First, forested areas are often cleared to provide agricultural land, thereby reducing the local role of the biosphere in removing atmospheric CO_2 during photosynthesis. Because the biomass that is removed is normally burned, the carbon stored in the trees is returned to the atmosphere. This process, therefore, leads to an overall increase in atmospheric CO_2 concentrations. Additionally, land clearance and conversion (e.g., from forest to pasture or row crops) cause changes in both reflectivity and local evapotranspiration, which can have varied effects on albedo.

Second, conventional agriculture strongly depends on the use of nitrogen fertilizers, the production of which requires fossil fuels such as natural gas as a feedstock and is highly energy intensive. For example, human fixation of nitrogen (principally for agriculture) exceeds that of all natural sources. Beyond the large CO_2 emissions associated with fertilizer production, nitrogen-based fertilizers also undergo chemical reactions, which result in nitrous oxide (and other oxides of nitrogen), which is a potent GHG. In addition to the effects on global warming, fertilizer production has also resulted in a serious and growing nitrogen contamination problem in the oceans and biosphere.

Livestock production represents a third and highly important source of GHGs. Obviously the growth of food crops for livestock (principally maize) and the transportation of animals and animal products represent a large portion of livestock-associated GHG production. Perhaps more important though, livestock production results in very large methane emissions. Methane is a highly effective GHG and is produced during enteric fermentation (the breakdown of carbohydrates into sugars in the digestive process of livestock). The recent trend in North America of switching cattle from grass to grain-based diets is believed to further contribute to enteric methane emissions. The decomposition of animal manure is also an important source of methane and nitrogen emissions. In addition to methane emissions by livestock, other agricultural methane sources, such as rice cultivation, contribute substantively to overall GHG emissions.

Finally, conventional agriculture results in substantial CO_2 emissions from general fuel use and energy used for irrigation or other needs. Although the agricultural sector is directly accountable for a relatively small percentage of CO_2 emissions, the processing, transportation, and especially refrigeration of food products is responsible for a much larger share of emissions.

Sustainable agricultural practices have the potential to considerably reduce the negative impact of agriculture in climate change and even make a positive contribution. Because sustainable agriculture is based on returning organic matter to the soil, it can increase soil organic carbon, and thereby increase carbon sequestration. Increases in soil organic carbon also increase soil fertility and water retention. These may be particularly attractive soil qualities in a climate characterized by warmer temperatures and more variable precipitation.

The sustainable replacement of nitrogen-based fertilizers with natural alternatives such "green manures" (e.g., nitrogen-fixing cover crops) or organic animal manures could also largely remove the agricultural contribution to CO_2 and nitrous oxide emissions. Although modern organic growers do not use nitrogen-based fertilizers, the transportation of large quantities of plant materials (e.g., compost) or manure still results in significant CO_2 emissions. In addition, the processing and transportation of organic foods has a similar and in some cases larger carbon footprint.

Sustainable, as opposed to conventional or even organic, livestock production typically relies on grazing rather than use of external animal feed (e.g., grains). Because external animal feeds usually require fossil fuel inputs and are often transported long distances, the sustainable approach should result in fewer associated emissions. The effects of sustainable agriculture on methane emissions are less clear. Although changes in animal diet can reduce enteric fermentation, use of organic manures, which promote aerobic microorganisms in soil, can increase methane emissions, and even sustainable or organic rice cultivation will result in methane emissions.

Sustainable food production and consumption clearly depends on the reduction of GHGs through a variety of mechanisms. Certainly one of the largest effects consumers and society could have on the production of GHGs in general, is through the reduction of meat consumption. Given the tremendous amount of fossil fuel resources that go into meat production, it is unlikely that any other consumer food choice would have as large a positive influence. Another method to reduce GHGs associated with food production is to decrease the processing and transportation of food products. Increased consumption of locally produced, or whole food products would have a large and immediate impact on GHG emissions.

Further Reading

Burroughs, William James. *Climate Change: A Multidisciplinary Approach*. 2nd ed. Cambridge, UK: Cambridge University Press, 2007.

Ruddiman, W. F., *Plows, Plagues, and Petroleum: How Humans Took Control of Climate*. Princeton, NJ: Princeton University Press, 2005.

Solomon, S., D. Qin, M. Manning, Z. Chen, M. Marquis, K. B. Averyt, M. Tignor, and H. L. Miller, eds. "Climate Change 2007: The Physical Science Basis. Contribution of Working Group I to the Fourth Assessment Report of the Intergovernmental Panel on Climate Change." New York: Cambridge University Press, 2007.

See Also: Environmental Issues; Fossil Fuel Use; Vegetarian Diet

Justin T. Schoof and Matthew D. Therrell

Cloning

"Cloning" is a term used to refer to a form of human-induced asexual reproduction of plants or animals. The technical term for animal cloning is "somatic cell nuclear transfer" (SCNT).

During the cloning process, a differentiated somatic cell (such as a cell from the ear of an existing animal) is introduced into an embryonic cell that has had its nucleus removed. Generally, an electric shock is applied, and the new cell containing the introduced genetic material replicates and forms a zygote. The zygote is implanted into a female animal (surrogate dam), where it develops into a fetus.

Although plants have been cloned through the use of cuttings and tissue culture for hundreds of years, animal cloning is a relatively new technology that is complex, technically demanding, and inefficient. There is no method that is universally employed. Different labs and cloning companies use different techniques, with varying levels of success.

In the 2007 "Draft Risk Assessment," the U.S. Food and Drug Administration (FDA) stated that only 4 to 7 percent of attempted clones result in healthy, live animals. Although some clones may develop into healthy animals, the low success rate is likely associated with the inability of clones to successfully reprogram the donor cell to become a properly developing embryo.

Many people falsely believe that clones are the same as identical twins, because, in theory, they carry the same sets of DNA as the parental stock. There are significant differences between cloned animals and identical twins. Identical twins originate from a single zygote, or fertilized egg. DNA from the father combines with DNA from the mother in the nucleus of the zygote. Once the zygote has undergone the first division, it is referred to as an embryo. In the case of twins, early in pregnancy, the zygote divides into two parts. The two parts develop into separate individuals who have the same genetic markup. Twins are formed through the union of egg and sperm carrying DNA from male and females animals. The DNA mixes, and then the cells divide. A clone, on the other hand, is the result of asexual reproduction, and carries the DNA of a single living or dead, male or female animal, but not both. In addition, identical twins develop from natural and spontaneous cell division, and no electrical or chemical stimulation is needed to reprogram the embryo.

Research has shown that animals involved in the cloning process (i.e., cattle and sheep surrogate dams) are at increased risk of adverse health outcomes compared to noncloned animals. Cows and ewes used as surrogate dams for clone-derived pregnancies appear to be at increased risk of late gestational complications. There is an increased risk of mortality and morbidity in young calf and lamb clones, compared with calves and lambs produced without cloning. In cattle and sheep, the increased risk appears to be primarily related to large offspring syndrome.

There are inherent uncertainties associated with the release of cloned animals into the agricultural environment. Although sheep and cattle are unlikely to interbreed with a local wild population, goats and swine are more likely to escape and interbreed with wild populations. Cloned animals may have compromised immune systems and could serve as vectors for pathogens for both animals and humans.

It is well known that a population with a narrow gene pool can collapse when animals encounter unanticipated diseases, parasites, and other threats. The narrowing of genetic diversity of animals, inherent to the cloning process, threatens to increase the susceptibility of animals to opportunistic diseases, parasites, and changing environmental conditions.

Despite ecological concerns and documented compromises to animal health and welfare, the FDA has ruled that the meat and milk from cloned cattle, hogs, and goats, and products from the offspring of clones, are allowed for human or livestock consumption, without further testing, tracking, or labeling. The United States Department of Agriculture (USDA) has determined that cloned animals, their products, or progeny may not be used in organic production.

Further Reading

Center for Food Safety. Not Ready for Prime Time: FDA's Flawed Approach to Assessing the Safety of Food from Animal Clones. 2007. www.centerforfood safety.org/Policy.cfm.

Rudenko, L., J. C. Matheson, S. F. Sundlof. "Animal Cloning and the FDA—The Risk Assessment Paradigm under Public Scrutiny." *Nature Biotechnology* 25(1) 2007: 39–43.

U.S. Food and Drug Administration. "Animal Cloning: Risk Management Plan for Clones and Their Progeny, January 15, 2008." 2008. www.fda.gov/cvm/Cloning RA_RiskMngt.htm.

U.S. Food and Drug Administration. 2008. Animal Cloning: A Risk Assessment— FINAL, January 15, 2008. www.fda.gov/cvm/CloneRiskAssessment_Final.htm.

See Also: Factory Farming

Jim Riddle

Coffee, Organic

Coffee is a perennial crop that needs at least four to five years to start producing berries. Organic coffee is grown using a production system intended to replenish and maintain soil fertility, without the use of toxic and persistent pesticides and fertilizers, and build biologically diverse agriculture. Coffee has been produced in Africa without synthetic fertilizers and agrichemicals for more than 400 years of the bean's 500-year history as an active product in global trade. During the last 200 years, coffee production expanded in Latin America and Asia as millions of indigenous small-scale farmers incorporated coffee into diverse organically managed farming systems. However, most of these systems were not certified organic until the 1990s.

In contrast to organic farming, conventional coffee production is dependent on petroleum-based fertilizers and agrichemicals. Agencies backed by the U.S. government promoted conventional coffee as part of an effort to spread the green revolution and sell more agrichemicals. The larger farms in Costa Rica, Colombia, and parts of Brazil were the first to adopt many of these technologies. These same green revolution farms cut down most, if not all, of the shade trees grown together with the coffee. The shade trees sustain a great reserve of biodiversity within the coffee farms. The rapid expansion in consumer demand for *certified* organic coffee has emerged in part as a counter move to avoid negative consequences of conventional coffee production.

Organic certification standards regulate the production process and require a separate chain of custody throughout different processing stages in the value chain. Most organic standards require ecological management of farms, soil conservation practices, and require intensive on-farm record-keeping, among many other criteria. Likewise, organic standards prohibit the use of non-approved synthetic fertilizers, synthetic pesticides, and genetically modified crops. Farms are certified organic by third-party inspectors who follow an international code for each crop.

As of 2007, North American coffee drinkers spent over one billion dollars on organic coffee. Many organic coffees have won top prizes for their quality and taste. Although some consumers may prefer organic coffee for its taste and perceived human health benefits, others seek it for the environmental benefits generated in producing countries. Mexico exported the first certified organic coffee. Today, the leading certified organic coffee exporting countries include Ethiopia, Peru, and Mexico. Many of the same small-scale farmers and cooperatives producing organic coffee are simultaneously selling Fair Trade certified coffee. Although many indigenous small-scale farmer cooperatives have contributed to the rapid spread of certified organic agriculture in the tropics, the price premiums are often not enough to compensate for the higher labor, training, and administrative costs associated with these more environmentally friendly production systems.

Further Reading

Bacon, C. M., V. E. Méndez, S. R. Gliessman, D. Goodman, and J. A. Fox, eds. *Confronting the Coffee Crisis: Fair Trade, Sustainable Livelihoods and Ecosystems in Mexico and Central America.* Cambridge, MA: MIT Press, 2008.

Bray, B. D., J. L. Plaza Sanchez, and E. C. Murphy. "Social dimensions of organic coffee production in Mexico: Lessons for Eco-Labeling Initiatives." *Society and Natural Resources* 15(6): 429–46.

Liu, Pascal, Alice Byers, and Daniele Giovannucci. "Value-Adding Standards in the North American Food Market: Trade Opportunities in Certified Products for Developing Countries." FAO Trade and Markets Division, 2008. Available at SSRN: http://ssrn.com/abstract=1107382.

Martinez-Torres, M. E. *The Benefits and Sustainability of Organic Farming by Peasant Coffee Farmers in Chiapas, Mexico.* Athens: Ohio University Press, 2005.

Perfecto, I., R. A. Rice, R. Greenberg, and M. E. Vand der Voort, "Shade Coffee: A Disappearing Refuge for Biodiversity." *BioScience* 46(8) 1996: 598–609.

See Also: Fair Trade

Christopher M. Bacon

Community-Supported Agriculture

Community-Supported Agriculture (CSA) is a local food marketing system that creates a direct link between consumers and farmers. But CSA is much more meaningful than a simple exchange of money for vegetables. CSA farmers and their members commit to each other in a way that is seen nowhere else in agriculture and rarely in American society.

CSA member selects his weekly share of vegetables. Courtesy of Leslie Duram.

The basic premise of CSA is that consumers and farmers need one another and should share the risks as well as the bounty of growing food. In a CSA system, a consumer agrees to buy a share of a farm's harvest, and a farmer agrees to provide a fair amount of high-quality food over a specific length of time. The benefits to both parties are enormous.

From the farmer's perspective, CSA ensures that his or her products are presold before they leave the farm. Farmers are freed of most marketing tasks and can devote their time to growing. They don't have to stand at farmers markets or drive around to restaurants and supermarkets, for example. Revenue is guaranteed, too, which gives the farmer a measure of financial security and allows for better planning. In the purest form of CSA, members pay before the first seed is planted, so farmers have their production costs covered. Farmers also enjoy getting to know their customers, and the immediate feedback they get from members helps them make better decisions about what and how much to grow.

From the member's perspective, belonging to a CSA ensures steady access, at a convenient location, to fresh, high-quality produce and other local foods. CSA members in most cases also gain access to the farm itself, where they can enjoy special events, pick their own produce and flowers, spend time outdoors, share the experience with other like-minded people, and teach their children about farming. Usually, the food is less expensive than in a supermarket.

CSA farms are proliferating throughout the world. In the United States, there are now estimated to be more than 2,000 CSA farms. They range in size from those with just a few members to those with several thousand members.

HISTORY

The first CSAs were in Europe beginning in the 1950s and were closely associated with Rudolf Steiner's biodynamic agriculture movement. In the United States, the model of a producer-consumer alliance sprang up almost simultaneously in 1986 on two farms, Indian Line Farm in Massachusetts and Temple-Wilton Farm in New Hampshire.

At Indian Line Farm, Robyn Van En heard about the concept from Jan VanderTuin, who had recently returned from working on biodynamic farms in Switzerland. In 1985, Van En and several neighbors sold thirty shares in an apple harvest, and distributed 360 bushels of apples plus cider and vinegar. The next year, the group agreed to start the CSA gardens, and to use land on Indian Line Farm. Van En later produced a video *It's Not Just about Vegetables*, and started CSA North America (CSANA), a nonprofit clearinghouse to support CSA development. In 1997, Van En died of an asthma attack at age forty-nine. Her son was forced to sell the farm, but it was eventually purchased by a land trust and the Nature Conservancy, to be held in perpetuity as farmland.

At Temple-Wilton Farm, a similar idea was sparked by Trauger Groh, who had farmed in Germany. Rather than selling shares, though, the farm presented interested members with a budget and asked for pledges. Once the budget was funded, members were free to take as much food as they wanted from the harvest. Groh went on to write the book *Farms of Tomorrow* with coauthor Steven McFadden.

The CSA idea spread rapidly, and by 1990, there were sixty CSA farms in the United States. A decade later, the number was estimated at 1,700 and by some accounts has now passed 2,000. No one is tracking them officially, though, so the exact number of CSA farms and members is not known.

FORMS

The first two CSAs, at Indian Line Farm and Temple-Wilton, are known as community farms, in which the entire production of the farm is shared among the members. Today, fewer than 25 percent of CSA farms follow that model. Instead, most CSA farms also sell at farmers markets, to supermarkets, and other local markets.

Many people recognize a distinction between CSA and subscription farms. In CSA, customers pay in advance for a share to the full season. In subscription, they pay as they go and can drop out at any time. However, there is a great deal of overlap between the two, and both depend on an ongoing relationship of trust and commitment between the farmer and the eater.

CSA variations are endless in the details, but most operate under these general guidelines:

1. Sustainable farming. Almost all CSA farms are organic farms. Some are certified organic under USDA regulations, but most CSA farmers don't feel the need to certify, because their customers trust them and are familiar with their farming practices.

2. Types of food. The vast majority of CSA farms are vegetable farms, so most of the food distributed through CSA is fresh produce. However, many CSA farms recognize the desire of customers to have access to other local foods and may offer bread, cheese, milk, eggs, meat, and other food from other local farms. These may be sold as a separate add-on shares or may be available for sale at distribution.

3. Core group. Most CSA farms have a group of members who are deeply involved in the management and marketing of the program. In some cases, a core group comes together in the community to start a CSA and then looks for a farmer to grow the food. In other situations, the farmer initiates the CSA and recruits friends and regular customers to be the core group. Not all farms have a core group.

4. Share price. Most farms set a price for a full share and a half share that is comparable or slightly less than the price a member would pay to buy the same food in a store. The share price can be paid in full before the start of the season, which is most helpful to the farmer, or can be paid in installments.

5. Weekly pickup or delivery. In the summer, most CSAs distribute shares once a week. In winter, they may offer a share of storage vegetables once or twice a month. Usually, the main CSA season is summer only; winter shares are usually priced and paid for separately.

6. Food choices. Some CSA farms pack a box of whatever produce is available that week, giving members little, if any, choice about what they receive. Other farms put out their harvest supermarket style and post signs telling members how much of each item they may take, offering choices such as "Take up to 8 of tomatoes, cucumbers, and peppers." Members are then free to choose their favorite things and to leave things they don't like.

7. Work shares. Most CSA farms offer some kind of work opportunity. At a few farms, members must commit to working a certain number of hours per season. The work can be on the farm, helping with harvest, washing produce, weeding, and so on. It can be done at distribution, helping greet members and keeping produce bins stocked. Or it can be done by managing the business end of the CSA, keeping the books and maintaining mailing lists. Other farms do not require members to work but some will reduce the share price in exchange for work.

8. Farm access. Some CSAs have pickup on the farm, whereas others may have distribution points at churches, natural foods stores, and members' houses. It depends largely on how far the farm is from most of its members. Almost all CSAs have opportunities for members to visit the farm: work days, parties, and potluck dinners are an essential component of CSA.

Farmer proudly surveys his rows of compost. Courtesy of Leslie Duram.

dangerous human pathogens and other contaminates may be transferred to food crops. Thoughtful use of this resource is necessary to maintain safety and sustainability. In certified organic operations, there are specific requirements for the use of raw manure, depending on the type of crops to be grown. For vegetable crops, manure cannot be applied within 120 days of harvest when edible portions of the crop come into contact with soil. There is a 90-day limit for crops in which the edible portion does not come in contact with the soil. For nonorganic production, guidelines suggest avoiding the spread of manure after planting, and recommend fall applications or use with cover crops. Knowing the nutrient content of the particular manure being used and knowing soil test results is imperative for sustainable fertility management. When over applied, manures can cause water quality problems as excess nutrients are washed away; significant amounts of nitrogen can be lost by leaching and conversion to ammonia gas (ammonification), thereby losing much of its fertility value to the crop. When manure is mixed with large amounts of bedding material, such as straw, woodchips, or saw dust, the high carbon to nitrogen ratio (C:N) causes mineral nitrogen to be temporarily unavailable to the crop, and may interfere with plant growth. The C:N ratio is important because of the microbial processes at work—microorganisms that decompose organic matter need both carbon and nitrogen to grow and multiply. When carbon is in great supply, the competition for nitrogen becomes elevated and any available mineral nitrogen will be taken up into microbial biomass (nitrogen immobilization); when there is no longer sufficient mineral nitrogen, a portion of the microbial biomass begins to die and decomposes until the C:N ratio is reduced to below 25:1 (25 grams of carbon for every 1 gram of nitrogen). At this ratio, the nitrogen is sufficient for microbial growth, allowing nitrogen to be released into the soil solution and available for plant uptake (nitrogen mineralization). When manure mixed with carbon rich materials is applied to soil, a period of nitrogen depression is expected, therefore either supplemental nitrogen should be applied or planting should be delayed for several weeks until the C:N is reduced enough to allow nitrogen mineralization.

There are many advantages to composting manure; it reduces weight, volume, and moisture content, transforming manure into a substance that can be more easily stored without fly or odor problems. The nitrogen content is reduced, but converted to a more stable organic form that is less susceptible to leaching and ammonification. Composting also lowers the C:N ratio of manures mixed with high carbon materials, and the heat produced from decomposition destroys many pathogens and weed seeds. The finished product has many applications, such as a soil amendment, mulch, or bedding for livestock and poultry; compost can be safely applied directly to growing vegetable crops.

The right starting materials and proper production technique is the key to creation of quality compost. Common composting materials include manure, kitchen and garden waste, and carbon-rich plant matter such as straw, wood chips, sawdust, and corn stalks. Woody inputs are more difficult to decompose, and may take longer to compost. A C:N ratio between 50:1 and 25:1 is recommended for efficient composting (organic regulations specify a range between 40:1 and 25:1). Aerobic microorganisms decompose the starting materials, consuming oxygen and breaking down organic compounds to fuel their metabolic processes, releasing water, carbon dioxide, and heat. By maintaining conditions favorable to the activity of these decomposing microorganisms, compost production becomes faster and more consistent. Oxygen must be continually introduced into composting materials to maintain the aerobic activity of microorganisms. Aeration can be achieved by turning the compost pile on a regular basis, such as in a windrow system, where the composting materials are set out in long rows so they can be turned mechanically. Compost can also be produced in aerated static piles or containers that can be rotated. Sufficient moisture is required to maintain microbial activity—between 40 and 65 percent is optimal. The heat produced by aerobic decomposition can reach temperatures high enough to kill pathogenic microorganisms and weed seeds. For organic compost production, a temperature of 131 to 170°F must be maintained for a minimum of three days in a static pile, or fifteen days in windrow, which must be turned a minimum of five times. Eventually, this phase of "active" composting will give way to a curing period in which decomposition occurs at a slower rate, consuming less oxygen and producing less heat.

Although decomposition never fully stops, compost can be considered finished when proper temperatures have been reached for an appropriate amount of time, the consistency is uniform, little or no original starting material can be identified, no strong odors are present, and the desired C:N ratio has been achieved. The length of time it takes to produce finished composts varies from months to years, depending on the method and starting material.

Safe, successful, and sustainable use of compost and manure depends on their proper use and knowledge of their nutrient contents, as well as that of the soil it is being applied to; regular testing is recommended.

Further Reading

Kuepper, George. "Manures for Organic Crop Production." National Sustainable Agriculture Information Services (ATTRA) Publication #IP127. ATTRA, 2003. http://attra.ncat.org/attra-pub/PDF/manures.pdf.

Rynk, Robert, ed. *On-Farm Composting Handbook (NREAS-54)*. Ithaca, NY: NRAES Cooperative Extension, 1992.

USDA. "National Organic Program Standards," Section 205.203. www.ams.usda.gov/nop/NOP/standards.html.

See Also: Agriculture, Organic; Agriculture, Sustainable; Agroecology

Mark A. Williams and Audrey Law

Concentrated Animal Feeding Operations

Concentrated animal feeding operations (CAFOs) are industrialized livestock or poultry production systems that confine animals at high densities in open feedlots or enclosed structures. A CAFO may include multiple confinement structures or a single large-scale structure in which hundreds or thousands of animals, or in the case of poultry, in which even millions of animals are raised.

The federal Clean Water Act regulations define CAFOs as animal feeding operations that house and feed animals in a confined area for 45 days or more during any 12-month period. In addition, crops, vegetation, forage growth, or post-harvest residues are not sustained over any portion of the operation's lot or facility. The regulations define large-scale CAFOs as those with more than 1,000 animal units, which translates into at least 1,000 beef cattle; 700 mature dairy cows; 1,000 veal calves; 2,500 swine (over 55 lbs); 10,000 swine (less than 55 lbs); 500 horses; 10,000 sheep; 55,000 turkeys;

125,000 chickens (dry systems); 82,000 layers (dry system); 30,000 ducks (dry system); 30,000 chickens or layers (liquid system); or 5,000 ducks (liquid system). Any size animal feeding operation that discharges pollutants to the surface waters of the United States may be declared by the U.S. Environmental Protection Agency to be a CAFO requiring a Clean Water Act discharge permit that applies to the site where the animals are confined and land application sites under the CAFO owner or operator's control.

Within CAFOs, animals may be further restrained in cages, gestation crates for pregnant and lactating swine, or crates for veal calves. The most extreme confinement is that of laying hens. Many CAFOs crowd six to nine hens into battery cages that give each hen floor space of about 8" × 10", less space than a standard piece of letter paper. Often these cages are stacked atop each other, with the waste from the hens above falling onto those below.

CAFOs generate huge quantities of animal waste. Agricultural operations produce an estimated 500 million tons of manure every year, which is three times the amount of waste produced by the human population of the United States. One beef cattle CAFO with 140,000 head of cattle could produce over 1.6 million tons of manure annually, more than the almost 1.4 million tons of waste generated by the more than two million residents of Houston, Texas. Unlike human waste, however, livestock and poultry waste is treated only minimally or not at all. The common practice at dairy and hog CAFOs is to store animal waste in manure pits or mix it with water in waste lagoons and then spread it onto the land. Animal waste and litter from poultry houses is often stored in large open piles before land application.

The large amounts of waste generated by CAFOs pose a number of public health and environmental risks. In many CAFOs, animals are routinely treated with subtherapeutic amounts of antibiotics to stimulate their growth under crowded and stressful conditions. Water pollutants associated with manure-related discharges at CAFOs include antibiotics and pathogens, such as parasites, bacteria, and viruses, which may have developed resistance to the antibiotics. Many of the antibiotics used by the CAFO industry are the same or similar to antibiotics used in human medicine. The use of antibiotics by CAFOs is contributing to the problem of antibiotic resistance, resulting in fewer drugs available to medical doctors to treat human infections and disease.

Animal waste discharges also contain high levels of ammonia, nitrogen, and phosphorus. At high concentrations in drinking water, these pollutants pose risks to human health. The health effect of most concern is methemoglobinemia or "blue-baby" syndrome, in which the oxygen-carrying capacity of a baby's blood is reduced by ingested nitrates. At lower concentrations, these nutrients can stimulate the growth of algal blooms and other microorganisms. These blooms can degrade the quality of drinking water and foul its taste. Large algal blooms can deplete surface waters of oxygen when the blooms die and decay. This can result in fish kills and other adverse effects on both freshwater and coastal ecosystems. CAFO waste also contains other oxygen depleting organic matter that can degrade surface water quality. CAFO pollutants can end up in groundwater that serves as drinking water, especially rural wells, posing health risks.

Many CAFOs use pesticides to control flies and other animal pests and treat their animals with growth hormones. Even when animals are not given synthetic hormones, the levels of hormones in animal waste may be quite high, especially from dairies, laying operations, and swine farrowing CAFOs where female animals are producing young and lactating. Researchers have associated CAFO waste discharges with hormone-related changes in downstream fish. CAFO waste may also contain trace elements such as arsenic and copper which can contaminate surface and ground water and possibly harm human health.

Large-scale CAFOs can release enormous quantities of toxic air pollutants, including ammonia and hydrogen sulfide from decomposing manure, in amounts comparable to air pollution from the nation's largest manufacturing plants. For example, Threemile Canyon Farms in Boardman, Oregon, reported that its 52,300 dairy cow operation emits 15,500 pounds of ammonia per day, more than 5,675,000 pounds per year. That is 75,000 pounds more than the nation's number one manufacturing source of ammonia air pollution, according to the 2003 Toxics Release Inventory for the United States. These toxic emissions can harm the health of people living within a few miles of CAFOs. In addition to health problems, residents in areas near CAFOs often report nuisances, such as constant noxious odor and flies. CAFO workers subject to toxic pollutants can develop chronic and acute respiratory illnesses and infections from antibiotic resistant pathogens.

Further Reading

Centers for Disease Control and Prevention. "Environmental Hazards and Health Effects: Concentrated Animal Feeding Operations (CAFOs)." www.cdc.gov/cafos/about.htm.

General Accountability Office. *Concentrated Animal Feeding Operations: EPA Needs More Information and a Clearly Defined Strategy to Protect Air and Water Quality from Pollutants of Concern.* Report No. GAO-08-944. (Sept. 4, 2008). www.gao.gov/new.items/d08944.pdf.

Pew Commission on Industrial Farm Animal Production. "Putting Meat on the Table: Industrial Farm Animal Production in America." Pew Charitable Trusts Web Site. 2008. www.pewtrusts.org/uploadedFiles/wwwpewtrustsorg/Reports/Industrial_Agriculture/PCIFAP_FINAL.pdf.

Union of Concerned Scientists. *Impacts of Industrial Agriculture* (multiple articles). www.ucsusa.org/food_and_agriculture/science_and_impacts/impacts_industrial_agriculture/.

See Also: Factory Farming; Family Farms

Martha L. Noble

Consumers

Consumers are responsible for the sustained growth in organic retail sales, increasing at 15 to 20 percent annually for more than a decade. Retail sales of organic food and beverages in the United States were estimated at $20 billion in 2007, about 2 percent of total food and beverage sales. Studies indicate that 50 to 60 percent of U.S. consumers purchased organic foods in 2007, and at least 26 percent purchased organic beverages. Businesses and consumers in the organic market select food not only for its taste and appearance, but also for the social, health, and environmental benefits it delivers. Prices and household budgets weigh against the perceived benefits of organic food in purchase decisions.

The typical organic food consumer was characterized as recently as 2002 as Caucasian, female, affluent, well educated, and concerned about health and product quality. More recently, surveys have revealed growing diversity among consumers of organic food. Gender, income, and ethnicity are no longer considered reliable predictors of who purchases organic food. A slightly higher percentage of women than men purchase organic food, and a larger percentage of higher income consumers report buying organics. However, half of those who frequently buy organic food have incomes below $50,000, and are people of color, particularly Asians and Asian Americans. Hispanic consumers are more likely than average to use organic beverages. Purchasers of organic foods skew toward younger ages (18 to 24 years) but consumers in their 40s are also significant buyers of organic foods. College education is a positive predictor of organic purchase. Some surveys have also shown that parents of young children or infants are more likely to purchase organic foods.

When consumer characteristics are more finely delineated, personal lifestyle choices and attitudes about food are revealed as explanatory factors in purchase decisions. Beliefs that reflect concern for animal and farm worker welfare, environmental protection, personal health, and dietary restrictions related to choices such as vegetarianism or conditions such as food allergies have been shown to drive organic food purchases. Consistent with organic consumption are lifestyle choices such as membership in a fitness club, shopping in a health food store or food cooperative, deriving enjoyment from cooking, and having no religious affiliation. These characteristics suggest a strong interest in healthy lifestyles among organic consumers, who commonly express the perception that organic food offers the safest, healthiest alternative.

Consumers do not all buy the same products, but a general pattern of purchase decisions has been identified. More highly processed items are purchased as consumers gain more familiarity with and express more support for organics. Upon introduction to organic products, consumers typically buy fresh fruits and vegetables, dairy products, nondairy products (such as soy milk), and baby food. These are the most widely available organic items in conventional food stores and have the largest total sales

nationally. These foods are consumed most heavily by vulnerable populations such as children and pregnant women, and purity of these products
are of greatest concern to consumers. With more experience, juice, single
serving beverages, meat, poultry, seafood, cold cereal, and snacks are added
to the shopping basket. These foods are processed by both organic and
mainstream companies and are often available side-by-side with conventional versions in the same retail outlets. Next, consumers expand their purchases to include frozen foods, breads, pasta sauces, salsas, and canned
tomatoes. These items are perceived to have less of a health advantage over
conventional versions because they are highly processed, and so are
accepted later. Other organic canned goods and bulk goods are usually next
purchased, with nonfood products (textiles, personal care products, and pet
foods) the last organic items consumers try. For these items, the perception
of a link between the organic version and better health is weaker, resulting
in later acceptance.

From a marketer's perspective, target groups may be statistically identified based on clusters of characteristics that include observable demographic factors, such as income, education, and geographic location, as well
as products the groups are most likely to purchase. This type of grouping
makes it possible to introduce organic food items with some assurance that
a market for the products will exist. The clusters change over time as product availability and consumer preferences change, but national surveys are
able to produce snapshots of the classes of individuals who buy organic
foods.

One such study in 2000 identified four buying demographics representing 31 percent of U.S. consumers who are multiproduct organic users.
Specialty Foodies, moderate income households of two forty-somethings
living in the Pacific region, make up 29 percent of the organic buying segment and purchase many stereotypical health foods outside the mainstream
shopping experience, such as soy and rice milk, herbal teas, tofu, ethnic
meals, and organic soft drinks. Pacific Produce Pickers, similar to Specialty
Foodies except with incomes over $75,000, make up 35 percent of the seg-
ment, and buy primarily fresh fruits and vegetables, coffee, and nuts. The
Miss American Pie demographic, comprised of single women over thirty-
five, living in the South Atlanta region, is 20 percent of the multiproduct
segment, and purchases products deemed typical of the American experience, including condiments, ice cream, meats, sauces, and dairy products.
Topper Shoppers, similar to Specialty Foodies except living in the South
Atlantic region, are only 16 percent of the segment, but buy on average of
forty-nine different products. These consumers purchase primarily condiments, sauces, and dairy products.

Although studies are aimed at finding out why consumer purchase
organic foods, they are also useful in identifying barriers to purchase. Health
concerns are the primary motivation for purchasing organic food, offering
a label that reduces exposure to pesticides, antibiotics, growth hormones,
and genetically modified organisms (GMOs). The environment has declined

in significance for consumers over time as taste, freshness, and food safety have become more important. Price and lack of availability are the two main reasons consumers cite for not buying organic foods.

Organic foods are typically more expensive, costing on average 10 to 30 percent more than their conventional counterparts. Surveys indicate mixed results about consumer response to higher priced organic food. A typical survey in 2003 indicated that 73 percent of consumers believe organic food is too expensive, confirming many other studies indicating that price is a barrier. Higher prices are less likely to reduce purchase probability if the organic benefits are perceived to be greater, such as for fresh produce or baby food. Econometric analysis indicates that organic and conventional milk, coffee, and baby food are substitutes, so that increases in the price of the conventional product result in an increase in the quantity of the organic products purchased.

Lack of availability for preferred products is still cited as a barrier to purchase despite widespread distribution of organic foods through mainstream channels. Approximately 38 percent of all organic food was sold in mainstream supermarkets in 2006, with 44 percent sold in natural product supermarkets and independent stores, and mass merchandisers and club stores accounted for an additional 12 percent. Several studies have indicated that up to 35 percent of consumers would not purchase organic foods even if the prices and availability were the same as for conventional foods. The market is not yet saturated for organic foods, so expanded availability will increase sales up to this limit. As basic attitudes about food and goals such as social justice and environmental protection evolve, this resistance to organic purchases could be eliminated.

There is evidence that consumers are being more thoughtful about the social and environmental aspects of their food purchase decisions. Values-based food purchases are made by consumers looking for purer products, locally grown foods, production systems that pay fair wages and are environmentally sustainable, products that are ecologically benign and packaged in recycled and recyclable containers. This interest has resulted in sales of natural products, of which organics is a legally defined subset, of $62 billion in 2007. The key driver behind the natural products market is a concern for the environment and for the way in which consumers treat and react to it. Cause-related marketing by manufacturers such as Ben and Jerry's, Newman's Own, and Burt's Bees supports a range of causes such as animal welfare, small-scale farming, and community development while delivering a product that is minimally processed and contains limited, if any, artificial ingredients. About 50 percent of consumers say that cause-related advertising has a positive effect on their purchase decision.

Growth in organic and natural products sales may slow with economic stagnation or recession in 2008 through 2010 if the price differential compared to conventional products does not decline. Many natural products stores such as the Whole Foods grocery retailer offer discounts and shopping hints to obtain the most for the consumer's food dollar. These approaches, combined with introductions of 1,000 to 1,500 new organic products and 1,600 to 2,000 new natural products per year, are likely to maintain consumer interest in these types of foods. As supply chains for organic foods and nonfood products become more mature and legislators take aim at lowering farm subsidies, the price differential with conventional foods will continue to fall, insuring continued, if slowed growth.

Further Reading

Dimitri, C., and L. Lohr. "U.S. Consumer Perspective on Organic Food." In *Organic Food: Consumers' Choices and Farmers' Opportunities*, M. Canavari and K. D. Olson, eds., 157–70. New York: Springer, 2007.

Hughner, R. S., P. McDonagh, A. Prothero, C. J. Shultz, and J. Stanton. "Who Are Organic Food Consumers? A Compilation and Review of Why People Purchase Organic Food." *Journal of Consumer Behaviour* 6(2007): 94–110.

Stevens-Garmon, J., C. L. Huang, and L. Biing-Hwan. "Organic Demand: A Profile of Consumers in the Fresh Produce Market." *Choices* (22) 2007: 109–15.

Zepeda, L., and J. Li. "Characteristics of Organic Food Shoppers." *Journal of Agricultural and Applied Economics* 39(2007): 17–28.

See Also: Food Deserts; Food Security; Food Sovereignty; Local Food; Marketing, Direct; Organic Consumers Association; Organic Labels; Appendix 4: Recent Growth Patterns in the U.S. Organic Foods Market, 2002

Luanne Lohr

Conventionalization

Conventionalization refers to the process by which the practices of organic farming become closer to those in conventional (nonorganic) farming. Organic farmers who conventionalize tend to follow an input substitution model of organic farming, meaning they avoid materials restricted by organic rules and regulations but otherwise rely on farming and marketing practices common to industrial agriculture, such as monocropping and selling by contract. Conventionalization has also come to mean that traditional agribusiness firms have become involved in organic farming and marketing.

The claim that organic farming is not as distinct from conventional farming as many imagine it to be was first made in a 1997 article by Daniel Buck, Christy Getz, and Julie Guthman in the journal *Sociologia Ruralis*. Later dubbed the "conventionalization thesis," the claim has sparked significant debate among scholars of alternative agriculture. Several articles, many published in the same journal, went on to test the thesis empirically or dispute the tenor of the characterization. Some studies designed to test the conventionalization thesis have found a "bifurcation" of the organic sector, such that old guard, small-scale, social movement–driven organic farmers seem to farm and market much closer to agroecological ideals, whereas new, large-scale, commercially driven producers seem to be more conventionalized. Other studies have found the organic sector to be more complex, constituted by farmers with a variety of motivations, practices, and operating scales, albeit often distinguished by crops grown, regional agrarian structures, and national regulatory contexts. In general, however, conventionalization seems to be most apparent in organic crops designed for processed foods or export markets, where large buyers often prefer to deal with conventional firms.

Whether conventionalization is a good thing is also a subject of debate. Some argue that conventionalization has allowed organics to become more widely available and accessible to consumers of all incomes and has allowed more farmers to convert to organic farming; others argue that the meaning of organics has become watered down and it no longer poses a formidable challenge to industrial farming.

Further Reading

Buck, Daniel, Christina Getz, and Julie Guthman. "From Farm to Table: The Organic Vegetable Commodity Chain of Northern California." *Sociologia Ruralis* 37(1) 1997: 3–20.

Guthman, Julie. *Agrarian Dreams: the Paradox of Organic Farming in California*. Berkeley: University of California Press, 2004.

Lockie, Stewart, Geoffrey Lawrence, and Kristen Lyons. *Going Organic: Mobilising Networks for Environmentally Responsible Food Production*. Oxfordshire, UK: CAB International, 2006.

See Also: Agriculture, Conventional; Agriculture, Organic

Julie Guthman

Cooperatives, Grocery

Food cooperatives are businesses that are owned and governed by their customers (consumer cooperatives) or by their workers (worker cooperatives). They share a commitment to providing high quality, affordably priced grocery products, to supporting their local communities, and to education. Food cooperatives may take the form of retail food stores, which are a legal entity and have a physical storefront location; buying clubs or preorder co-ops, usually an informal group pooling its resources to purchase and share food; or cooperative food warehouses, which supply food to retail co-ops, buying clubs, or other stores.

CONSUMER COOPERATIVES

The feature that most distinguishes consumer food co-ops from other grocery retailers is that co-ops are owned by the people who use them— their members. Member-owners are invited to participate in the co-op's

Cooperative groceries are now full-service markets that compete with larger chain supermarkets and natural food stores. Neighborhood food co-op in Carbondale, Illinois. Courtesy of Leslie Duram.

ownership structure (through investment) and decision-making process. This participation helps to ensure that co-ops are successful economically while meeting the needs of those members. Cooperative values and principles guide organization and decision-making.

Consumer cooperatives are governed according to principles agreed on by the international cooperative community. Based on guidelines first laid out in 1844 by members of a store started by mill workers in Rochdale, England, the International Co-operative Alliance (ICA) adopted seven cooperative principles, most recently revised in 1996. The first is open and voluntary membership, which prevents member discrimination regardless of gender, race, social position, or political or religious affiliation. An often-diverse membership organizes, patronizes, invests in, and runs the cooperative business.

Democratic control, the next principle, means that most cooperatives are run on the basis of one-member, one-vote. Active participation in decision-making, policy setting, and governance of the cooperative is encouraged, and co-ops hold annual and other membership meetings. Members also elect the board of directors, which oversees the growth, direction, and success of the co-op business.

Member economic participation, the third principle, means that members own and control the assets of the business. They contribute to the business capital (often with investment) and provide funds to ensure the financial stability of the cooperative. A few food co-ops also require or provide an option of working in the store as a means of meeting or offsetting operating costs. When the co-op thrives, member-owners may receive dividends of cash or equity (in the form of patronage rebates) in proportion to the amount of business they did with the cooperative that year. In general, the more a member patronizes the co-op, the greater the benefits to the individual, both monetarily and otherwise.

The fourth principle states that cooperatives must maintain their member control, ensuring autonomy and independence, even when entering into agreements with other groups, including the government, lenders, or suppliers.

As stated in the fifth principle, cooperatives are committed to educating their members and employees as well as the public about cooperative business. Food co-ops also typically provide information about products and food issues. Education is often accomplished via member meetings, newsletters and other publications, signage, training sessions, classes, and other presentations.

Cooperation among cooperatives, the sixth principle, means that cooperatives increase their effectiveness and strength (and so the cooperative movement worldwide) by pooling people power, knowledge, and assets. Cooperatives band together locally, regionally, nationally, and internationally.

And, finally, cooperatives make decisions that are in the best interests of their communities, the seventh principle. Food cooperatives pride themselves on being good citizens, actively contributing to and working for the health and sustainable development of their neighborhoods.

Although there are benefits of membership, not all shoppers at a consumer grocery cooperative are necessarily members of the co-op; at most food co-ops, everyone is welcome to shop.

WORKER CO-OPS

In addition to consumer cooperatives, worker cooperatives are found in the food industry. Consumers don't have ownership in a worker cooperative; instead, employees (and only employees) own shares in and run the business. The worker co-op generates benefits (wages and other) to meet the needs of its employee-owners. Workers own many natural food businesses, such as food cooperatives and restaurants.

Like consumer co-ops, worker co-ops are driven by values and principles. Workers invest in and own the business, and decision-making is democratic, with workers electing a board of directors, primarily from within its membership. (Outside directors and advisors serve on some worker co-op boards.) Workers have a commitment to the sustainability of the business and the community, including the local economy.

The management structure of worker cooperatives varies as in other businesses. In some, the workers are allowed active involvement in management decisions, whereas others use a more traditional and hierarchical style of management. There is often a probationary period of a few months or more for joining a worker co-op. In some worker co-ops, members vote on acceptance of new members.

As with consumer co-ops, worker co-ops return profits to workers—in the form of patronage rebates—based on hours worked, salary, seniority, or position.

The most well-known system of worker cooperatives in the world is in Spain's Mondragon region. What started as a worker cooperative that manufactured stoves now has hundreds of co-ops, with 23,000 member-owners in the region. It has expanded to offer cooperative insurance, education, health care, and banking. In the United States, the United States Federation of Worker Cooperatives, founded in 2004, represents the interests of worker cooperatives nationally. Some of the more well-known worker co-ops in the food industry include Equal Exchange, which insures that growers in developing nations are paid a fair living wage for their products, and Alvarado Street Bakery.

HISTORY OF FOOD CO-OPS

Many of today's food co-ops were first organized in the United States in the 1970s. The median year co-ops were founded was 1975, according to National Cooperative Grocers Association (NCGA). But the co-ops of the 1970s are actually considered a resurgence of a movement that started much earlier.

In fact, in 1828, *The Cooperator,* a monthly periodical of cooperative philosophy and advice about running a shop using cooperative principles, was published in England by William King. In 1844, industrial weavers in England who were unhappy with their neighborhood merchants formed the Rochdale Equitable Pioneers Society to provide food cooperatively. The group actively encouraged others to form cooperatives and wrote down the principles for running a successful cooperative. In the United States, as early as 1865, Michigan passed what is thought to be the first law in recognition of the cooperative method of buying and selling. In 1895 the International Cooperative Alliance (ICA) was established. An independent, nongovernmental association, today ICA represents over 800 million people worldwide, from 88 nations and over 200 national cooperative organizations.

BENEFITS OF SHOPPING AT FOOD CO-OPS

The benefits of shopping at a food co-op include those available to all co-op shoppers: high-quality food, reliable information regarding food issues, and the opportunity to support ethical product sourcing and business practices. The benefits of being a co-op member include having a voice in the governing of the co-op, receiving timely information via newsletters and meetings, and, in some cases, better prices (nonmembers may be assessed a surcharge on purchases or members may receive a discount).

Additionally, many shoppers choose to shop at their local food co-op because their purchases support values that are important to them, such as increasing local and organic food production and supporting sustainable agriculture. And many consumers enjoy the sense of community that co-op membership and shopping provides.

Natural, healthy, high-quality food has been the mainstay of grocery cooperatives, which many consider their year-round farmers market. For example, co-ops generally offer food that has been raised by local farmers whenever possible; this enables them to provide the freshest, most nutrient-rich food for its members. Purchasing locally also supports the local economy, including farming families. In addition to saving travel miles, local food is often grown with respect for the local land and local residents. Local farmers are more likely to grow their food organically and sustainably than are industrial agricultural producers.

In many communities, food co-ops are the go-to source for organics—products that have been grown without the use of synthetic pesticides, herbicides, fertilizers, synthetic growth hormones, antibiotics, or genetic engineering (GE). Co-op concern for sustainable food systems is demonstrated in many products purchased on behalf of their shoppers, from meats raised without antibiotics or growth hormones to eggs from free-range chickens to sustainably harvested seafood.

Food for those with special dietary needs (such as food sensitivities and allergies to milk or wheat, for example) or dietary practices (such as vegetarianism and veganism) are also commonly offered by grocery cooperatives to help meet the needs of their members. Most food co-ops also offer a selection of foods in bulk. These bulk items might include herbs and spices, grains, beans, nuts and seeds, and cereals. Shoppers are able to weigh and purchase the exact quantities they desire of these items. Bulk buying is economical, because most products are much less expensive per ounce than their prepackaged versions, and because consumers needn't pay for more product than they can use. Some of the earliest food cooperatives sold almost entirely from bulk bins. Most food co-ops today offer both bulk and prepackaged products.

In addition to food, grocery co-ops typically offer environmentally friendly household and personal care products, including cleaning supplies, soaps, shampoos and toothpastes, herbal remedies, and dietary supplements. Other products typically found at a grocery cooperative include handmade items like pottery, candles, soaps, and cards; and cookbooks and other food-related and wellness-related publications.

Food co-ops take education seriously, and they serve as a source of information about a wide array of food-related issues. Via publications and signage, co-ops let shoppers know where and how food is produced. They educate shoppers about the health benefits of specific foods—from standard fare like whole grains to foods in the trend spotlight, like pomegranate or the acai berry—and the importance of Fair Trade products, the partnership between consumers and producers that supports farmers, artisans, and other workers by guaranteeing that they receive a fair wage for their work and ensuring that neither they nor their environments are exploited.

Food cooperatives highlight food safety and sustainability issues (like organic standards and other agricultural and environmental concerns) and many offer cooking classes and workshops (from basic bread making to gourmet ethnic fare). Co-ops have also led the market in unit pricing and nutritional labeling, as would be expected in stores owned by the shoppers. Information is often provided about other wellness products, such as toiletries, cosmetics, cleaning supplies, and paper products. Co-ops offer shoppers flyers or workshops on topics of concern, such as recycling, chemicals in cosmetics, and making your own cleaning products.

Because co-ops are locally owned, they make decisions that are in the best interests of the community, and they give back to their communities. They seek local resources to provide goods and services, typically cultivating supportive relationships with local and regional suppliers, including farmers and artisans. They provide local jobs and are generally very generous contributors to local and community events, nonprofit organizations, and initiatives.

NATIONAL COOPERATIVE GROCERS ASSOCIATION

The National Cooperative Grocers Association (NCGA) is an example of a business services cooperative; it is owned and served by organizations (rather than individuals) with similar needs, goals, and interests—in this case, food co-ops. Started in 1999, NCGA has over 100 cooperative members that operate over 130 storefronts across the country. The smallest of these generates less than $1 million in annual sales, whereas the largest has over $100 million in annual sales, eight storefronts, and over 35,000 member-owners.

By joining together as members of NCGA, grocery cooperatives are able to strengthen purchasing power, obtaining the best prices on quality products for their members. Combined annual NCGA member sales total over $1 billion. In addition to increasing purchasing power, NCGA membership unifies food co-ops to optimize operational and marketing resources.

NCGA's concerns and the issues it advocates for reflect those of consumer grocery co-ops in general and its co-op members in particular and include sustainable food systems, fair treatment of people, and a healthy environment.

Further Reading

Consumer Federation of America. "Consumer Cooperatives." www.consumerfed
.org/coops.cfm.

Cooperative Grocer. "Food Cooperative Directory." www.cooperativegrocer
.coop/coops/.

Cooperative Grocers' Information Network. "How to Start a Food Co-op." www
.cgin.coop/public/food-coop-info/start-a-food-coop.

Fair Trade Labelling Organizations International. "Standards." www.fairtrade
.net/standards.html.

International Cooperative Alliance. "What Is a Co-operative?" www.ica.coop.

National Cooperative Business Association. "About Cooperative: Food Cooperatives." www.ncba.coop/abcoop_food.cfm.

National Cooperative Grocers Association. "About Us: Mission." www.ncga.coop/about.

National Cooperative Grocers Association. "Profiles of Cooperatives in Various Industries, Developed for Co-op Month." www.go.coop.

U.S. Federation of Worker Cooperatives. "Farther Faster Together." www.usworker .coop.

See Also: Farmers Markets; Processors, Local Independent

National Cooperative Grocers Association

Corn Syrup

Consumers are increasingly aware of the extensive use of high-fructose corn syrup (HFCS) in most processed foods. This consumption prompts health concerns among many. In addition, HFCS is typically produced from genetically modified corn, which concerns other consumers. Genetically engineered products are not allowed in organic foods, and because non-GE sources of corn are difficult to find, most organic products simply avoid HFCS and use organic sugar instead.

HFCS is a sweetener made from corn that has existed only since 1970. Using a method developed by Japanese researchers, corn starch is processed to make long-chained glucose molecules that are then treated with enzymes to convert some of the glucose into fructose. The enzyme alpha-amylase breaks the long chain of glucose into shorter chains and then glucoamylase breaks them into single glucose molecules. The third enzyme, glucose-isomerase, changes the configuration of half the glucose to convert them to fructose. The enzymes alpha-amylase and glucose isomerase are genetically modified to make them more stable.

HFCS is slightly higher in fructose content (55% fructose, 45% glucose) compared to table sugar (50% fructose, 50% glucose) which is made from sugar cane. Conversely, corn syrup contains no fructose at all, and should be thought of as different than HFCS, despite the similarity in names. It is the higher percentage of fructose that makes HFCS the sweetest of these sweeteners.

Pure fructose is only metabolized by the liver, whereas glucose can be metabolized by all cells. However, fructose is typically combined with glucose in foods and is not used in the pure form. Consumption of high amounts of HFCS has been implicated in some adverse health conditions, but scientists and corn producers continue to debate the issues. Current research topics with regard to HFCS and health include its effect on feelings of fullness, its impact on the ability to produce insulin, its relationship with the rise of obesity in the past thirty years, and its effect on blood triglycerides and uric acid levels.

HFCS is 20 percent cheaper than table sugar and well accepted by food companies because it is liquid, is easier to transport, mixes well with many foods, and improves shelf-life of processed foods. As a result, HFCS is found in soft drinks, jam, jelly, baked goods, and candy, as well as bread, breakfast cereal, English muffins, soup, ketchup, yogurt, pasta sauce, breakfast meats, mustard, peanut butter, and coffee creamer.

The major sweetener of soft drinks in the United States is HFCS. The amount of HFCS consumption by an average American has increased from .5 pound per year in 1972 to 41 pounds in 2007. However, during that same time period, consumption of refined sugar decreased from 72 pounds per person in 1972 to 44 pounds in 2007. Thus, refined sugar intake decreased with the increased consumption of HFCS.

There are some packaged snacks and beverages that do not contain HFCS. For example, Goose Island Soda, Jones Pure Cane Soda, and Jolt Cola use cane sugar, and Steaz sodas and energy drinks use only organic cane sugar. Soups from Amy's, Pacific, or Imagine contain no HFCS and are made with 100% organic ingredients. Country Choice and Back to Nature cookies are made with organic cane sugar. Foods labeled as 100% organic should be free of HFCS.

Further Reading

Corn Refiners Association. "The Facts about High Fructose Corn Syrup." 2008. www.hfcsfacts.com.

Johnson, J. Richard, and Timothy Gower. *The Sugar Fix: The High Fructose Fallout That Is Making You Fat and Sick.* New York: Rodale, 2008.

Weston A. Price Foundation. "The Murky World of High Fructose Corn Syrup." www.westonaprice.org/motherlinda/cornsyrup.html.

See Also: Diet, Children's; Health Concerns

Sharon Peterson

Country-of-Origin Labeling

The U.S. food system puts hundreds of thousands of food products on store shelves around the country with around 10,000 new product introductions each year. This long list of foods embodies an even longer list of attributes. Country-of-origin labeling is one such attribute that informs consumers where their food may be grown and produced. For example, farmers, food manufacturers, and retailers can designate the United States, Canada, Mexico, Chile, or mixed origin as the country of origin. However, all labels must be truthful, accurate, and not misleading. A food product labeled from the United States as the country of origin can be designated in this manner only if the food is exclusively a U.S. product.

Some consumers are willing to pay more for the U.S. label to ease their food safety concerns, others may be interested in supporting U.S. producers, whereas still others may prefer U.S. foods because they think they are of higher quality. They may also use the U.S. label as a way of buying foods that may not have to travel over long distances and thus, may have a smaller carbon footprint relative to imported foods.

Some producers and food companies may also have an interest in labeling their foods based on origin. Sensory and some credence attributes (product characteristics consumers individually cannot verify, such as organic, fair trade, or origin) are thought by some suppliers to influence consumers' purchasing decisions, and thereby potentially raise profits. Food companies advertise and label such products as Danish hams, New Zealand lamb, and French wines to increase sales and to attain price premiums for what they perceive as high-quality products.

INFORMING CONSUMERS

For many food products, suppliers are required or may want to inform consumers of the country of origin, although unambiguously doing so is often problematic. Our globally integrated food system blurs distinctions between foods produced in the United States, those produced in another country, or those of mixed origin. A discussion on identifying where food originates may not be that much different than the more familiar question, "Where was this automobile produced?" Cars can contain parts manufactured in many countries, and may be assembled in multiple countries as well. It would not be unusual for a "dressed" hamburger to be made from a meat patty composed of lean meat from Australia and trimmings from the United States, with Florida tomatoes, California lettuce, and Mexican onions on a bun fabricated from U.S. wheat, and with Mexican sesame seeds.

Country of origin labeling for meat and fish may be particularly complicated. Retailers can designate the United States as country of origin only if the food is exclusively a U.S. product. For meat, this means that it is derived from an animal born, raised, and slaughtered in the United States. For wild fish, the product needs to be harvested in U.S. waters or by a U.S.-flagged vessel and processed in the United States or aboard a U.S.-flagged vessel. Meat or fish are considered to be of mixed origin when the final production step occurs in the United States but one or more production steps occur outside the United States. For example, pork that is derived from a hog imported from Canada and raised and slaughtered in the United States would be of mixed origin.

BENEFITS AND COSTS

Evidence is limited on how important the country-of-origin label attribute is to the majority of consumers and how much they are willing to pay for the information. Foods compete primarily on the basis of taste, price, and quality; marketing efforts generally emphasize these attributes. Consumers are most likely to choose domestic over imported foods (or the reverse) when they can distinguish price or sensory differences. Even if some consumers do favor domestic over imported products, labeling costs may outweigh the benefits from increased demand. The costs of country-of-origin labeling depend on the number and difficulty of activities that suppliers must undertake. Costs depend on the extent to which new activities differ from current production and marketing practices and how easily firms can adjust their practices. Firms may incur three main costs in meeting country-of-origin legal requirements—labeling, recordkeeping, and operating (identifying, segregating, and tracking the food products through the supply chain). To limit costs, the 2008 U.S. Farm Bill does not require records of country of origin other than those maintained in the normal course of business for the food products covered under the legislation.

SUMMARY OF CURRENT REQUIREMENTS

Federal law—the Tariff Act of 1930 as amended, the Federal Meat Inspection Act as amended, and other legislation—requires most imports, including many food items, to bear labels informing the "ultimate purchaser" of their country of origin. Ultimate purchaser has been defined as the last U.S. person who will receive the article in the form in which it was imported. The law requires that containers (e.g., cartons and boxes) holding imported fresh fruits and vegetables, for example, must be labeled with country-of-origin information when entering the United States. If produce in the container is packed in consumer-ready packing and sold to the consumer, then that item must already be labeled as well. Consumer-ready packages, such as grapes in bags or shrink-wrapped English cucumbers, although they are packed in a box, must have country-of-origin labels on each consumer-ready package. Most recently, in the 2002 and 2008 Farm Bills, legislation was passed requiring that country-of-origin information be provided to consumers by label, stamp, mark, placard, or other clear and visible sign on the commodity or on the package, display, holding unit, or bin for muscle cuts of beef, chicken, goat, lamb, pork; ground beef, ground lamb, and ground pork; farm-raised fish and shellfish; wild fish and shellfish; perishable agricultural commodities (fresh fruits and vegetables); and ginseng, peanuts, pecans, and macadamia.

Further Reading

Carter, Colin, Barry Krissoff, and Alix Peterson Zwane. "Can Country-of-Origin Labeling Succeed as a Marketing Tool for Produce? Lessons from Three Case Studies." *Canadian Journal of Agricultural Economics* 54(4) December 2006: 513–30.

Krissoff, Barry, Fred Kuchler, Kenneth Nelson, Janet Perry, and Agapi Somwaru. "Country-of-Origin Labeling: Theory and Observation." United States Department of Agriculture WRS-04-02. January 2004. www.ers.usda.gov/publications/WRS04/jan04/wrs0402/wrs0402.pdf.

See Also: Policy, Agricultural; Policy, Food

Barry Krissoff and Fred Kuchler

Crop Rotation

Crop rotation is the practice of planting different types or species of crops on the same field, in a particular sequence. This practice has been employed for many centuries for the purposes of keeping the land productive, ensuring a constant supply of food, fiber, or feed for farm animals; improving soil fertility, planting leguminous crops, which can fix nitrogen in the soil; enhancing soil organic matter content, by planting crops with extensive root systems or that will leave more residues in the soil after harvest; and disrupting the life cycles of pests and pathogens that may otherwise establish themselves in the field if only one type of crop is grown every year.

Crop rotations vary in the number and the type (species) of crops that are planted one after the other on the same piece of land. Simple rotations involve two crops that are grown in alternating years; other rotations include three or more crops, hence requiring more years for the rotation scheme to be completed. Crop rotation systems vary widely from field to field and also from grower to grower, because of differences in environmental conditions (e.g., soil type, altitude, climate), grower preferences, farm resources (labor supply, equipment, etc.), and market conditions (demand for the produce, selling price, etc.).

There are, however, some general guidelines for ensuring good yields, maintaining healthy soils (e.g., high fertility, high organic matter content), and reducing pests and diseases (i.e., nematodes, insects, fungi, bacteria, weeds) through the employment of crop rotation. These guidelines include the following:

1. Planting crops that require higher amounts of nitrogen (such as corn) right after leguminous crops (such as soybean), to take advantage of the increased nitrogen levels
2. Refraining from growing one species of annual crop more than once in a rotation to avoid the build-up of its associated pests and diseases
3. Refraining from planting closely related crop species (such as squash and zucchini) right after one another to break the life cycles of pests and pathogens that attack related crop species
4. Including weed-suppressive plant species (such as sorghum, sunflower, or barley) to reduce weed problems in the field
5. Including crops that have extensive root systems or those that will leave significant amounts of residues in the soil in the rotation, to increase soil organic matter content.

Some examples of crop rotation schemes include: corn, soybean (two-year, two-crop rotation); corn, soybean, winter wheat (three-year, three-crop rotation); and sugar beets, beans, wheat or a cover crop, corn (four-year, five-crop rotation). Some crop rotation systems involve growing a mixture of several related species (i.e., from the same plant family) or crops that have similar cultural requirements (such as irrigation and fertilization) in one season instead of only one species of crop at a time.

Further Reading
Magdoff, Fred, and Harold Van Es. *Building Soils for Better Crops*, 2nd ed. Washington, D.C.: Sustainable Agriculture Network, 2000.

See Also: Agriculture, Organic; Agriculture, Sustainable; Integrated Pest Management; Soil Health

Camilla Yandoc Ables

Cuba

Unlike the isolated sustainable agriculture movements that have been developed in most countries, Cuba developed a massive movement with wide, popular participation, where agrarian production was seen as the key to food security for the population. Still in its early stages, the transformation of agricultural systems in Cuba has mainly consisted of the substitution of biological inputs for chemicals and the more efficient use of local resources. Through these strategies, numerous objectives of agricultural sustainability have been serendipitously reached. The persistent shortage of external inputs and the surviving practices of diverse production systems have favored the proliferation of innovative agroecological practices throughout the country.

The enormous economic, ecological, and social crisis that was unleashed at the beginning of the 1990s was the result of the high level of dependency reached in Cuba's relationship with Eastern Europe and the Soviet Union. Many studies demonstrate the depth of the crisis and almost all agree with the conclusion that it would have been much worse if there had not been the will to change to centralized planning of material resources and to work toward an equitable social structure. Government assistance, together with its encouragement of innovation, the high educational level of the population, and the exchange of resources and knowledge among the people, allowed the creation of a sustainable agriculture movement and its implementation at a national scale.

THE ORGANIC MOVEMENT

In 1992, the promotional group of the Cuban Association for Organic Agriculture (ACAO) emerged in the Higher Institute for Agricultural Science of Havana (ISCAH). A group of people, mainly technicians and professionals from different institutions interested in supporting the process of change and the redesign of Cuban agriculture on an agroecological and sustainable basis, started to work together towards organic agriculture. The first attempts were related to how to apply the organic concepts in education and research as an important way of influencing the productive sector. ACAO worked for about six years, with a committee group and an executive board. The main objectives of ACAO were: (1) to raise national awareness on the need to carry out agriculture in harmony with nature and man, as well as to produce, in an economically viable way, sufficient and healthy food for the population; (2) to implement projects for the development of an ecological agriculture, including the education and training of human resources to lead with this new approach of sustainable rural development; (3) to stimulate agroecological research and teaching, as well as the recovery and learning of the basic principles of traditional production systems; (4) to establish pilot models of farmer self-management, for their reassessment and use within the new production systems; (5) to coordinate the advising of producers, government, and nongovernment organizations in the establishment of organic and natural production systems; (6) to facilitate the exchange of experiences with foreign organizations (with emphasis on tropical and sub-tropical Latin America) and individual actors working on sustainable agriculture and rural development; and (7) to stimulate the production, certification, and marketing of organic products.

As a result of widespread activities in the country, and the results achieved in the promotion and application of organic agriculture, in November 1999 ACAO was honored with the Right Livelihood Award (also called the Alternative Nobel Prize) by the Swedish Parliament. Some months later, the ministry of agriculture decided to give the mandate carried out by ACAO until then, of the development of organic agriculture in the country to the Cuban Association of Agriculture and Forestry Technicians (ACTAF). Currently, ACTAF continues editing the *Agricultura Orgánica* magazine, organizes the National Meetings on Organic Agriculture, and carries out projects and other research and development activities, which have been the principal means to promote organic agriculture in Cuba. Its current membership exceeds 20,000 members, from all over the entire country.

Unlike many countries in the world, in Cuba the market has not been the driver for organic agriculture. There is no national certification and control system for organic productions, and this has been a great limitation for the recognition and monitoring of the developments in the sector. Few projects have pursued control and certification of organic products. Such projects have been developed by foreign certification bodies, to export Cuban products to the European market. However, most of organic production is not being certified due to the lack of differentiation for food quality in the agricultural markets.

Cuba has been labeled as the country that carried out the greatest conversion of its agriculture towards organic or semiorganic systems. Among the most successful programs based on organic and sustainable farming are the National Program for Urban Agriculture and the Agroecological Farmer to Farmer Program of the National Association for Small Farmers (ANAP). Other programs with a nationwide effect include the cultivation of rice in small plots in all areas suitable for this purpose, also called popularization of rice production, the promotion of integral farms within all sectors and branches of agriculture, and the National Program of Local Innovation of the Institute of Agricultural Sciences (INCA). The creation and dissemination of Centers for Artisanal production of Entomophagus and Entomo pathogenous (CREE) has have been successfully applied to the ecologization of Cuban agriculture.

Cuban agricultural systems applying organic farming practices have been motivated by three main factors: the lack of chemical inputs at the beginning of the 1990s made it necessary to find other alternatives and adaptation to the new scenario (input substitution), environmental conservation and the use of locally available resources, and positive results in organic systems' efficiency and productivity.

Although certified organic agriculture has increased between 1998 and 2008, the volume of organic products likely would have experienced a substantial decrease in the same period. Although the practices and methods of organic farming were massively applied in the 1990s (the use of biopesticides, organic soil fertilization, the use of animal traction, the use of polycropping and other diversified crop arrangements, and the use of local feeds for animal nutrition), the recent recovering of the Cuban economy has favored the return to conventional practices and there is an increasing access to chemical inputs. Such trends put the growing capacity of the organic sector at risk.

Organic production systems in Cuba are strongly linked to the agroecological concept that combines productive, ecological, economic, and social components of agricultural sustainability. Due to the high heterogeneity, diversity, and dynamism of organic systems on the island, it would not be possible to characterize a typical Cuban model of ecological production. The lack of standards makes organic agriculture conceived as an attitude towards the rational use and conservation of locally available resources. Hence, in practice it is difficult to precisely determine which proportion of the Cuban land could be considered to be in organic use. Many farmers and agriculture officials have misunderstood ideas about organic agriculture. Most of them consider themselves organic simply because they don't utilize chemical inputs (by default). Studies have demonstrated that self-accredited organic farmers would use chemical inputs if available. The common situation currently is a mixture of organic and conventional practices.

ACHIEVEMENTS AND CHALLENGES

The transformation that occurred in the Cuban countryside during the last decade of the twentieth century is an example of a large-scale agricultural conversion from a highly specialized, conventional, industrialized agriculture, dependent on external inputs, to an alternative input–substitution model based on principles of agroecology and organic agriculture. Numerous studies of this conversion attribute its success to both the form of social organization employed and the development of environmentally sound technologies.

However, further steps—indeed, profound changes—are necessary in Cuban agriculture. Although innovation has been present in all branches of agriculture and the scientific institutions have tested environmentally sound technologies on a large scale, these efforts have tended to focus on the substitution of inputs, and there remains a disjunction between the biophysical and socioeconomic aspects of agricultural development. If this newest stage in Cuban agriculture, characterized by the emergence of diverse agroecological practices throughout the country, is to progress further, neither the conventional pattern nor that of input substitution will be versatile enough to cover the technological demands and socioeconomic settings of the country's heterogeneous agriculture. Therefore, it is now necessary to develop more integrated, innovative, and locally oriented solutions rather than continuing to solve specific problems from the top down.

Despite the existence of an organic movement in Cuba for about fifteen years, many of its original goals have not been reached. The golden opportunity for Cuba to make a consistent transition from a large-scale, monoculture, and centralized conventional agriculture to a small-scale, diversified, and decentralized organic model is being threatened. As in most countries, organic agriculture in Cuba faces challenges and goals that will not be reached in the short term. Many see that more innovative and adaptive initiatives and innovations need to be continuously created to achieve sustainable agriculture. Cuba is probably in better condition than any other country to make the large-scale transition possible.

Further Reading

Funes, Fernando, Luis García, Martin Bourque, Nilda Pérez, and Peter Rosset, eds. *Sustainable Agriculture and Resistance: Transforming Food Production in Cuba.* Oakland, CA: Food First Books, 2002.

Funes-Monzote, Fernando R. "Farming Like We're Here to Stay: The Mixed Farming Alternative for Cuba." PhD thesis, Wageningen University, The Netherlands, 2008.

Rosset, Peter, and Medea Benjamin. *The Greening of the Revolution: Cuba's Experiment with Organic Agriculture.* Melbourne, Australia: Ocean Press, 1994.

Wright, Julia. "Falta Petroleo! Cuba's Experiences in the Transformation to a More Ecological Agriculture and Impact on Food Security," PhD thesis, Wageningen University, The Netherlands, 2005.

See Also: Agriculture, Organic

Fernando Funes and Fernando R. Funes-Monzote

D

Development, Sustainable

Although the ideas behind sustainable development are not solely modern, the term itself became a popular liberal idea in the 1980s. It emerged from the more radical "ecodevelopment" of the 1970s and has recently been largely supplanted by the broader term "sustainability."

EARLY HISTORY

Renaissance philosopher Nicholas Machiavelli wrote in *First Book of the History of Florence* that "the people which live north-ward beyond the Rhine and Danube, being born in a healthful and prolifick Clime, do many times increase to such unsustainable numbers, that part of them are constrained to abandon their native, in quest of new Countries to inhabit." This observation recognized the fundamental limits of natural resources, which economists call the primary economy and which satisfy people's basic physical needs. It typically includes all natural resource exploitation, such as farming, fishing, forestry, and mining (including oil, sand, gravel, etc.), though solar energy is rarely specifically mentioned.

In 1798, economist Thomas Malthus published his *Essay on the Principle of Population*, stating that although growth in resource exploitation could occur only in a straight-line, linear fashion, population grows in a geometric, exponential way, so that population is bound to outrun resource availability. In doing so, Malthus foreshadowed the idea of what today would be called the "ecological footprint" or larger ecological effect of a population's life activities. Like many still today, however, he did not refine his theory by including the factor of technological efficiency, whereby larger populations may be sustained by more ecologically sustainable practices.

In the nineteenth century, the growing industrial revolution, in the secondary economy of adding value to natural resources through manufacturing, raised a number of serious questions, both about the effect of resource limitations on social development as wealth disparity grew quickly and about how long such economic growth, in the forms it took, could last. The easy availability of petroleum, however, along with its fluid form and extremely high energy content, only added to the modern surge in economic activity. This in turn has resulted in both the extreme wealth of today's richest countries and the commensurately powerful resource flows away from the poorest countries.

THE 1970s AND ECODEVELOPMENT

Ecodevelopment was a development strategy popular in less-developed countries that eschewed economic growth as a primary consideration, forgoing even the concept of national boundaries as a useful development guide in favor of an ecoregional approach.

The term "ecodevelopment" was used as early as 1973, by Maurice Strong, then Secretary General of the United Nations Environment Programme. In 1975 UNEP defined ecodevelopment as "development at regional and local levels . . . consistent with the potentials of the areas involved, with attention given to the adequate and rational use of the natural resources, and to applications of technological styles . . . and organizational forms that respect the natural systems and local sociocultural patterns."

Ecodevelopment was further promoted by the International Research Center on Environment and Development in Paris (CIRED), especially by Ignacy Sachs. It was suggested for national development plans in both Africa and Latin America, including El Salvador, Costa Rica, the Dominican Republic, and Mexico. The UN's Peter Bartelmus was still strongly advocating for ecodevelopment in 1986, and the term was used as late as 1990 in a paper on Costa Rica.

Concerns included the global loss of agricultural land, whether by development, loss of soil quality, or simple erosion. Ecodevelopment cast the question as one of ecological stability, applied to human development patterns. Since this ecological focus did not maximize the economic valuation or exploitation of natural resources, mainstream development professionals sought a different path.

THE 1980s AND SUSTAINABLE DEVELOPMENT

In 1981, the Worldwatch Institute's Lester Brown published *Building a Sustainable Society*, focused more on satisfying broader social needs in their ecological context rather than narrower economic demands. He noted growing food insecurity, downward land productivity trends since 1970, and typically presciently, the use of grain for feed and food-fuel competition as major problems to be solved. He called for soil security, stable biological systems, and preserving the web of life as major strategies for sustainability.

Counter to this effort, the World Commission on Environment and Development (WCED), chaired by Gro Harlem Brundtland, was convened in December, 1983 as a result of a United Nations General Assembly resolution. Its mandate included studies of agriculture and food security, among other topics. In 1987, the WCED published its report, Our Common Future, giving equal weight to considerations of environment and development.

Whether this was in response to the success of the more radical ecodevelopment model and the perceived need to compromise with business interests has yet to be demonstrated with finality. At the least, the surprisingly vigorous publicity surrounding the work of the World Commission on Environment and Development may be seen as an attempt to draw attention away from the growing acceptance of ecodevelopment in less-developed countries.

Whatever its intent, the new term was widely promoted and taken up, in some cases by progressive authors paradoxically unsympathetic to its "three-legged stool" approach, now the "triple bottom line," which incorporates economic, social, and environmental concerns as more or less equal factors. By 1990, "sustainable development" was a term used in ecology, geography, economics, politics, international relations, regional and landscape planning, and, notably, the design professions of landscape architecture and architecture.

One of the reasons for the decades-long success of the term "sustainable development" has been its usefulness to both those who would draw attention to the ecological limits of the current globally dominant development path and to those who approve of the current development path and would like to make it permanent.

Nonetheless, the introduction of the economy as a co-equal partner with the environment and society, however unwise, did allow the development of a number of resources that are stepping stones to a more sustainable human ecology. The choice of the word "development" was certainly preferable to the term "sustainable growth." Many critics had stressed that "development" could contain a much richer meaning than "growth," as growth was and continues to be seen in terms defined exclusively by the dominant economic path. Yet the noun remained "development," as opposed to the current trend toward developing sustainability.

THE 1990s

As a follow-up to the WCED, the Earth Summit was held in Rio de Janeiro in 1992, resulting in Agenda 21, a program for promoting sustainability for the twenty-first century. The results were largely ignored by the outgoing Bush administration but adopted enthusiastically by the incoming Clinton administration.

The structure of President Bill Clinton's President's Council on Sustainable Development (PCSD, 1993–96) emphasized the Brundtland vision of not interfering with business as usual, being co-chaired by Jonathan Lash, president of the World Resources Institute, and David Buzelli, vice-president of the Dow Chemical Company.

If the WCED had compromised the meaning of sustainability by insisting on the sustainability of the current economic system, the PCSD embraced business as usual. Its definition of sustainable agriculture was "agriculture that combines modern technological innovation with proven resource conservation . . . food and fiber production practices that protect environmental quality and maintain and enhance profitability. . . ." The abridged version of the Council's reports is very diplomatically written, mentioning neither genetic engineering nor organic agriculture, but clearly referring to both while explicitly emphasizing only the need to minimize pollution and enhance conservation.

Following the lead of Agenda 21 and the PCSD, Canada and other nations followed in the "sustainable development" mode. Since this formulation allowed business as usual, though, the basic patterns of which were seen by many as part of the problem, theorists soon began to shift the debate toward the sustainability of the global human ecological system.

THE TWENTY-FIRST CENTURY AND SUSTAINABILITY

In the early years of the twenty-first century, both the popular and professional discourse moved to considering the global transition to sustainability. Given an increased global awareness, any single nation's sustainable development was no longer a satisfying or even relevant goal.

There are still both stronger and weaker meanings of "sustainable," with most focusing on the primary, natural resources economy; some considering the secondary, manufacturing economy (e.g., linking the benefits of renewable energy production to jobs, foreign policy, and climate change), and very few proposing a whole-system approach. One notable exception includes the relatively new field of ecological economics.

The human and ecological costs of maintaining the current dominant global development path are great, and the more delay there is in addressing these, the more wrenching the ultimate transition will be. Already there are "environmental refugees" fleeing both drought and flooding caused by greenhouse gases released from petroleum and its by-products. While State, municipal, and local efforts slowly gain momentum, nation-states lag, and the easily apparent brittleness of the current narrow path is creating global anxiety. The fundamental question remains whether the current formulation of sustainability suffices for a transition to an ecological and social future that can provide for the continued evolution of human activity.

Further Reading
Bartelmus, Peter. *Environment and Development*. Boston: Allen & Unwin, 1986.
Brown, Lester. *Building a Sustainable Society*. New York: W. W. Norton, 1981.
Sitarz, Daniel, ed. *Sustainable America: America's Environment, Economy, and Society in the 21st Century*. Carbondale, IL: Earth Press, 1998.

See Also: Fair Trade; Food Sovereignty; Globalization; Green Revolution; Political Ecology of Food; Shiva, Vandana (1952–)

Thomas W. Hutcheson

Diamond v. Chakrabarty

Ananda Mohan Chakrabarty, a molecular scientist working for General Electric, requested a patent on a bacterium (derived from the *Pseudomonas* genus) that was designed using molecular (but not rDNA) techniques to dissolve crude oil and was intended to treat oil spills. His request was rejected by a patent examiner, because living organisms were not legally defined as patentable materials.

Chakrabarty appealed to the Board of Patent Appeals and Interferences, but it agreed with the original decision. Chakrabarty then appealed to the United States Court of Customs and Patent Appeals and it ruled in favor of Chakrabarty's position. The Appeals court held that living organisms were like any other invention. Sidney A. Diamond, Commissioner of Patents and Trademarks, appealed to the Supreme Court. The Supreme Court case was argued on March 17, 1980 and decided on June 16, 1980. In a 5–4 ruling, the court ruled in favor of Chakrabarty, and upheld the patent, holding that a "live, human-made micro-organism is patentable subject matter under [Title 35 U.S.C.] 101. Respondent's micro-organism constitutes a "manufacture" or "composition of matter" within that statute."

The majority focused on language in the original patent act that seemed to provide extremely wide coverage; in addition, as late as 1952 Congress had confirmed this interpretation by decreeing that patents could be granted for "anything under the sun." That is, any device or technology "made by humans" could be patented if it does not occur naturally and is a product of human ingenuity. Chakrabarty's engineered bacterium met the definition of a human-made technology.

The minority argued in its dissent that Congress had specifically taken up this issue through the establishment of the Plant Patent Act of 1930 and the Plant Variety Protection Act of 1970. If the original patent act had covered living organisms, then Congress would not have been required to enact separate legislation for agricultural innovations such as improved crop varieties. Furthermore, the 1970 Act had specifically excluded bacteria from matter or organisms that could be patented.

The decision was an extremely critical building block for the biotechnology or life sciences industry. The expansive ruling assumed no central role for Congress in shaping the development of the emerging life sciences industry. The majority took the position that the necessary legislative deliberations and authority in regards to patenting living organisms had already taken place and been established. A narrower ruling or one that held existing legislation was insufficient to patent living organisms would have placed the issue of technology assessment and the future direction of the life sciences industry under the aegis of Congress. Critics of the nature of the majority's opinion have argued that the court would have been on firmer legal ground if it had decided that the nature of human invention had changed sufficiently such that Thomas Jefferson's original notion of intellectual property and patentable matter was hopelessly out-of-date. Due to this ruling, corporations have been obtaining patents on many things in nature (i.e., trees, plants, bacteria) that will then be the basis of genetic modifications, which will be sold at a high cost to bring profit. Critics argue that nature should be available to all people in perpetuity.

Further Reading
Diamond v. Chakrabarty. 447 U.S. (303) 1980, 306.
Jasanoff, S. *Designs on Nature: Science and Democracy in Europe and the United States.* Princeton, NY: Princeton University Press, 2008.
Jasanoff, S. *Science at the Bar: Law, Science and Technology in America.* Cambridge, MA: Harvard University Press, 1995.

See Also: *Bacillus thuringiensis* (Bt) Corn; Genetically Engineered Crops; Roundup Ready Soybeans; Substantial Equivalence; Terminator Gene

Rick Welsh

Diet, Children's

As babies develop they are gradually introduced to solid foods, such as rice cereal, between four and six months of age. Gradually, pureed vegetables and fruits can be given to them. There are many organic baby cereals and baby foods available in today's grocery stores to provide infants with food that is free of pesticides, antibiotics, genetically modified ingredients, and other chemicals. It is not very difficult to prepare organic baby food at home, even if the rest of the family doesn't have organic food practices. However, buying organic foods can result in relatively higher grocery bills when compared to buying for a nonorganic diet. Parents and the medical community are particularly concerned about children's diets because kids

Children tend to make smart choices about the food they eat if given healthy options. Courtesy of Leslie Duram.

eat a much higher percentage of their body weight than adults each day. Indeed, it is generally accepted that young children eat four (or more) times as much food per pound body weight as average adults. This increases their risk of exposure to pesticides and other toxins.

When infants approach nine to ten months, they can be offered drinks in sippy cups instead of bottles and can be introduced to 100% organic fruit juices. Large amounts of apple and pear juice can cause diarrhea in children less than a year old because these juices contain Sorbitol, a sugar alcohol that is poorly absorbed. Foods with added sugar and high-fructose corn syrup should be avoided to ensure a nutrient-rich diet for infants.

Although honey is 100% natural and better than sugar because it is not processed or refined, foods made with honey can cause botulism poisoning in children less than a year old. Honey can contain spores of *Clostridium botulinum* that can germinate in the body to produce toxins that cause weakness, respiratory problems, and even paralysis.

After an infant reaches one year of age, organic whole milk can be given instead of formula or breast milk. Organic milk comes from cows reared organically, receiving no hormone or antibiotic injection, and fed only 100% organic feed. As toddlers reach two years of age and become familiar with a wide variety of foods, the new "MyPyramid" (www.mypyramid.gov) can be used as a guide. A colorful, customized, printable meal plan can be created.

For example, MyPyramid recommends 1 cup of fruit and 1 cup of vegetables per day for a two-year-old boy who exercises less than 30 minutes a day. However, 1.5 cups of fruit and 2 cups of vegetables are recommended for a six-year-old boy who exercises 30 to 60 minutes per day.

MyPyramid also recommends grain servings for children. A four-year-old girl who exercises more than 60 minutes per day should aim for 5 ounces of grains. When recommended amounts of whole grains, fruits, and vegetables are eaten, it should provide appropriate levels of fiber, vitamins, and minerals. Fiber supplements are not recommended for young children. Likewise, vitamin and mineral supplements are usually not necessary for children who eat recommended amounts of organic produce.

In addition to being free from toxic chemicals and pesticides, organic produce usually has higher nutrient content than conventionally grown produce. For example, researchers have found that one organic tomato had the antioxidant equivalent of four conventionally grown tomatoes. Children have smaller stomachs and therefore need frequent small meals and nutritious snacks, so organic foods can provide higher quality nutrition for young children.

Parents sometimes think the amount of produce eaten is more important than the source of the produce. Many people might not know that produce imported from South America has high levels of pesticide residues, including some pesticides that are banned in the United States. According to the National Research Council, the major source of exposure to pesticides for infants and children is through their diet. Pesticides can cause neurologic, dermatologic, and respiratory problems and even cancer, miscarriage, and birth defects. Organic fruits and vegetables consistently have one-third as many pesticide residues as those grown conventionally. For example, researchers have found a dramatic decrease in the level of organophosphorus pesticide residue in a group of children when they were given an organic diet.

The United States Environmental Protection Agency (EPA) had planned to ban a pesticide commonly known as DDVP (Dichlorvos) on several occasions during the 1970s for its potential carcinogenic effects and dangers to the central nervous system, especially in children, but never banned it. Information such as this often causes many parents to choose organic baby foods, but the better taste and higher nutrient content of organic produce is also a factor.

As children grow and become more active, their calorie and protein requirements increase with age. For example, a two-year-old boy needs about 1,000 calories and 2 ounces of protein per day, whereas a six-year-old boy needs about 1,600 calories and 5 ounces of protein per day. The highest recommended sources of dietary protein are meats, fish, poultry, dairy products, and legumes. Organically reared farm animals are treated differently than conventionally raised animals because they are fed only high-quality organic feed and given no hormones or antibiotics. MyPyramid recommends 2 cups of milk per day and 2 to 5 ounces of meat/beans per day for young children.

When it comes to dietary fat intake, MyPyramid recommends 3 to 4 teaspoons of fat/oil per day for young children. Approximately 70 percent of children in the United States consume higher amounts of dietary fat (including saturated fat) than recommended. To prevent future elevated blood cholesterol and heart disease, a lower-fat diet with a variety of nuts and no added trans-fat can provide children with adequate dietary fat that is necessary for normal growth and development. Nutritionists recommend that about 30 to 40 percent of total calories come from dietary fat for one- to three-year-olds, and 25 to 35 percent of total calories come from dietary fat for older children.

Adequate dietary calcium is also needed to develop strong, dense bones to reduce risk of having osteoporosis later in life. The daily recommended calcium allowance is 500 to 800mg for children two to eight years of age. Up to 20 percent of U.S. children do not consume adequate amounts of calcium despite the public relations efforts to encourage kids to consume 2 cups of milk/dairy products per day. Because of increased demand for hormone-free and antibiotic-free dairy products, organic milk, cheese, and yogurt are increasingly offered in mainstream grocery stores.

The most common nutrient deficiency in children is iron deficiency anemia, which can affect a child's stamina, ability to learn, intellectual performance, and resistance to illness. Fortified grains and cereals, lean red meat, eggs, and raisins are promoted as some good sources of dietary iron for children. Iron supplements are only prescribed with diagnosed anemia and only for as long as it takes to replenish iron levels. Iron overdose from supplements is a major cause of poisoning deaths among children under six years of age. Children may be able to avoid iron deficiency anemia by eating iron-rich organic foods, which are often higher in iron compared to conventionally grown iron-rich foods.

Children should not consume too much salt. The adequate intake of sodium/day for children one to five years old is 1,000 mg. Canned soups and most processed foods contain large amounts of sodium. When reading food labels, one should keep in mind that 140 mg or less per serving is considered a "low-sodium" product. It is also recommended that children's consumption of soft drinks be limited or eliminated, because of the high sodium and high-fructose corn syrup content.

Five and a half to seven cups of water are recommended for children one to eight years of age, and 80 percent should come from drinking water and other fluids and 20 percent from watery foods. Children who are physically active need to drink more water, especially in the summer. Organic sports drinks that are made with pure organic cane sugar are recommended for children who will be exercising for more than an hour at a time.

To ensure adequate nutrition for children, some believe that it is important to provide foods with a variety of colors from every food group and to limit the intake of artificial preservatives, food colors, added salt, sweeteners, saturated fat, and trans-fat. Sometime it is difficult to ensure adequate nutrition for children because they go through phases where they will only eat certain foods and reject everything else. The general guideline is to keep offering a "new food" with the foods the child already eats. It is estimated that a "new" food may need to be offered up to ten times before a child will accept it.

Creativity is important when feeding young children and helps parents incorporate healthy ingredients into recipes to improve the nutritional value of meals. For example, organic vegetables and meats can be added to soups, sandwiches, lasagna, pasta sauce, or pizza. Organic milk can be added to soups or puddings, and organic fruits to breakfast cereals and milk shakes. If a parent really wants to make sure a child's food is from a source that is free of chemicals, pesticides, hormones, antibiotics, and genetically-modified ingredients, the only way to do this is to "buy organic."

Further Reading

Grosvendor, Mary, and Lori Smolin. *Nutrition: Everyday Choices*. Hoboken NJ: Wiley and Sons, 2006.

ShapeFit. "Organic Foods—Natural Foods with No Preservatives or Fertilizers. http://shapefit.com/organic-foods.html.

Weston A. Price Foundation. "The Double Danger of High Fructose Corn Syrup." www.westonaprice.org/modernfood/highfructose.html.

See Also: Corn Syrup; Farm to School; Fast Food; Gardens, Children's; Health Concerns; Organic Foods, Benefits of

Sharon Peterson

E

Education and Information, Organic

The state of education and information related to organic food broadly reflects the current status of organics in the overall food industry. Sales of organic foods constitute approximately 3 percent of the U.S. food industry, or about $20 billion in annual sales as of 2007. Yet, the organic market has consistently outpaced the conventional food economy, with a 20 percent sales increase per year since 1990.

Such consistent, rapid growth indicates a pattern that goes beyond a mere trend; a pattern that indicates consumers are becoming more educated and informed about organic products.

However, the vast majority of promotional expenditures and consumer exposure to food stems from the conventional food industry, which spent approximately $7 billion in advertising in 1997. According to a report by the United States Department of Agriculture's Economic Research Service (USDA ERS), most of the $7 billion centered on highly processed and packaged goods. Advertising dollars spent on meat, fruits, and vegetables were statistically negligible.

In contrast to money spent on commercial advertising, the USDA spent more than $300 million on nutrition education, evaluation, and demonstrations—a mere fraction of dollars devoted to advertising for coffee, soft drinks, beer, candy, and cereal.

In the classroom, more formal channels of education on dietary guidelines in general are through primary and secondary educational institutions. Yet there is nothing specifically tailored to organic or local food.

In postsecondary education, organic programs are slowly being integrated into major universities. For example, Michigan State University, University of Florida, Washington State University, Colorado State University, and University of Guelph (Canada) offered a major or a certificate program in organic agriculture.

The bulk of information promoting, educating, and informing consumers about organic and local food is through the Internet, grassroots organizations and programs, land grant extension agencies, and the increasing amount of exposure in chain grocery stores devoted to natural and organic products.

Information about organic products faces several barriers to being understood by consumers. Organic food is perceived as being higher priced; it also suffers from a "granola" or elitist stigma. However, health concerns related to industrial food production are creating more awareness about diet-related diseases and processed food. As a consequence, information about organic food is starting to take hold in the mainstream mentality as a healthier, safer alternative to conventionally produced food.

Information about organic farming has been established and formalized by the USDA through Organic Standards that were established in 2002. In 2007, the USDA established rules for pastured, or grass-fed, beef, which comes from cattle raised using environmentally sustainable standards and practices.

University extension services sometimes offer organic certification training, but this varies by state. Overall, critics complain that there is a lack of education among conventional growers who are accustomed to industrial, large-scale growing methods. Efforts to inform conventional growers about the opportunities of switching from traditional to organic methods are lacking.

Issues of climate change, unstable energy markets, and environmental practices are also leading to a greater awareness of the benefits of organic and local foods from an ecological perspective.

Further Reading

Iowa State University, Agricultural Extension. http://extension.agron.iastate.edu/organicag/.

Midwest Organic and Sustainable Education Service. www.mosesorganic.org.

Northeast Organic Farming Association. www.nofa.org.

University of California, Santa Cruz, Center for Agroecology & Sustainable Food Systems. http://casfs.ucsc.edu/education/.

USDA, Sustainable Agriculture Research and Education. www.sare.org.

See Also: Research

Jerry Bradley

Endocrine Disruptors

Endocrine disruptors and endocrine-disrupting chemicals (EDCs) are both naturally occurring in specific plants and human-made (synthetic) chemicals. These compounds or substances interrupt or impede the communication signals of the human body, specifically the hormonal messaging that occurs within the endocrine system. Endocrine disruptors interfere with the body's hormone regulation, which typically requires messages to travel to distant, specific target organs via the body's "ductless system" of endocrine glands. For example, androgen is the male sex hormone that must turn on within early embryonic development. Endocrine-disrupting chemicals prevent the necessary communication by attaching to a specific receptor, and the message "the baby is a male" is prevented from being delivered.

Endocrine-disrupting chemicals are known to act at extremely low levels. Pesticides, some natural plants, and drugs like Diethylstilbestrol (DES) have the ability to mimic natural estrogens. Several species ranging from wildlife to humans can develop mutations after exposure to certain pollutants. Hazardous substances enter the body through ingestion, inhalation, and contact with the skin. Humans are top predators, and with the consumption of food they risk the consequences that occur from biomagnification and bioaccumulation. Biomagnification is the increase in concentration of a pollutant from one link (organism) in a food chain to another link within the food chain. Bioaccumulation is the difference between an exposure to a chemical and the uptake, storage, and degradation (ability to break down the exposure) within an organism. Perhaps the greatest threat from food consumed by humans occurs from bioconcentration. Bioconcentration is when a chemical has the tendency to increase in concentration levels as it moves upward in the food chain. A pollutant from the environment, such as a pesticide used to increase crop production, enters the food chain from direct application to the crops. Additional sources of pesticide exposure include aerial drift and pesticide runoff into nearby streams that can enter the aquifer. Typical health concerns are generally thyroid disruption from exposures and the tetrogenic or transgenerational effects of EDCs. Tetrogenic or transgenerational effects occur when the offspring of a species are at the greatest risk for disease and not the primary person exposed.

Sustainable food production using concepts from sustainable agriculture could lower the exposure to EDCs. Riparian buffer systems are streams that are lined with trees, shrubs and grasses. Riparian buffers could improve water quality and reduce the impact to adjacent land from agriculture runoff. Suggestions that may lower EDCs while maintaining food production practices include: minimizing pesticide use, spraying pesticides only in problem areas, and switching to organic farming. Commodity and price support programs could be restructured to allow farmers to realize the full benefits made possible through alternative agriculture practices. Maintaining food production without increasing the amount of EDCs that are already in the soil, water, and air is the goal. Sustainable foods attempt to incorporate environmental health, economic profitability, and socioeconomic equity. When applied properly, this concept may be a preventive tool or damage control for the current crisis of EDCs.

Further Reading

Carson, Rachel. *Silent Spring.* Boston: Houghton Mifflin Co., 1962.

Colburn, T., D. Dumanoski, and J. P. Myers. *Our Stolen Future.* New York: Penguin Books, 1996.

Damstra T., S. Barlow, A. Bergman, R. Kavlock, and G. Van Der Kraak. Global Assessment of the State-of-the-Science of Endocrine Disruptors. IPCS (International Programmee on Chemical Safety). 2202. Geneva, Switzerland: World Health Organization. www.who.int/ipcs/publications/new_issues/endocrine_disruptors/en/.

"Endocrine Disrupting Chemicals." Center for Bioenvironmental Research at Tulane and Xavier Universities. http://e.hormone.tulane.edu.

Wingspread Consensus Statement. 1991. http://www.ourstolenfuture.org/Consensus/wingspreadimmune.htm.

See Also: Buffers; Health Concerns

Annette M. Hormann

Environmental Issues

Agriculture uses 38 percent of the world's lands and is the greatest effect that humankind has on the environment. The five billion hectares in agriculture exceed the area in woodlands or forest and each year more than thirteen million hectares in natural ecosystems are converted to agricultural use. The most dramatic example of the effects of agriculture on the ecosystem is the loss of tropical rainforests in the Amazon, Congo, and Southeast Asia to export crops and livestock.

Agriculture deliberately maintains ecosystems in simplified, disturbed, and nutrient-rich condition. In many instances, agricultural monocultures replaced natural ecosystems that contained thousands of plant, insect, and vertebrate species. This simplification of ecosystems caused a loss of biodiversity, the extinction of species, and the dominance of vast tracts of land by once rare cereals (barley, maize, rice, and wheat), legumes (soybean), and tubers (potato). Although there were great gains in yield, the genetic diversity of crop plants has narrowed. Current agricultural productivity relies on control of pests through chemical pesticides and managing soil fertility via chemical fertilizers and irrigation. Sustainable agricultural practices are being adopted that reduce the environmental impact.

The conversion of natural ecosystems to agricultural uses impacts soils and the global carbon cycle. Compared to monocultures of annual crops, many natural ecosystems are net carbon sinks buffering global inputs of carbon dioxide from industrialization. Also, cultivation through impacts on soil organic matter alters the global carbon cycle. Agricultural systems have reduced carbon inputs into the soil and accelerated decomposition of soil organic matter. In tropical regions, within the first fifty years of cultivation, 40 to 70 percent of original soil organic matter is lost. Cultivation also contributes to soil erosion, the loss of soil at a faster rater than it is deposited. Erosion removes the most productive topsoil, leaving behind subsoils that are less fertile and less able to produce crops. Global topsoil loss is estimated at fourteen million metric tons annually. Soil particles from eroded soils can clog rivers, lakes, and reservoirs. This reduces water quality, increases flooding, lowers reservoir capacity, and alters aquatic habitats.

Agriculture is a major user of surface and ground water and effects water quality. In the United States, agriculture accounts for 80 percent of the comprehensive water use. Worldwide, during the last thirty-five years, there was a doubling of food production, with an 1.68-fold increase in irrigated cropland. Agricultural water use conflicts with needs for ecosystem services, human consumption, and industrial activities. For example, in the western United States, water use has been shifted from agriculture to support freshwater fisheries and urban population growth. Agriculture depletion of groundwater has become a serious problem in areas as diverse as the high plains of the United States and the coastal areas of Bangladesh.

Agriculture also has reduced water quality through riparian ecosystem destruction, soil erosion, contamination with animal wastes, and nutrient and pesticide pollution. Confined livestock production leads to large amounts of waste that contribute human pathogens and plant nutrients to water. This water contamination can make drinking water unacceptable for human use and require extensive treatment before consumption. Agriculture harms aquatic foodweb structure through loss of biodiversity, outbreaks of nuisance species, shifts in structure of food chains, and impairment of fisheries.

Biological productivity in most waters is controlled by nutrient (nitrogen, phosphorus, and iron) supply in surface waters. Nitrogen, phosphorus, or iron fertilizer contamination can cause excessive algae growth. As the algal cells die and decompose, the dissolved oxygen in water is decreased, leading to death of fish and other aquatic organisms in the world's oceans.

A large portion of oceans in the tropics and subtropics has extreme nitrogen depletion and is acutely vulnerable to nitrogen pollution. In many other aquatic ecosystems, phosphorus is the limiting nutrient.

Supplementing crop nutrient requirements with mineral fertilizers is necessary in current high-yielding agricultural systems. The doubling of world food production has required a 6.87 and a 3.48-fold increase in nitrogen and phosphorus fertilization, respectively. The United Nations estimates that on a global scale, by 2020, 70 percent of plant nutrients will come from mineral fertilizers. Commonly, crops absorb less than 50 percent of applied fertilizers. Environment problems from mineral fertilizers are due to imbalance or lack of closure of nutrient cycles. Fertilizers are significant contributors to the greenhouse gases involved in global warming. Nitrous oxide and ammonia originating from fertilizer decomposition or animal agriculture can contribute to acidification of rain and surface waters. The addition of a limiting nutrient (such as nitrogen or phosphorus) can cause weedy species to dominate ecosystems. For example, addition of nitrogen fertilizers can cause prairie grasslands to become virtual monocultures of otherwise rare agricultural weeds.

High crop yields in monocultures is dependent on pest control with synthetic pesticides. In 2008, the global pesticide market was $52 billion and was anticipated to average an 8 percent annual increase over the next five years. Improper use of pesticides can kill or injure nontarget organisms. For example, broad-spectrum insecticides can directly kill predators and parasites of insect pests, allowing pest populations to explode. Pests can develop resistance to pesticides. Declines in amphibian populations have been blamed on some pesticides having hormone-like activity. People can be poisoned when applying pesticides or being exposed to them. Chronic effects (such as cancer) may occur from long-term, cumulative exposure to some pesticides in foods, water, or the environment. Some pesticides (such as DDT) persist in the environment and bioaccumulate, causing declines in predator populations.

Agriculture production contributes to greenhouse gases and global warming. Conversion of vast areas of natural ecosystems to agricultural lands has eliminated their ability to sequester carbon. Manufacturing agricultural inputs (fertilizers, pesticides, and farm equipment), crop and animal production, food processing and global movement of agricultural products contributes large amounts of carbon dioxide to global warming. Large-scale animal production (especially production using high-protein diets) results in large amounts of methane being emitted from the livestock.

Agricultural practices have been developed to minimize environmental damage. Integrated Pest Management (IPM) uses a range of practices to control pests while minimizing use of commercial pesticides. For example, IPM sets pest thresholds, monitors and identifies pests, uses preventative measures, and practices mechanical and cultural controls (like trapping or hand-weeding). Conservation tillage (reduced or no-till) grows crops with minimal tillage, avoiding soil compaction and leaving plant residues on the soil surface to reduce erosion. Sustainable and other ecological approaches to agriculture have goals to maintain environmental health, economic profitability, and social and economic equity. Research and education hold promise to resolving the tension between agricultural production, resource conservation, and environmental well-being.

Further Reading

Ayoub, A. T. "Fertilizers and the Environment." *Nutrient Cycling in Agroecosystems* 55 (1999): 17–121.

Butler, S. J., J. A. Vickery, and K. Norris. "Farmland Biodiversity and the Footprint of Agriculture." *Science* 315 (2007): 381–84.

Food and Agriculture Atlas. www.fao.org/es/ESS/chartroom/gfap.asp.

Porter, Keith S., Nancy M. Trautman, and Robert J. Wagenet. "Modern Agriculture: Its Effects on the Environment." Natural Resources. Cornell Cooperative Extension. http://pmep.cce.cornell.edu/facts-slides-self/facts/mod-ag-grw85.html.

Robertson, G. P., and S. M. Swinton. "Reconciling Agricultural Productivity and Environmental Integrity: A Grand Challenge for Agriculture." *Frontiers in Ecology and the Environment* 3(1) 2005: 38–46.

Tilman, D. "Global Environmental Impacts of Agricultural Expansion: The Need for Sustainable and Efficient Practices." *Proceedings of the National Academy of Sciences* 96 (1999): 5995–6000.

U.S. Geological Survey. "Investigating the Environmental Effects of Agriculture Practices on Natural Resources." 2007. http://pubs.usgs.gov/fs/2007/3001.

See Also: Agroecology; Climate Change; Fossil Fuel Use; Geographic Information Systems and Global Positioning Systems; Gulf Hypoxia; Vegetarian Diet; Water Quality.

John Masiunas

Environmental Protection Agency

The U.S. Environmental Protection Agency (EPA) is a federal agency charged with the implementation and enforcement of environmental laws and the development of environmental policy related to land, water, and air.

The legislative history of the EPA is rooted in the National Environmental Policy Act (NEPA) established by Congress and signed by President Richard Nixon in early 1970. In accordance with the provisions of the Act, President Nixon established the EPA through Reorganization Plan No. 3, which was sent to Congress in July of 1970. The EPA began operating on December 2, 1970.

The agency is divided into ten regional offices, coordinated through a central office in Washington, D.C. The EPA's top official is the administrator, a noncabinet position appointed by the U.S. president. As a result, its policies are typically influenced by the political ideology of the executive branch.

The EPA, along with the state departments of agriculture, is charged with registering or licensing pesticides for use in the United States through its Office of Pesticide Programs and Office of Prevention, Pesticides, and Toxic Substances. The agency also has a role in evaluating and regulating genetically modified organisms (GMOs) that contain or produce pesticides (such as Monsanto's Bt Corn).

In registering new pesticides, the EPA is charged under the Federal Insecticide, Fungicide, and Rodenticide Act (FIFRA) with ensuring "a reasonable certainty of no harm to human health" and an absence of "unreasonable risks to the environment" when the product is used according to label directions. The terms "reasonable" and "unreasonable" in these requirements leave regulations open to a significant level of interpretation. Furthermore, in making these determinations EPA requires and reviews "more than one hundred different scientific studies and tests," yet the tests and studies are provided, and often conducted, by the applicants themselves. The EPA also sets "tolerances" (maximum pesticide residue levels) for the amount of the pesticide that can legally remain in or on foods.

In 2003, the EPA issued a final rule clarifying two specific circumstances in which a Clean Water Act (CWA) permit is not required to apply pesticides legally registered under FIFRA to or around water. They are the application of pesticides directly to water in order to control pests and application of pesticides to control pests that are present over or near water, where a portion of the pesticides will unavoidably be deposited to the water in order to target the pests.

The EPA also has a role in dealing with water pollution associated with Confined Animal Feeding Operations (CAFOs) through the National Pollutant Discharge Elimination System (NPDES) Permit Regulation for CAFOs (40 CFR Part 122), and the Effluent Limitations Guidelines and Standards (ELGs) for CAFOs (40 CFR Part 412).

Permits for direct discharge from large CAFOs into U.S. waterways set discharge limits, require recordkeeping, and contain provisions requiring the use of technologies for limiting discharge. Permits for "small" and "medium size" CAFOs (those with fewer than 1,000 cattle other than mature dairy cows or veal calves, 2,500 swine over 55 lbs., or 125,000 chickens other than laying hens) do not set specific effluent limits, but are instead set by the "best professional judgment" of the permitting authority.

Further Reading
Environmental Protection Agency. www.epa.gov.

See Also: Policy, Agricultural; Policy, Food; Water Quality

Taylor Reid

European Union Organic Regulation

In June 1991, Council Regulation (EEC) No. 2092/91 established the first government rule for production, labeling, and inspection of organic plant production and wild product collection in the European Union (EU). The regulation was implemented in all EU countries in 1992. In July 1999, Livestock and animal husbandry were regulated in Council Regulation (EC) No. 1804/1999. Several countries in Europe and a number of private organizations had established standards prior to the passage of the EU regulations. Without a unifying regulation, countries could deny equivalency of certifiers, impeding trade across national and even regional boundaries. Obtaining multiple certifications to enable a firm to move its products across several jurisdictions was prohibitively expensive for all but the largest companies, and multiple certification emblems on packages were confusing to consumers. Claims of fraudulent organic packaging could be neither substantiated nor denied because all standards were equally valid (or invalid, depending on the viewpoint). The EU organic regulations were a compromise that established an equivalent, minimum set of rules for production, labeling, and marketing designed to improve the flow of organic products.

The EU standards were rooted in the International Federation of Organic Agriculture Movements (IFOAM) Basic Standards. IFOAM, an umbrella organization of more than six hundred affiliate organizations in more than one hundred member countries, developed basic standards for organic production in the 1980s directly from existing practices and philosophy of organic farmers. IFOAM created an accreditation program in 1992, which verifies that certifying organizations of governmental and nongovernmental members inspect and certify operations in their jurisdiction in compliance with the IFOAM Basic Standards. For years, the IFOAM Basic Standards were the only international regulations in the organic market. Even now, certifications from IFOAM accredited bodies are usually granted equivalency in the EU. Equivalency means a country accepts a foreign certifier's rules and inspections as equivalent to its own and permits marketing of the products certified by the foreign body within its boundaries. This facilitates EU trade with many developing countries where federal organic laws have not been established, but an IFOAM accredited body is conducting certifications.

The EU governments realized that the IFOAM Basic Standards

International Federation of Organic Agriculture Movements (IFOAM)

IFOAM's mission is "leading, uniting and assisting the organic movement in its full diversity." Interestingly, the goal of IFOAM is the "worldwide adoption of ecologically, socially and economically sound systems that are based on the principles of Organic Agriculture." What began as a small movement in Europe has now become a global organization with member organizations located in more than one hundred nations worldwide. Members include farmer organizations, certification agencies, and marketing groups.

www.ifoam.org

and accreditation program provided a valuable bridge in international trade of organic products. However, as a nongovernmental body, IFOAM has no authority to enforce its standards, other than through revoking accreditation. In countries without regulations, "organic" can still be claimed on a label without being certified. Further, as an accreditor of organizations rather than a certifier of production, IFOAM does not label products, so consumers have to know whether a certifier is accredited to be sure the product meets IFOAM Basic Standards. IFOAM standards are stricter than many national standards, making it more difficult for countries to pursue target goals for conversion to organic production.

Relying on the IFOAM Basic Standards, the EU crafted a minimum rule for all countries to follow. The EU regulations govern production and processing inputs and activities, emphasizing soil health and animal welfare attained in an ecologically sound way that contributes to biological diversity and responsible use of natural resources. At least 95 percent of the product ingredients must be organic. Genetically modified organisms and ionizing radiation are prohibited. The underlying principle for production and processing is that inputs must be of natural organic origin if available and, if not, the use of alternatives, particularly synthetics, must be minimized.

Under the EU regulation, each country interprets and implements the rules, including enforcement, monitoring, and certification. The EU recognizes a "competent authority" designated in each member state as the body that can certify organic products as complying with EU law. The competent authority may delegate some authority for certification to "control authorities" within the member state. Although the competent authorities may enact stricter regulations than the EU standards, they should not impede international trade for any products properly certified as complying with the EU rules. Non-EU countries or entities may apply for "third country" status, which is granted by the EU if equivalency between the foreign standard and the EU rule is verified. However, this is a tedious and lengthy process, so most organic imports from countries outside the EU are granted "import derogation" or permission to enter the EU on a case-by-case basis.

In practice, EU countries and control authorities have enforced varying degrees of strictness in their organic standards, resulting in a tiered structure in which not all certificates are equally acceptable. This has resulted in difficulty exporting to certain countries, particularly where the domestic organic industry is heavily encouraged. Since the number of control bodies varies by country according to the decision of the competent authority, exporters meet with variable success. From anecdotal evidence, U.S. exporters have found the following countries to be among the easier ones to enter: the Netherlands (1 control body), Denmark (1), Sweden (2), the United Kingdom (1), Germany (22), and Spain (20). Italy (13 control bodies), Switzerland (1), and France (6) are considered very difficult to enter due to strictness of the regulation and protectionist attitudes. Consumer loyalty to certification labels has been heavily courted, and national pride encourages some consumers to seek out labels issued by control authorities in their own countries.

In an attempt to strengthen recognition of the precedence of EU regulations and to stimulate demand for EU-based products, a logo marked "Organic Farming—EC Control System" was introduced in 2000. The voluntary logo could be used on any product derived at least 95 percent from EU origins. There was not significant adoption of the logo because firms were concerned that the addition to existing certifier labels was confusing to consumers, costly to implement, and unnecessary for loyal consumers who knew which certifier labels were EU based.

The EU organic regulation was integral to the development of policies supporting the growth of organic production and markets in Europe. The EU passed an agri-environmental program, the 1992 Common Agricultural Policy (CAP) reform (EC Regulation 2078/92), which created the policy for member states to support organic farming through direct subsidies paid to farmers for converting to organic farms beginning in 1994. Financial support measures were included in the CAP reform program (Rural Development Regulation No. 1257/99) implemented from 1999 to 2001. Most EU countries pay both existing and converting farmers for the environmental benefits generated for society from organic farming. In 2001, about €500 million were spent on organic farm subsidies under these two policies. Success in converting land to organic management has stimulated organic action plans by several EU countries that identify target levels of organic conversion, such as 20 percent of cropland certified organic by 2010. Without the EU organic regulation, policies such as these that advance social goals of organic farm production would not be possible.

After fifteen years, the European Commission agreed on the need to revise the organic regulations to account for new products (seaweed, yeast, aquaculture, wine), clarify and standardize rules, and add flexibility. In June 2007, Council Regulation (EC) No. 834/2007 was passed, repealing the old organic rule as it established new standards to take effect in January 2009.

The new regulation is more uniform in its application of the principles and objectives of organic farming throughout the stages of production and processing. For example, livestock must be of organic origin and fed with organic feed or milk from birth. Incidental contamination from genetically modified organisms now precludes sale of the product as organic, whereas before only intentional contamination did so. The new rule will increase trade flows with third-world countries by allowing accreditation of both government and private certifiers in these countries. Better information will be available to consumers by permitting organic ingredients to be specified in the ingredient list for a product and by making the EU organic logo mandatory. The EU logo indicates whether the product is composed mainly of ingredients of EU or non-EU origin. Consumers looking for particular organic ingredients or concerned about supporting EU farmers can use this information in product selection.

Article 22 of the new regulation is devoted to flexibility in production rules. This is likely to be the most controversial portion of the new regulation in application. Requiring only that they be kept to a minimum and for a limited time as set by the commission, exceptions to the rules may be granted when necessary. For example, exceptions may be granted to initiate or maintain organic production under climatic, geographical, or structural constraints; to aid production when organic versions of farm inputs, food additives, feed additives, or processing ingredients are not available on the market; and to solve specific livestock management problems. A similar attempt to introduce flexibility in the U.S. regulation, the commercial availability clause, has generated controversy in determining what constitutes "availability." In particular, an ingredient that is available on the market may not be economically obtained. Of greater concern for organic importers buying from the EU is the potential for abuse invited by the exception allowed for production constraints. Temporarily permitting nonorganic inputs or ingredients on the rationale of introducing or sustaining production in a region may significantly damage consumer trust in the integrity of organic labels.

Obtaining agreement on EU regulations is difficult in the face of national and regional interests and may reduce the risk of the flexibility article. This is apparent in the recent amendment adopted in September 2008, Council Regulation (EC) No. 967/2008, which pushes the date of mandatory EU labeling to 2010 to give the commission sufficient time to design a logo that does not look like existing designation of origin labels. Organic regulations are still evolving, and will undoubtedly become more unified globally over the next twenty years.

Further Reading

"Council Regulation (EC) No 834/2007 of 28 June 2007." *Official Journal of the European Union.* 2007. http://eur-lex.europa.eu/LexUriServ/LexUriServ.do?uri=OJ:L:2007:189:0001:0023:EN:PDF.

Dimitri, C., and L. Oberholtzer. *Market-Led Versus Government-Facilitated Growth: Development of the U.S. and EU Organic Agricultural Sectors.* WRS-05-05. Washington, D.C.: U.S. Department of Agriculture, 2005.

European Commission Directorate-General for Agriculture and Rural Development. "Organic Food: New Regulation to Foster the Further Development of Europe's Organic Food Sector." *Agri-Newsletter.* April 4, 2007. http://ec.europa.eu/agriculture/publi/newsletter/2007/04/04_en.pdf.

See Also: 1990 Farm Bill, Title XXI; USDA National Organic Program

Luanne Lohr

F

Factory Farming

The term "factory farm" is most commonly used as a pejorative to refer to a facility in which large numbers of genetically similar livestock (e.g., chickens, turkeys, cattle, or swine) are confined to control their diet and environment to maximize production. The management regime often includes the administering of synthetic versions of growth hormones, as well as antibiotics to maximize growth and prevent disease transmission exacerbated by the close proximity of animals. A closely related term is "large concentrated animal feeding operation" (large CAFO). The U.S. Environmental Protection Agency defines large, medium, and small CAFOs based on the number and type of animals. For example, a large CAFO operation would contain at least: 1,000 beef cows, 700 mature dairy cows, 2,500 swine weighing at least 55 pounds each, or 125,000 broilers where a non-liquid manure handling system is employed.

Combining the two words, "factory" and "farm," purposefully implies a contradiction or conflict from the imposition of industrial practices onto a production process associated traditionally with family ownership arrangements and managerial control and reliance on biological and regenerative processes and inputs to sustain yields.

The term can also be used more generally to refer to large-scale agricultural production that tends to be highly mechanized, employs a large number of wage laborers, and uses standardized practices of management and production. In addition, industrially produced inputs such as transgenic or genetically engineered crops, commercial fertilizers, and pesticides of various types are used in place of cultural management practices such as diverse crop rotations, green and animal manure application, biological controls, and tillage.

Further Reading

Environmental Protection Agency. "Concentrated Animal Feeding Operations Industry: 2008 CAFO Regulations." www.epa.gov/guide/cafo/.

Food and Agriculture Organization of the United Nations. "Avian Influenza Glossary." 2009. www.fao.org/avianflu/en/glossary.html.

Goodman, D., B. Sorj, and J. Wilkinson. *From Farming to Biotechnology*. Oxford: Basil-Blackwell, 1987.

Hinrichs, C. C., and R. Welsh. "The Effects of the Industrialization of U.S. Livestock Agriculture on Promoting Sustainable Production Practices." *Agriculture and Human Values* 20 (2003): 125–41.

Lyson, T. A. *Civic Agriculture: Reconnecting Farm, Food and Community*. Medford, MA: Tufts University Press, 2004.

McWilliams, Carey. *Factories in the Field: The Story of Migratory Farm Labor in California*. Berkeley: University of California Press, 2000.

See Also: Agriculture, Conventional; Agribusiness; Animal Welfare; Antibiotics and Livestock Production; Cloning; Concentrated Animal Feeding Operations; Free-Range Poultry; Production, Treadmill of; rBGH

Rick Welsh

Fair Trade

Fair trade is simultaneously a critique of the unfairness associated with the commercially focused free trade system and a proposition that an alternative fairer trade system is possible. Fair trade advocates cite the persistent poverty among small-scale agricultural commodity producers, workers, and artisans in the global south to demonstrate the need for a different type of trading system. Fair trade organizers have worked to create and expand an alternative trade system that starts with a focus on human development since the late 1940s. An alliance among several international fair trade associations states that "Fair trade is a trading partnership, based on dialogue, transparency and respect, which seeks greater equity in international trade. It contributes to sustainable development. Fair trade organizations (backed by consumers) are engaged actively in supporting producers, awareness raising and in campaigning for changes in the rules and practice of conventional international trade."

Fair trade is a global movement and an expanding market. The twin strategies for implementing fair trade principles are to build an alternative market that prioritizes small-scale producer organizations and offers better prices, and to provide support to producer partners through international development nongovernmental organizations (NGOs). Fair trade organizations have established an international labeling system (Fair Trade Labeling Organizations International) as well as international associations of alternative trade organizations (IFAT) to build demand; develop standards; coordinate standards, certifications, and inspections; and create more alternative trade policies. Although fair trade accounts for 1 to 5 percent of the global trade in the specific agricultural commodities and handicrafts, the markets have expanded rapidly. As of October 2006, the global fair trade–certified network included 586 producer organizations and 1.4 million farmers, artisans, and workers in 58 developing countries from Latin America, Asia, and Africa. Consumers worldwide spent U.S. $2.024 billion on fair trade–certified products. The expanding list of fair trade products includes coffee, cocoa, tea, fruits, wine, sugar, honey, bananas, rice, crafts, and some textiles.

The global fair trade movement finds its roots in international partnerships, which extend into many lives, organizations, and landscapes. The primary partnership consisted of impoverished small-scale producers organizations and artisans selling high-quality goods to northern volunteers, NGOs, and later, businesses. Producers and consumers shared the risk: farmers and artisans sometimes provided their products months or even years before receiving full payment after volunteers and alternative trade organizations sold their goods into distant and uncertain markets. On the other hand, fair trade organizers from the north provided producers with loans that would otherwise be unavailable and bought crafts and coffee before they had established demand in their home markets.

The first alternative trade organizations emerged around handicrafts, often connecting religious and politically motivated northern groups with groups of female artisans in impoverished communities. Examples of these pioneering organizations include the Mennonite Central Committee in Pennsylvania, which started buying quilts directly from seamstresses in Puerto Rico in the late 1940s and a decade later created Ten Thousand Villages, an alternative trade organization that as of 2006 connected to some one hundred artisan groups and had annual sales in excess of $20 million; SERVE International (Sales Exchange for Refugee Rehabilitation and Vocation), and a campaign by Oxfam, UK, called the Helping-by-Selling project. The first Worldshop opened in the 1950s, and by 2005, there were more than 2,800 Worldshops throughout western Europe selling alternatively traded products.

Coffee led the rise of certified fair trade agricultural products, which followed an organizational development trajectory similar to crafts until alternative trade organizations united with other NGOs and traders to create an international certification system in 1988. The emergence of a product certification system allowed the participation of more conventional companies, expanded fair trade markets, and shifted the ratio of global fair trade goods from crafts to foods and beverages.

In response to the problem of how to expand fair trade product demand and distribution without compromising the values, benefit flows, and consumer trust created through decades of solidarity-based alternative trade, Dutch Jesuit priest Franz Vander Hoff partnered with an indigenous coffee cooperative in southern Mexico, first to create the alternative-trade coffee network and then to start exploring ways to expand larger market segments. Vander Hoff returned to Holland and recruited several advocates and business leaders to create a product certification and labeling system called Max Havelaar. In 1988, Max Havelaar united with European and North American NGOs and in 1997 created Fairtrade Labeling Organizations International (FLO). FLO is an international nonprofit multi-stakeholder association that seeks to establish fair trade standards; support, inspect, and certify disadvantaged producers; and harmonize the fair trade message across the movement.

The rapidly expanding retail sales figures tell us little about the ability of fair trade to deliver on its stated empowerment and sustainable development goals. A vital consideration in assessing fair trade impacts is the fact that as of 2006 only 20 percent of the agricultural goods produced by fair trade–certified organizations are sold according to generally accepted fair-trade terms. The remaining 80 percent of products are generally sold into domestic and international markets under less favorable terms. An important percentage is often consumed within the household or traded locally. Most scholars agree that small-scale producers linked to fair trade are better off than producers that lack these connections. Many cooperatives use their sense of collective empowerment by building stronger organizations, and they have also conserved biological and cultural diversity through their farming practices. However, the combination of fair trade sales and additional support from allied international development NGOs is not a panacea for eliminating poverty or stopping outmigration even within fair trade organizations. The minimum coffee prices are especially important when conventional market prices fall. Artisans have generally been able to sustain their crafts and cultures and partially support their livelihoods even as they have improved their organizations' business capacities through direct connections to better markets for their products.

Fair trade has emerged as a potentially useful tool for southern producer organizations to strengthen their collective market influence and for northern social justice activists, such as those within United Students for Fair Trade, to coordinate campaigns to convince their college campuses to sell fair trade, local, and organic foods and beverages. However, fair trade advocates have yet to make significant contributions to reforming the international free trade and development policy agendas. In fact, some certification agencies seem to have dropped this agenda altogether. Small-scale producer organizations have also used their participation in fair trade to strengthen their alliances, increase their visibility, and expand their negotiating power. For example, the Latin American and Caribbean Network of Small-Scale Fair Trade Producers (CLAC), which represents more than 200,000 producer families, has used its participation in fair trade to win a seat on the FLO board of directors, to gain partial ownership of the certified fair trade system, and to advocate for minimum prices that keep up with inflation and cover the costs of sustainable production.

A core paradox within the certified fair trade system is that it sets out to achieve social justice and environmental sustainability within the same market that many believe impoverished small producers in the first place. The fair trade system is also experiencing increased competition from a rapidly expanding array of sustainable product certification programs, such as the Rainforest Alliance and Utz Certified. Many of these programs have lower social and environmental standards. Certainly, these challenges have made it difficult for fair trade advocates, conscious consumers, producer organizations, NGO administrators, and business leaders. However, movement is spurred forward to individual stories of impact and change as well as the societal needs to create a fairer and more sustainable trade system.

Further Reading

Bacon, C. M., V. E. Méndez, S. Gliessman, D. Goodman, and J. A. Fox, eds. *Confronting the Coffee Crisis: Fair Trade, Sustainable Livelihoods, and Ecosystems in Mexico and Central America*. Cambridge, MA: MIT Press, 2008.

Barratt-Brown, Michael. *Fair Trade: Reform and Realities in the International Trading System*. London: Zed, 1993.

CLAC. *Estudio de Costos y Propuesta de Precios para Sostener el Café, las Familias de Productores y Organizaciones Certificadas por Comercio Justo en América Latina y el Caribe*. Domincan Republic: Assemblea de Coordinadora Latinoamericana y del Caribe de Pequeños Productores de Comercio Justo, 2006.

DeCarlo, Jacqueline. *Fair Trade: A Beginner's Guide*. Oxford, U.K.: One World, 2007.

Fairtrade Labelling Organizations International. *Shaping Global Partnerships: FLO Annual Report 2006/2007*. Bonn, Germany: FLO International, 2007.

Hernández Navarro, Luis. "To Die a Little: Migration and Coffee in Mexico and Central America." Special Report, Americas Program. Silver City, NM: Interhemispheric Resource Center, 2004. Assessed December 21, 2008, at www.americaspolicy.org/reports/2004/0412coffee.html.

Jaffee, Daniel. *Brewing Justice: Fair Trade Coffee, Sustainability and Survival*. Berkeley: University of California Press, 2007.

Krier, J-M. *Fair Trade in Europe 2005: Facts and Figures on Fair Trade in 25 European Countries*. Brussels, Belgium: Fairtrade Labeling Organizations International, International Fair Trade Association, Network of European World Shops, and European Fair Trade Association Fair Trade Advocacy Office, 2005.

Leclair, M. S. "Fighting the Tide: Alternative Trade Organizations in the Era of Global Free Trade." *World Development* 30(6) 2002: 949–58.

Oxfam UK. "Oxfam in Action." 2007. www.oxfam.org.uk/oxfam_in_action/index.html.

Raynolds, L. T., D. Murray, and A. Heller, A. "Regulating Sustainability in the Coffee Sector: A Comparative Analysis of Third-party Environmental and Social Certification Initiatives." *Agriculture and Human Values* 24(2) 2007: 147–63.

Renard, Marie-Christine. "Quality Certification, Regulation, and Power in Fair Trade." *Journal of Rural Studies* 21 (2005): 419–31.

See Also: Coffee, Organic; Development, Sustainable; Farm Workers' Rights; Free Trade; Globalization

Christopher M. Bacon

Family Farms

"Family farms" have no single, universally accepted definition. The United States Department of Agriculture (USDA) identifies all farms owned by individuals, partnerships, or family corporations as family farms. This excludes only those farms owned by nonfamily corporations. Using the USDA definition, 98 percent of all farms in the United States were identified as family farms in the 1994 Census of Agriculture. This definition may be useful for political purposes but it is of little use in understanding or describing the basic structure or character of U.S. agriculture.

Historically, people probably assumed that a family farm consisted of a husband, wife, and their children all living and working full-time on a farm they owned and managed. However, this arrangement has not been the reality of U.S. agriculture in many decades, as most families must supplement their farm income through off-farm employment. In fact, less than half of all farmers identified in recent USDA census reports indicate that farming is their primary occupation, meaning at least half of the farmer's hours of work were spent farming. In addition, throughout the 1990s and early 2000s, farm families on average earned approximately 90 percent of their total family income from off-farm sources. Furthermore, with increased reliance on mechanization and agrichemicals, there are few opportunities for the children of farm families to be significantly involved in working on the farm. Regardless, most U.S. agricultural policies are still defended politically as being necessary to preserve the family farm and the traditional way of farm life. Most tax payers probably still think of the traditional hardworking farm family as the primary beneficiaries of government farm programs. Contrary to popular belief, there are very few traditional family farms left in the United States, with the exception of some Amish and Mennonite farms.

A more useful and relevant definition of a family farm is a farm that is not only owned and operated by a family but also one in which the family is involved in all important management decisions and provides most of the farm labor. Such farms need not employ the whole family full-time or be solely reliant on family labor, but the family needs to be actively involved in the day-to-day operation of the farm. This definition excludes farms that are owned by families or family corporations that are operated by someone other than the owner. A farm that represents a family investment, rather than a

Ryne and Karen Tharp, of Terrebonne Farm, pasture-raise a flock of some forty sheep with their young son Rhett on a rolling patchwork of fifty-five acres near Draper's Bluff, just outside the tiny town of Lick Creek in southern Illinois. Courtesy of Jerry R. Bradley.

family occupation, is excluded from this definition as well. It also excludes contract production, such as most concentrated animal feeding operations (CAFOs), where all of the important management and marketing decisions are made by someone other than the owner or operator. This definition essentially excludes nonfarm corporations, not only from ownership but also from significant involvement in true family farming operations.

From the perspective of understanding how a farm operates and the consequences of its operation, a family farm represents a distinctive farm way of life, which includes farming as a way of making a living. From a quality of life perspective, a family farm means a farm on which the farm and the family are inseparable. The family considers the needs of the farm—the land, plants, animals, wildlife—as well as the needs of the family, in making all management decisions. The farm, with a different family farming it, would be a fundamentally different farm and the family on a different farm would be a significantly different family. So, on this type of farm, the farming operation evolves as the family changes, and each family member plays a different role as they mature. The farm evolves to accommodate the members of each new generation of the family who choose to become farmers. The farm, in its appearance and operation, is a reflection of the family. The family, in its relationships and expressions, is a reflection of the farming operation. Their identities are inseparable.

This quality of life definition fits most traditional family farms of the past but it also fits a rapidly increasing number of farms today. Many of today's quality-of-life family farmers are identified by labels such as organic, biodynamic, permaculture, ecological, practical, innovative, or holistic. The "families" on these farms may or may not be married couples with children, but the people who farm the land together have a familylike commitment to each other. They are not just business partners, they are partners in life.

They typically market their livestock and crops into specialized markets, which include natural, grass-based, free-range, without hormone and antibiotics, without genetic modification, as well as natural and organic. Many of these new family farmers market directly to local customers through farmers markets, roadside markets, or community-supported agriculture organizations (CSA). Increasingly, these new family farmers work with independent food retailers—restaurants, food cooperatives, smaller shops—to gain access to larger numbers of like-minded customers. These new family farms are defined by the same characteristics as traditional family farms; the farms and the families are inseparable. To the farmers' most loyal customers, the fact that their food is grown on a family farm is an important reason for their choice of food and for their loyalty.

Many people question whether even these new family farms are sustainable. Most may not be truly sustainable, in the sense of being able to maintain their productivity and value to society indefinitely, at least not under existing conditions using existing know-how. Many have made great strides in recent years, but economic stability is often the most difficult challenge in sustainable farming. Regardless, these new family farms clearly offer the current best hope for sustaining agriculture in the future. The industrial farming systems that have displaced most traditional family farms are very productive and efficient in the short run, but they quite clearly are *not* sustainable over the long run. Industrial agriculture is critically dependent on nonrenewable fossil energy, which is being rapidly depleted. Industrial agriculture may be sustainable for a few more decades, or perhaps another fifty or even one hundred years, but as energy becomes more limited and costly, it will lose its productivity. In meeting its economic needs today, it is degrading and depleting the natural and human resources of the earth, leaving nothing with which to meet the needs of future generations.

The new family farms are the best hope for national food security. A nation is food secure only when it is able to meet its basic food needs in a time of crises. In times of crisis, global markets for food and for fertilizers, chemicals, seed, and other farm inputs may be disrupted for weeks, months, or years. A nation that lacks sufficient domestic food production to survive nutritionally for an extended period of time is not food secure. During some future global crisis, even an economically powerful nation such as the United States might well be forced to rely on its own farmers for its very survival. If so, it will need to have a sufficient number of farmers on the land who know how to work with nature, using organic, biodynamic, holistic, and other sustainable approaches to agriculture, to produce sufficient quantities of safe and healthful food without relying on imported commercial inputs, such as nitrogen fertilizer. (The last nitrogen fertilizer plants operating in the United States were closed in 2007, leaving U.S. farmers dependent on nitrogen fertilizer imported from the Middle East and Russia.)

Equally important, the new family farmers are committed to farming in their home countries. Unlike the multinational corporations that increasingly dominate global food markets, family farmers are real people. They have families, friends, and national citizenship. They are rooted in the places where they grew up and would like their children to continue living and farming in those places in the future. They are not going to renounce their citizenship and leave this country just because they could earn more profit farming in some other country, nor are they going to sell their products elsewhere when there are hungry people at home. Ultimately, a nation's food is no more secure than are the relationships among its people and their relationships with the land. For most people, their only meaningful relationship with land is through a family farmer.

Further Reading

Berry, Wendell, *What Are People For?* New York: North Point Press, 1990.

Ikerd, John. *Small Farms Are Real Farms: Sustaining People through Agriculture.* Austin, TX: Acres USA, 2007.

Levins, Richard. *Willard Cochrane and the American Family Farm.* Lincoln: University of Nebraska Press, 2000.

Strange, Marty. *Family Farming: The New Economic Vision.* 2nd ed. Lincoln: University of Nebraska Press, 2008.

See Also: Agriculture, Alternative; Agriculture, Sustainable; Agritourism; Concentrated Animal Feeding Operations; Grassroots Organic Movement; Local Food; Marketing, Direct; Sustainability, Rural; Transition to Organic

John Ikerd

Farm to School

School settings provide excellent teaching opportunities about nutrition and healthy foods. Research is clear that good nutrition can improve student learning. Farm to School includes many types of programs and experiences for children to learn about food production, such as farm tours, nutrition education, local food tastings, and school gardens. Teachers, district staff, parents, community members, and students work together to provide these opportunities. Goals of Farm to School vary, depending on the type of program or experience. Typical goals are educating students about food items and sustainable food production, retaining food dollars in the community, reducing carbon emissions, and improving health. Concerns about obesity rates and environment have influenced districts to offer Farm to School programs to help students develop healthy decision-making skills.

BACKGROUND

The National School Lunch (NSL) Act of 1946 formalized a national child nutrition policy to "safeguard the health and well-being of the Nation's children, and to encourage the domestic consumption of nutritious agricultural commodities." The USDA administers nearly twenty nutrition programs, including School Breakfast Program (SBP) and After-School Snack Program. All programs must meet nutritional standards. Child Nutrition Programs (CNP) can include locally grown foods as part of the meal, a la carte, or through snacks programs.

RECENT FEDERAL INITIATIVES

Two recent and important programs affect school food projects. First, the Local Preference for School Food Purchases regulatory reform allows school districts participating in NSL or NSB programs to specify the geographic origin of products. Prior to this change, USDA Programs' Procurement Guidelines did not allow preferences or exclusion of prospective vendors of district purchases. Second, the Fresh Fruit and Vegetable Snack Program (FFVSP) provides funding to each state for purchase of at least one daily fresh fruit or vegetable snack with a focus on low-income school districts. Funding is indexed to inflation.

TYPES OF PROGRAMS

Farm to School programs include locally grown items in school meals, farm tours, school fundraisers (sell pumpkins rather than candy bars), nutrition education, food tastings, Ag in the Classroom, and school gardens. Through FFVSP, Farm to School encourages children to select fresh fruits or vegetables. Nutrition education research has found that children have more favorable attitudes about foods if there is engagement—such as growing vegetables in a school garden or tasting product directly on the farm.

PROCUREMENT METHODS

There are three procurement methods for school lunch programs:

1. *Direct from the Farmer.* Food service director and farmer communicate directly about needs for the district: product specifications (amounts, size, packaging, ordering, delivery, payment procedures, etc.).
2. *Farmers' Cooperatives.* School food service director works with a consortium of food growers at the beginning of the planting season to contract amounts and prices or during the harvest season.
3. *Department of Defense.* In certain states, the department provides a distribution network.

CHALLENGES

Farm to School presents challenges. Because fresh produce is often served raw, communication between district buyers and growers should define specifications and on-farm practices to ensure safety of food. Small-to medium-sized farms do not typically invest in mechanical equipment, so growing and harvesting products is labor-intensive and costs can be higher. School districts often lack the additional kitchen staff to prepare fresh foods. Yet opportunities abound: producers benefit from the stable institutional market provided by a school distract and school districts benefit from the educational opportunities provided by local farms, as well as the fresh food they provide.

Further Reading
National Farm to School. www.farmtoschool.org.

See Also: Campus Programs; Diet, Children's; Gardens, Children's; Gardens, Community

Catherine H. Strohbehn

Farm Workers' Rights

Farm workers' rights are the rights of workers to adequate working and living conditions. Struggles concerning these rights center on issues such as adequate wages, health and safety measures, and even protection of workers' lives. The rights of farm workers are disdained by conventional agriculture in the drive for capital accumulation. Organic, sustainable, and local food movements have also largely neglected farm workers' rights, a disregard shared with conventional agriculture.

"Farm workers" are those workers who toil in the field. They face unique labor issues and are especially disadvantaged in their organizing efforts. Farm workers differ from both farmers and growers. Although farmers own the means of production, they employ few additional laborers, and much of the farm work is done alone or with family. Alternatively, growers employ large numbers of workers and rarely do farm work themselves. Rather, they purchase the labor power of farm workers, who have no direct access to the means of production. Farm workers, most often immigrants and people of color, are exploited and then discarded in a never-ending drive to extract surplus value.

Contractors often mediate the relationship between growers and farm workers. These "gang bosses" buffer growers from unruly workers and absolve growers of any obligation for workers' rights. Gang bosses also control farm laborers' working and living conditions, which are among the more deplorable in the United States. Poor working conditions often lead to accidents resulting in physical injuries that disfigure farm workers and may lead to cumulative trauma disorder as a consequence of repetitive work. Farm workers also face serious health dangers from exposure to toxins in the fields that can cause cancer, reproductive disorders, developmental impacts, and neurological and lung damage. As one of the most dangerous jobs in the United States, farm work is often a matter of life and death.

The dependence of agricultural production on organic processes necessitates a migratory workforce, one that follows crops and moves with the harvest. At the same time, this need for geographically mobile labor creates conditions whereby workers can move in search of better work opportunities. The contradiction of the need for labor mobility and the need to hinder it is a concern for growers, who often employ extra-economic means to maintain control over the labor process.

To address this contradiction, many growers turn to an immigrant labor force. As a socially isolated group that often lacks access to the most basic of social services, immigrant farm workers face a variety of additional workers' rights abuses, including the threat of deportation, illegally low wages, and fear of job loss. These abuses of power add another obstacle in the struggle for farm workers' rights.

BRIEF HISTORY OF FARM WORK AND WORKERS' STRUGGLES

The unique characteristics of farm work have meant that workers are historically marginalized and manipulated, facing a multitude of difficulties in mounting successful struggles. Despite these obstacles, farm workers have fought for their rights throughout history.

Farm worker picking eggplant in Florida. (AP Photo/Lynne Sladky)

By the end of World War I, the United States had a fully developed agricultural economy. Many different ethnicities were working alongside native-born laborers in the fields of the United States, including Chinese, Filipino, Italian, Japanese, Mexican, Portuguese, and Punjabi immigrants. Expansion of large-scale fruit and vegetable production ushered in the use of seasonal hired labor, a persistent characteristic of agricultural production. This harvest labor system emerged within a labor market that favored workers, as the demand for farm workers exceeded supply, providing some leverage for workers to make claims for improved work conditions. The Industrial Workers of the World (IWW) seized the opportunity afforded by the favorable conditions and organized workers around living conditions. The uprising failed as the growers quickly responded with violence. This period of worker struggle colored future organizing as growers became steadfast in their refusal to allow workers any gains.

Farm worker agitation flared up again during the 1920s out of anarchist and communist politics. With support from the American Communist Party, workers in California went on strike. Once again farm workers were met with violence, but a new wave of farm worker struggle was under way and would continue to grow through the 1930s.

The Great Depression stirred racist tendencies and immigrant farm workers were deported or refused admittance to the United States. Growers turned to recruiting small-scale family farmers reeling from the Dust Bowl. The new pool of wage laborers, white "Okies" escaping the destroyed lands of Arkansas, Texas, Missouri, and Oklahoma, were easily exploited, willing to work for low wages, and soon filled the ranks vacated by displaced immigrant workers. However, the tightening of the immigrant labor pool, coupled with wage suppression brought on by the Depression, provided fertile ground for worker organizing.

The early-1930s witnessed a number of strikes across California, involving tens of thousands of farm workers. Farm worker agitation was successful in gaining national attention through the publication of Carey McWilliams's bestseller *Factories in the Field* (1939) and John Steinbeck's novel *The Grapes of Wrath* (1939). Again, the growers responded with a heavy hand. Soon thereafter, growers united in an all-out class war by joining with producers and others beyond the farm gate to proactively thwart worker struggles.

When World War II siphoned farm workers for the war effort (both as soldiers and urban industrial workers), growers pressured the federal government for support. The government responded by creating the Bracero Program, a contract labor program established in 1942 to provide cheap, abundant labor to the U.S. agricultural sector, a move that proved to be a serious setback for farm workers' rights.

The Bracero Program brought indentured workers from Mexico to the United States. Workers were under contract to the government, and their movements were closely controlled: they lived in work camps, were bused to and from the fields, and were sent back to Mexico after harvest. The program served growers well, providing them with a ready labor pool that they leveraged to lower farm wages and push back against the organizing of the 1920s and 1930s. Despite the stated need to replace workers lost to the war effort, the Bracero Program lasted until 1964, peaking in 1956 when almost half a million *braceros* worked in U.S. fields.

Following the Bracero Program, the "green revolution" brought about the mechanization of agriculture, transforming post-1960 food production into a more energy- and labor-intensive industry. These shifts were rooted in crop hybridization along with implementation of resource-dependent technologies such as synthetic fertilizers, pesticides, herbicides, and machinery. Unlike the labor saving impacts of mechanization on crops such as corn, cotton, and soy, the expanded output in fruits and vegetables necessitated more workers for harvest, which is still mainly done by hand.

Again, growers turned to immigrants, many of whom are undocumented Latinos from Mexico and elsewhere in Latin America, to fulfill the bulk of hired farm work, ensuring a steady stream of exploitable labor. Even though the agricultural sector enjoys strong revenues, farm wages remain extremely low, and farm workers' rights are some of the most neglected rights of workers across the country.

In the 1960s and 1970s, another era of farm worker struggles coalesced. A new leader emerged in Cesar Chavez, who, in the early 1960s, organized workers in the fields throughout California. Chavez formed the United Farm Workers (UFW) in 1969 with support from the American Federation of Labor and Congress of Industrial Organizations (AFL-CIO). The UFW successfully fought for farm workers' rights, instituting contracts for thousands of workers. Chavez built alliances with sympathetic consumers in U.S. cities through a national boycott of lettuce and grapes. Around the same time, the Farm Labor Organizing Committee (FLOC) emerged in Ohio to help organize tomato pickers.

This successful period of worker activism lasted until 1980. Growers fought back and have succeeded in undermining many of the gains of the previous two decades. However, new struggles have recently emerged, most notably the Coalition of Immokalee Workers (CIW) in Florida. Taking their cue from the UFW and FLOC, the Immokalee Workers have assumed leadership roles in the contemporary phase of farm worker agitation, with much of their recent success stemming from collaboration with students and other sympathetic groups. In a promising sign of future developments, the Immokalee Workers have recently partnered with sectors of the new food movements.

LABOR AND NEW FOOD MOVEMENTS

Despite this new partnership, new food movements for organic, local, and sustainable foods have failed to fully incorporate farm workers' rights as a central concern. Labor issues play second fiddle to matters such as personal health and environmental degradation, and there appears to be an overall sense that farm workers' rights can be ignored. Although this negligence is beginning to change, new food movements have largely reproduced the persistent, but primarily unsuccessful, history of farm workers' struggles.

In part, the new food movements' negligence can be attributed to their uncritical back-to-the-land romanticism. The consequence of this is a wholesale embrace of small-scale production, where farmers are all too likely to engage in a system of self-exploitation. This agrarian romanticism idealizes the rural landscape, whereby the blood and sweat of farmers and farm workers is obscured by the "lie of the land." Movements for organic, local, and sustainable food seem willing to accept this.

Further Reading

Cook, Christopher. *Diet for a Dead Planet: How the Food Industry Is Killing Us.* New York: The New Press, 2004.

Guthman, Julie. *Agrarian Dreams: The Paradox of Organic Farming in California.* Berkeley: University of California Press, 2004.

Kautsky, Karl. *The Agrarian Question.* Translated by Pete Burgess. London: Zwan Publications, 1988 [1899].

Magdoff, Fred, John Bellamy Foster, and Frederick Buttel, eds. *Hungry for Profit: The Agribusiness Threat to Farmers, Food and the Environment.* New York: Monthly Review Press, 2000.

Mitchell, Don. *The Lie of the Land: Migrant Workers and the California Landscape.* Minneapolis: University of Minnesota Press, 1996.

McWilliams, Carey. *Factories in the Field.* Boston: Little, Brown and Company, 1939.

Steinbeck, John. *The Grapes of Wrath.* New York: Penguin Books, 1939.

Walker, Richard. *The Conquest of Bread: 150 Years of Agribusiness in California.* New York: New Press, 2004.

See Also: Agriculture, Conventional; Fair Trade; Globalization

Evan Weissman

Farmers Markets

Farmers markets have provided a traditional outlet for the selling and buying of food products for centuries. Farmers bring their products to a central location and consumers browse among the products, buying directly from the producer. On a practical level, farmers markets provide easy access to a customer base, which allows consumers and producers to interact directly. Farmers generally receive a fair price for their product and consumers get fresh, quality food in return. Markets often become social events, with food and music in a central location. Farmers markets are becoming an important aspect of the local food movement and community development.

Farmers markets were commonplace and extremely important during the settlement of North America. Although farmers markets are a historically important method of selling food products, the rise of the industrial

Farmers markets are an increasingly popular venue to purchase local food. Courtesy of Jon Bathgate.

food system made farmers markets almost obsolete in many places in North America. As cities grew and agribusiness began to control food distribution, the importance of farmers markets diminished. Small farms were consolidated into larger farm enterprises, and farmers markets no longer had enough farmers or consumers to be viable retail outlets. Farmers markets were almost nonexistent for most of the twentieth century except during difficult economic times, when farmers markets flourished temporarily. This trend has been slowly reversing over the past two decades with the growth of an alternative food economy. Farmers markets are becoming an integral part of direct marketing and small to mid-scale food production. As consumer interest in alternative sources of food grows, the demand for "locally grown, fresh food with a story" increases, and farmers markets are the main outlet for these types of products. In 1976, Congress passed the Farmer-to-Consumer Direct Marketing Act. The act led to the development of many farmers market initiatives. Currently there are over four thousand markets in the United States alone.

MARKET VENDORS AND CONSUMERS

Farmers market vendors tend to have small to midsized operations and depend on the market for at least a portion of their total income. Markets are a low cost (although time consuming) way to get started in selling products directly to consumers. Often producers who use other direct marketing strategies use farmers markets as a way of networking with consumers and other producers to get the word out about their products. It is a way for farmers to avoid the intermediaries, which allows them to collect a premium price for their product. Most farmers who sell at a market also enjoy the social aspects of interacting with their customers. They appreciate the opportunity to share their knowledge about the products they have provided.

Consumers attend farmers markets with the expectation that they will find fresh, local, and often organic food. Most people associate farmers markets with fresh fruits and vegetables, but most markets incorporate a wide variety of products. These can include heirloom varieties of vegetables, specialty cheeses, meats, and preserves. There are often booths for artists who are selling crafts, pottery, jewelry, and other handmade items. In many ways, farmers markets are more than just places for the buying and selling of food products. They provide a particular experience that incorporates atmosphere, entertainment, and community gathering space in addition to the buying and selling. There are often musical performers, demonstrations booths, and space for sitting and eating.

Market location is very important, as consumers prefer to visit markets close to home, especially within walking distance. The face-to-face interaction at the market gives consumers a unique buying experience that cannot be replicated elsewhere. Consumers visit farmers markets for a variety of reasons. They make purchases at the market because they think the product is local. They also make purchases for philosophical reasons, such as supporting local farmers and the local economy, and because they believe the food is safer and healthier. Studies have also shown that consumers visit the market for social reasons almost as much as for access to fresh food. Studies on consumer expectations have found that most customers expect that food sold at the market to be fresher and generally of higher quality than the supermarket. Factors such as production methods, organic certification, and appearance tend to be less important, although many customers appreciate that they can discuss these factors directly with the farmer.

MARKET REGULATIONS

As farmers markets grow in popularity, the product ranges at the markets are expanding and more farmers are getting involved. As the number and size of markets grow, issues around market regulations, health and safety concerns, liability, and validity for product claims become increasingly important. Most vendors at the market must meet specific health and safety requirements as well as market level regulations. Currently,

USDA Agricultural Marketing Service: Farmers Market Locator

The USDA Agriculture Marketing Service Web site maintains a current listing of farmers markets across the United States. You can search by name, state, city, county, or ZIP code. The Web site even indicates whether each market accepts Women, Infants, and Children Program vouchers (WIC) or other benefit vouchers. This resource allows interested consumers to buy fresh food close to home.

Source: http://apps.ams.usda.gov/FarmersMarkets/

most markets are self-regulated by a market association, although additional regulations may be imposed as varying levels of government develop their own sets of regulations. Beyond health and safety concerns, the major issue in farmers markets is authenticity of the product and the producer. Market consumers trust the vendor to share information about where the product came from and how it was produced. The issue can become complicated when vendors at the market are not the growers or makers of the product.

Middlemen or resellers have been a concern at many farmers markets for as long as they have been in existence. Reselling at markets is a common practice for several reasons. One of the main reasons it occurs is to allow markets to maintain a wide product range year round. During the off-season, vendors can resell products imported from other regions to maintain a customer base and make shopping more convenient for consumers. Reselling enables vendors to supplement their own produce with products from neighbors who do not have a stand at the market. It enables vendors to carry a broader product range, include products that are not grown in the region, and extend the season on a variety of products. Reselling can be controversial as many people who shop at farmers markets believe that all the products were produced in the region where they are sold and often believe it is the farmer or a farm worker selling the items at the market.

BENEFITS AND CHALLENGES

Some of the challenges facing farmers markets include the need for producers to be present at the market, the limited variety of products, and the capacity of markets to meet demand (space, infrastructure, access). Another issue is how often markets are held. Many markets (especially in the United Kingdom) are only held monthly and cannot be considered a regular shopping source for consumers or a consistently reliable outlet for producers. This can make it more challenging to retain a regular customer base. It can also be challenging for farmers to ensure they have sufficient product available the week of the market and a separate market outlet for the products that need to be consumed before the market is held.

Farmers markets contribute to the alternative food economy and make up an important component of the local food movement. Proponents of farmers markets tout the economic, social, and environmental benefits of direct and local sales. Farmers markets encourage a local economy to develop based on trust and sense of community. The economic benefits of farmers markets often extend to the entire community. Consumers and vendors often visit shops and restaurants in the local community on the day of the market, putting extra money into the local economy. The social benefits include the sense of community that is fostered when local products are sold directly to consumers. The social atmosphere at a market encourages relationships between food producers and consumers. It also encourages food producers to interact with each other, providing important connections among local growers. Finally, farmers markets provide a good opportunity to educate consumers about the benefits of buying locally grown or raised food products.

Further Reading

Brown, Allison. "Farmers' Market Research 1940–2000: An Inventory and Review." *American Journal of Alternative Agriculture* 17(4) 2002: 167–76.

Byczynski, Lynn, ed. "Growing for Market." 2008. www.growingformarket.com.

Feagan, Robert, David Morris, and Karen Krug. "Niagara Farmers' Markets: Local Food Systems and Sustainability Considerations." *Local Environment* 9(3) 2004: 235–54.

Robinson, Jennifer Meta, and J. A. Hartenfeld. *The Farmers' Market Book*. Bloomington: Indiana University Press, 2007.

Smithers, John, Jeremy Lamarche, and Alun Joseph. "Unpacking the Terms of Engagement with Local Food at the Farmers' Market: Insights from Ontario." *Journal of Rural Studies* 24 (2008) 337–50.

United States Department of Agriculture, Agricultural Marketing Services. Wholesale and Farmers' Markets Web site. www.ams.usda.gov/farmersmarkets.

See Also: Agriculture, Alternative; Agriculture, Organic; Agritourism; Community-Supported Agriculture; Cooperatives, Grocery; Food Miles; Gardens, Community; Local Food; Sustainability, Rural; Urban Agriculture

Shauna M. Bloom

Fast Food

Quick-service or "fast food" restaurants make up the largest segment of the food service industry. Customers patronizing quick-service restaurants expect low-priced, quality food in a relatively short time frame. Customers want their food "fast" and will leave or get out of a line if the wait is too long. Competition is fierce in the fast food industry, since there is usually another quick-service restaurant nearby to meet consumer needs. Most customers expect fast, friendly service, but service expectations don't generally go beyond the basics of low price, quality food, and fast service.

There are a growing number of fast food outlets across the globe, typically by U.S. owned companies. For this reason, fast food items may come in a variety of different menus to appeal to a country's unique religious or cultural differences. Typical menu items at fast food outlets include fish and chips, sandwiches, hamburgers, fried chicken, French fries, chicken nuggets, tacos, pizza, hot dogs, and ice cream. However, all fast food outlets are noted for the unlimited supply of convenient, relatively inexpensive, and energy-dense food. Fast food appears to fulfill consumers' basic needs of providing quick and convenient breakfast, lunch, or dinner options at any time of the day.

Convenience is a major sales point for fast food operators. It is reported that "consumer spending on fast food has increased from $16 billion to 150 billion in 2004 and also the number of fast food restaurants has increased from 30,000 in 1970s to 280,000 in 2004."

Although there is no clear definition, "fast food" outlets have witnessed the most rapid expansion within the U.S. food distribution system. The National Restaurant Association reported that in "2006 the global fast food market grew by 4.8 percent and reached a value of 102.4 billion and a volume of 80.3 billion transactions. The average American eats out more than five times a week and often, one or more of those five times is at a fast food restaurant."

Fast food restaurants that depend principally on convenience foods are generally included in the low profit margin category. Critics claim that fast food chains do not provide a living wage to sustain their employees' families and their local community. Because of relatively lower menu prices, the percentage of food costs in the fast food restaurants tend to be higher than those in other restaurants. However, fast food restaurants tend to hire unskilled personnel, pay lower wages, and keep the number of employees at a minimum. This makes it possible for them to offset the high percentage of their food cost with their low percentage of labor cost.

Recent research into the increasing obesity trend has focused on the social and environmental factors that increase consumer expenditures on foods away from home. Increased food intake away from home may be one of the important factors in increased obesity. Fast food restaurants have been highlighted as a public interest concern due to the easy accessibility of these high-fat, high-calorie foods. Most people would agree that "fast" foods include the fare typically served in hamburger, pizza, taco, and fried chicken chain restaurants. Moreover, the high fat and salty foods provided by fast food operations, and their aggressive marketing strategies for increasing profits, have become major public issues.

Slow food or local food movements are in a sense a "backlash" to the fast food culture. The essence of slow food is to help individuals appreciate and respect the natural qualities found in traditional foods, without the industrial processing that our typical food supply travels through. Slow food recognizes the importance of local agriculture and the subsequent culture that inspires our cuisines. Slow food restaurants and events use local ingredients from farmers with respect to the environment, which in turn enhances the flavor and nutritional qualities inherent in the foods prepared.

Further Reading

National Restaurant Association. "Restaurant Industry Forecast." 2009. www.restaurant
.org/research/.
Schlosser, Eric. *Fast Food Nation.* New York: Penguin, 2002.

See Also: Diet, Children's; Restaurants; Slow Food

Sylvia Smith

Fish, Farm-Raised

Fish farming, or aquaculture, is the process of rearing fish (including shellfish) under controlled conditions for part or all of their life cycle. There are numerous ways to farm fish, ranging from simple methods like tending oyster beds on the ocean floor, to more complex, self-contained tank systems that recycle water and waste products.

Traditional fish farming began thousands of years ago in the carp ponds of China, and early native Hawaiians developed their own inventive fish pond system. More recently, a growing world population, combined with increased seafood demand and declining wild fisheries, has turned fish farming into a booming global industry. Aquaculture production has nearly tripled since 1990 and now accounts for almost 50 percent of edible seafood worldwide (at a value of $80 billion in 2006). It remains the fastest-growing method of food production, growing by almost 10 percent each year.

China still dominates the industry, accounting for approximately 70 percent of global production. By comparison, U.S. aquaculture contributes approximately 1 percent to the global total, comprising largely pond-raised catfish in the Southeast.

Globally, the most commonly farmed species are those found at or near the bottom of the food chain—kelp, oysters, carp, seaweeds, tilapia, and so on. The environmental impacts of such types of aquaculture are usually benign, as they often require little or no feed, and their harvest does not impact sensitive marine habitats.

For example, filter-feeding shellfish like mussels, clams, oysters, and scallops can actually help cleanse local waters. In addition, they can often be grown on ropes, rafts, or hanging nets suspended in the ocean, which precludes the need for dredging (a high-impact harvesting method that can severely damage the ocean floor).

However, a great deal of aquaculture development over the last thirty years has focused on carnivorous species like salmon and shrimp, which present severe environmental challenges. Although these high-value products are largely marketed to the United States, Europe, and Japan, they are often farmed in developing countries, where environmental regulations are lax or nonexistent. Shrimp are grown primarily in coastal ponds in Asia and Latin America, where large swaths of mangrove forest have been cleared to site shrimp farms. This not only destroys essential coastal habitat, it makes these areas more prone to greater erosion and storm damage.

Salmon farms offer their own suite of environmental concerns. Among them are the heavy use of wild-caught fish for salmon feed (on average, three to five pounds of wild fish are needed to produce one pound of farmed salmon). Also, salmon are raised in coastal netpens—floating cages that allow uneaten feed, fish waste, chemicals, diseases, and parasites to pass untreated to surrounding waters.

Finally, there is some confusion when it comes to organically raised farmed fish. Currently, there are no USDA standards, although organically labeled farmed fish from Europe are still available in the United States market. There are more than a dozen private certifiers—the two most prominent are Naturland (Germany) and the Soil Association (United Kingdom). Each organization is allowed to develop its own standards, meaning U.S. consumers can buy organic farmed salmon, for example, raised under several different sets of organic principles. In an attempt to stem such confusion, California recently ruled that no seafood can be sold as organic in the state until USDA standards are finalized.

Further Reading

Environmental Defense Fund. "The Promise and Perils of Fish Farming." www.edf.org/page.cfm?tagID=16150.

Food and Agriculture Organization of the United Nations. "The State of World Fisheries and Aquaculture 2006." www.fao.org/docrep/009/A0699e/A0699e00.htm.

U.S. Department of Agriculture National Organic Program. "Interim Final Report of the Aquaculture Working Group." Winter 2006. www.ams.usda.gov/AMSv1.0/getfile?dDocName=STELPRDC5062436&acct=nopgeninfo.

See Also: Fish, Line-Caught; Seafood, Sustainable; Tuna, Dolphin-Safe

Tim Fitzgerald

Fish, Line-Caught

Fish sold at restaurants and seafood markets may be marketed as "line caught." Most forms of line fishing are considered highly selective and a more eco-friendly alternative to fish caught with other gears, such as bottom trawls.

A number of different fishing gears fall under the category of line gear. In general, "line gear" refers to the use of a single fishing line to catch a target species. Depending on how it is configured, line gear can target both pelagic and benthic fishes. Some examples of benthic species that may be targeted with line gear include cod and haddock; some examples of pelagic species that may be targeted with line gear include tunas and mahi mahi (dolphinfish). Fisheries in different regions may have different specific names for the type of line gear that is used. Some examples of terms that are used to describe line gear include handline, pole and line, baitboat, troll, jig, and longline. With the exception of longlines, these gears are thought to result in minimal ecological impact due to the lack of accidental catch (bycatch) associated with them.

When targeting tuna, pole and line is also called baitboat fishing. For this gear type, fishers use a pole with fixed length line that has a barbless hook with either an artificial lure or live bait. In this way, fish are caught one at a time, and fishers can immediately throw back any unwanted catch. Trolling consists of towing artificial lures with barbless hooks behind the fishing vessel. Troll gear may also be called jig gear.

Pelagic longlines consist of a main horizontal fishing line that can be 50 to 65 nautical miles long, and are fished in the upper water column. Smaller vertical lines with baited hooks are spaced intermittently along the main line, and can be rigged to fish at various depths depending on the target species and fishing conditions. Pelagic longline gear accidentally catches protected and threatened species such as sea turtles, sharks, and seabirds. Bottom longlines are used to target benthic species, and in some fisheries may result in bycatch of seabirds.

Further Reading

Monterey Bay Aquarium. Conservation Research. http://www.montereybayaquarium .org/cr/research.asp.

See Also: Fish, Farm-Raised; Seafood, Sustainable; Tuna, Dolphin-Safe

Jesse Marsh

Food and Drug Administration

The history of the United States Food and Drug Administration (FDA) can be traced as far back as 1862 when a single chemist working in the United States Department of Agriculture (USDA) was responsible for regulation of food and food additives. It continued to grow as the Division of Chemistry until the passage of the Federal Food and Drugs Act (FFDA) in 1906, which was established to prohibit interstate commerce of adulterated foods, drinks, and drugs. Changes of titles and responsibilities continued until 1931 when the Food, Drug, and Insecticide Administration was renamed the Food and Drug Administration, still housed within the USDA but with new regulatory functions added to its mission.

In 1940, the FDA was moved to the Federal Security Agency (FSA). The FDA was transferred once again to the Department of Health Education and Welfare (HEW), and finally, in 1980 to the Department of Health and Human Services (HHS). From the inception of the FDA to its inclusion as an agency within HHS, confusion and at times conflict have marked its relationship with the USDA and other federal regulatory agencies. The division and discord between these institutions has often impeded progress toward a unified approach to the regulation of food, the applications of biologically active agents which are used in its production, and the regulation of substances which are ingested by U.S. consumers.

The 1938 Food, Drug, and Cosmetic Act (FDCA) codified many of the regulations the FDA was meant to enforce. Years of congressional hearings chaired by Rep. James T. Delaney of New York resulted in the Pesticide Amendment (1954), the Food Additives Amendment (1958), and the Color Additive Amendments (1960). These laws were unprecedented in their insistence that no substance could be introduced into the U.S. food supply without a prior determination that it was "safe."

In her seminal book *Silent Spring* (1962), Rachel Carson specifically called out the FDA for its failure to adequately protect consumers from exposure to specific chemicals that it had determined to be toxic, and ostensibly did not allow within the U.S. food supply. The impact of *Silent Spring* on public perception of the FDA's role and ability to protect consumers from toxins in their food was extremely significant. The fallout from *Silent Spring* and other critiques that followed throughout the 1960s and 1970s are largely credited with helping to establish the organic and sustainable agriculture movements. Yet the FDA currently has no role in assuring the "purity" or "quality" of organic foods. The regulation of organic as a "process-based" rather than "product-based" standard, implemented by third-party certifiers accredited by the USDA has been an enduring criticism of its effectiveness as an assurance of integrity.

In 1995, the Environmental Working Group (EWG) published a scathing report based on an analysis of FDA documents. It discovered that the FDA reporting process was deeply flawed and failed to accurately report many of its own findings on pesticide contamination. For instance, the rate of illegal pesticides on forty-two heavily consumed fruits and vegetables was found to be 76 percent higher than that reported by the FDA. And U.S. grown produce was more than twice as likely to be contaminated with illegal pesticides than reported.

In 1996, the Food Quality Protection Act (FQPA) was signed into law. It amended the FDCA and the Federal Insecticide, Fungicide, and Rodenticide Act (FIFRA), a 1947 law which established procedures for registering pesticides which was implemented by USDA, but did not regulate pesticide use. Enacted in large part to purge the zero-tolerance provisions of the Delaney legislation from federal law, FQPA set up a system for establishing "acceptable tolerance levels" for pesticides in food. As a result of this legislation, the Environmental Protection Agency (EPA) is now charged with setting tolerances for pesticide residue based on a "reasonable certainty of no harm"; the FDA is charged with monitoring and enforcing these tolerances in raw agricultural produce, fish, dairy products, and processed foods; and the Food Safety and Inspection Service (FSIS) of USDA is responsible for monitoring pesticide residues in meat, eggs, and poultry products. The FDA is also charged with establishing and enforcing many of the federal food labeling requirements.

In its role as a pesticide policing agent, the FDA is required to test the food products that fall under its auspices for their compliance with EPA's "reasonable" residue tolerances. Yet a 2007 Government Accountability Office (GAO) report cautioned that the fragmented nature of the U.S. food regulatory system is limited in its ability to assure the safety of U.S. food. It also drew attention to the fact that the understaffed and underfunded FDA lacks the capacity to test a significant portion of the U.S. food supply for harmful contaminants, as well as the fact that federal law does not stipulate the frequency of inspections the FDA is required to perform on foods that fall within its jurisdiction.

Testing and enforcement has become even more challenging in the era of globalization. In 2005, the United States became a net food importer for the first time in over 50 years. And monitoring pesticide residues on imported food is inherently more difficult than monitoring domestic supplies. The absence of unannounced visits to test produce, and the ability of importers to switch the lots to be tested, have been pointed to specifically by critics of the FDA's pesticide residue "sample" monitoring program. According to a 1986 GAO study, the "sample" monitoring program of the FDA had been ineffective in preventing the marketing of about 50 percent of imported fruits and vegetables that contained illegal levels of pesticide residues. As the 2006 spinach *E. coli* and 2009 peanut salmonella scares made clear, even if food dangers are identified, tracking their source can be extremely difficult, if not impossible, in a poorly regulated global food system.

Aside from pesticides, for which the establishment of premarket residue tolerances falls to the EPA, the FDA's activities mainly involve premarket control mechanisms aimed at assessing the safety of products before they are approved for use. In this role, the agency relies heavily on tests and reports conducted and provided by product manufacturers themselves. Critics and scholars have charged that this process has introduced the potential for significant conflicts of interest, and inadequate oversight. Many look specifically to what has come to be known as the revolving door between the food industry and the federal agencies charged with the regulation of food products.

This issue has been raised repeatedly with regard to the approval by the FDA of genetically modified (GM) food products, and the use of genetically modified organisms (GMOs) in food production. Critics specifically cite the case of the bovine growth hormone rBGH, a genetically engineered product used to enhance milk production in dairy cows. Michael Taylor, the FDA commissioner responsible for writing the labeling guidelines, had worked for seven years as a lawyer for Monsanto, the product's manufacturer. And Margaret Miller, then Deputy Director of the the FDA's New Animal Drugs Office, had been a Monsanto research scientist working on rGBH safety studies prior to her appointment.

The criticisms regarding the FDA's handling of GM foods do not end with rBGH. In 1992, the FDA ruled that GMOs were "substantially equivalent" to products developed through conventional breeding. As a result, it granted GMO technology "generally recognized as safe" (GRAS) status, which allows products to be commercialized without additional government testing. Federal law stipulates that GRAS status be provided only in cases where there is substantial peer-reviewed data (or the equivalent), and overwhelming consensus within the scientific community that a product poses no significant health risks. Critics charge that GM food products fail to meet either of these criteria, and that GRAS status was granted prematurely (even illegally) under pressure from the biotechnology industry.

In its 1992 policy statement regarding GMOs, the FDA established a voluntary consultation process through which it could review a developer's determination that their product met the criteria of "substantial equivalence" before it was marketed. Through this voluntary process, GM product developers meet with the FDA to discuss how they can establish "substantial equivalence" for their product. The FDA has yet to establish specific protocols for conducting scientific assessment regarding issues of toxicity or the potential for allergic reactions within GM products.

A 2003 study by the Center for Food Safety (CFS) analyzed documents related to fourteen data summaries reviewed by the FDA. It found that when the FDA requested additional information from companies, 50 percent failed to comply. And they concluded that "FDA performs a less-than-thorough safety analysis" of GM products.

Further Reading

Environmental Working Group. *Forbidden Fruit: Illegal Pesticides in the U.S. Food Supply*. Summary by Susan E. Richard. www.ewg.org/reports/fruit.

Food and Drug Administration. www.fda.gov.

Government Accountability Office. *Federal Oversight of Food Safety: High-Risk Designation Can Bring Needed Attention to Fragmented Food System*. 2007. http://frwebgate .access.gpo.gov/cgi-bin/getdoc.cgi?dbname=gao&docid=f:d07449t.pdf.

Gurian-Sherman, D. *Holes in the Biotech Safety Net: FDA Policy Does Not Assure the Safety of Genetically Engineered Foods*. Washington D.C.: Center for Science in the Public Interest, 2003.

United States Congress House Committee on Agriculture; Subcommittee on Domestic Marketing, Consumer Relations, and Nutrition. *Review of the U.S. General Accounting Office Report on Mismanagement of the 1987–88 Nationwide Food Consumption Survey*. Joint hearing before the Subcommittee of the One Hundred Second Congress, first session, October 16, 1991. Washington: U.S. Government Printing Office, 1992.

See Also: Health Concerns; Policy, Food; rBGH

Philip D. Reid and Taylor Reid

Food Deserts

A food desert is a large geographic area with no or distant "mainstream" grocery stores but often high concentrations of "fast" and other types of "fringe" food. Many stereotypes exist about food deserts. For example, many assume they exist in only poor, urban districts. However, a food desert can also be located in an urban or rural area. Furthermore, not everyone in the food desert is poor, although often residents of food deserts are indeed disadvantaged single heads of households and young children, populations that are especially sensitive to nutritional intake.

Often, food deserts have food available, but there is an imbalance of food choice, meaning a heavy concentration of food high in salt, fat, and sugar. Many fringe locations also offer "quick meals" that are highly convenient but cannot support a healthy diet on a regular basis. The study of food deserts is important for every type of community—urban, suburban, and rural—because findings reveal that residents of food deserts suffer worse diet-related health outcomes, including diabetes, cancer, obesity, heart disease, and premature death. These effects are independent from other contributing factors, such as income, race, and education. For diabetes and obesity, these relationships are statistically significant.

Whether an area is urban, rural, or suburban, an assessment of the local food should distinguish and quantify these types of food venues. For this reason, having a common definition for both mainstream and fringe food is very important.

A mainstream grocer is a place where people can support a healthy diet on a regular basis. A fringe food location is the opposite; it is not inherently bad, but if it were the primary food source, local diets and public health would likely suffer. Mainstream grocers need not be part of a major "full service" chain; total square footage is not important. Mainstream grocers can be independent or small food stores. The key defining factor is that they sell an assortment of healthy and fresh foods, such as produce, fruits, and meats.

Fringe food venues include fast food restaurants and convenience stores. However, they can also include gas stations, liquor stores, department stores, discount bakeries, pharmacies, and a multitude of other retailers that sell ready-made, fast, boxed, canned, and other types of food products but for whom fresh and healthy food is not the primary line of business. Again, these foods are usually high in salt, fat, and sugar and have very limited nutritional value.

Further Reading

Moore, L. V., and A. V. Diez Roux. "Associations of Neighborhood Characteristics With the Location and Type of Food Stores." *American Journal of Public Health* 96(2) 2006: 325–31.

Pearce, J., R. Hiscock, T. Blakely, and K. Witten. "The Contextual Effects of Neighbourhood Access to Supermarkets and Convenience Stores on Individual Fruit and Vegetable Consumption." *Journal of Epidemiology and Community Health* 62(3) 2008: 198–201.

Wrigley, Neil. "'Food Deserts' in British Cities: Policy Context and Research Priorities." *Urban Studies* 39(11) 2002: 2029–40.

See Also: Consumers; Food Security; Food Sovereignty; Political Ecology of Food

Mari Gallagher

Food Dollars

"Food dollar" refers to the price spreads from farmers to consumers for agricultural and food commodities, and specifically the percentage of the final consumer expenditures on food which is captured by farmers. That is, it is the difference between the farm value and consumer expenditures for foods at both food stores and restaurants. Thus, the marketing bill covers the processing, wholesaling, transportation, retailing costs, and corporate profits and the farm value of the food dollar is what remains. Over time the percentage of consumer expenditures captured by farmers has declined, and greater percentages have been captured by food manufacturing firms to cover processing and marketing costs such as labor, packaging, energy, transportation, and marketing costs. The Economic Research Service of the U.S. Department of Agriculture (ERS) has calculated that farmers received 19 percent of the final consumer dollar in 2006, 31 percent in 1980, 32 percent in 1970, and 41 percent in 1950. ERS also found that labor costs in the processing and marketing sectors is the largest single component of the marketing bill, accounting for 38 percent of total consumer expenditures on food in 2006.

The low and declining percentage of the consumer food dollar accounted for by farmers is viewed as a social problem and a symptom of our current industrialized food system. The majority of the food offered in restaurants and food stores is highly processed and is not produced in the region in which it is sold and consumed; this contributes significantly to the problem of the low and declining farm share of the food dollar. When food products travel hundreds of miles from the farm, through processing and marketing, to reach the final consumer, it negatively impacts the financial interests of farmers. Antidotes include purchasing food locally through farmers markets, community-supported agriculture operations, farm-to-school programs, reinvigorating and developing regional food processing and distribution systems, promoting fair trade and farmer collective bargaining organizations that retain a greater share of the food dollar for the farm-level workers, and eating less processed and more "whole" foods. These types of interventions benefit farmers, consumers, smaller-scale processing and food manufacturing firms, and the rural economy.

Further Reading

Lyson, T. A., G. W. Stevenson, and R. Welsh, eds. *Food and the Mid-Level Farm: Strategies for Renewal.* Cambridge, MA: MIT Press, 2008.

Sustainable Table. "What Is Local?" January 2009. www.sustainabletable.org/issues/buylocal/.

USDA ERS. "Price Spreads from Farmer to Consumer." May 28, 2008. www.ers.usda.gov/Data/FarmToConsumer/index.htm.

See Also: Production, Treadmill of; Sustainability, Rural; Appendix 3: U.S. Organic Farming Emerges in the 1990s: Adoption of Certified Systems, 2001

Rick Welsh

Food Miles

"Food mile" is a term used to describe the distance a food product travels from place of production to point of sale. Food mile has become a tool used to determine the sustainability of a food system. In particular, it assesses the carbon imprint of a particular food product, the cost of transporting a product from field to processor, distributor, and finally to a retailer, and the environmental impacts of the current global food system. The term was first used in a report released in 1994 by SAFE Alliance in the United Kingdom.

Food mile is typically calculated by using a weighted average formula, which can take into account factors such as distance traveled, energy used, and carbon emitted. These calculations are used to estimate the average distance of both a single source product and multiple source products (for example processed foods containing multiple products). The two most commonly used formulas are the Weighted Average Source Distance (WASD) and Weighted Average Emissions Ratio (WAER).

Studies done by the Leopold Center in Iowa have found that domestic produce in North America travels approximately 1,500 miles before reaching the consumer. The average distance traveled increases significantly when imported products are included, which is an increasingly common occurrence. One study found that transportation accounts for 14 percent of all energy consumption in the entire food system. Locally sourced products, on the other hand, travel approximately 50 miles and have significantly lower emissions and fuel usage than products sourced from the conventional food system. Other concerns that arise when food travels long distances include freshness, taste, and nutrient levels. Retailers and distributors, on the other hand, are concerned about shelf life, transportability, and appearance.

Food miles have increased significantly as the food system has become more global, with changes in distribution and retailing making importing and exporting food products more efficient and cost effective. This is particularly true for countries that have subsidized the fuel costs for transporting food. Food products have become more packaged and are selected for longer shelf life to match consumer preference for making fewer shopping trips.

The information gleaned through studies on the sustainability of long-distance shipping has encouraged many local food promotion groups to use food miles in their marketing and messaging. The actual impact of food miles is under debate and continuous study. The issue is complicated by multiple modes of transportation used, production methods, packaging, and concerns about local products increasing consumers' emissions and fuel consumption while they travel to multiple locations to find local products.

Further Reading

Halweil, Brian. "Home Grown: The Case for Local Food in a Global Market." World Watch Paper 163. 2002. www.worldwatch.org/system/files/EWP163.pdf.

Hill, Holly. "Food Miles: Background and Marketing." National Sustainable Agriculture Information Service. National Center for Appropriate Technology. 2008. www.attra.ncat.org/attra-pub/PDF/foodmiles.pdf.

Pirog, Rich, Timothy Van Pelt, Kamyar Enshayan, and Ellen Cook. "Food Fuel, and Freeways: An Iowa Perspective on How Far Food Travels, Fuel Usage, and Greenhouse Gas Emissions." Iowa State University, Leopold Center for Sustainable Agriculture, 2001. www.leopold.iastate.edu/pubs/staff/ppp/food_mil.pdf.

Pretty, J. N, et al. "Farm Costs and Food miles: An Assessment of the Full Cost of the UK Week Food Basket." *Food Policy* 30 (2005): 1–19.

See Also: Farmers Markets; Local Food; Appendix 4: Recent Growth Patterns in the U.S. Organic Foods Market, 2002

Shauna M. Bloom

Food Policy Council

Since the 1980s, food policy councils (FPC), also known as food system councils, have emerged as a means to address various food system issues at the local or state level. A council is typically an officially sanctioned entity with diverse representation across food system sectors, including production, processing, distribution, retail, and consumption. Government agencies often provide leadership to FPCs; however, in some regions a nonprofit agency provides administrative support. Food system councils also are materializing from grassroots networks.

On average, FPCs maintain twelve to fourteen members with a mix of stakeholders representing agriculture, health, food industry, antihunger, government, and nonprofit entities. Subunits such as ad hoc committees, task forces, or work groups, are formed to focus on specific issues. It is estimated that more than fifty FPCs have formed in North America in the past thirty years; however, the average life span of an FPC is less than ten years.

ROLES

The common roles of FPCs are to examine the existing food system, determine assets and gaps, and identify policies or programs to build the stability and resiliency of the food landscape. FPCs may be the only channel for communicating the needs of food assistance, agriculture viability, rural vitality, local food markets, ecological capacity, emergency preparedness, or food system sustainability to local or state governments.

Activities of FPCs are dependent on stakeholder composition, political climate, leadership, and funding. Although guided by innumerable values and visions, many councils provide policy and program recommendations to government officials. This is in addition to networking, facilitation, education, and research activities. Despite the label of food policy council, some entities do not engage in legislative or policy recommendations. However, a universal value of FPCs is that local or state governments assume responsibility to ensure that all eaters have regular access to healthy, fresh, and safe food at all times.

Examples of emerging trends of food policy and food system activities include bridging obesity prevention initiatives to farm-to-school programs; building rural economic development through diversified farming; and connecting low-income families to urban gardens. New stakeholders, such as local and state planners, architects, and chamber of commerce officials, are involved in food policy discussions.

FUTURE

FPCs have a critical role in shaping sustainable local and regional food systems; however, many challenges face the viability and endurance of councils—specifically, funding, staff, and government support. A significant challenge for councils is to close the gap among stakeholders who may or may not grasp the construct of food systems and how policy shapes the food system.

Characteristics of a successful FPC are sustained funding, vision, engaged and experienced leadership, stakeholder diversity and commitment, agency administrative support, identifying win-win solutions, government engagement, community connections, and visibility. FPC leaders extol that the success and longevity of FPCs is dependent on obtaining a level of integration or institutionalization within local or state government. In light of the current food crises, the need is greater than ever for FPCs to champion a vision that transforms the food system to one that is healthy, green, fair, and accessible.

Further Reading

Clancy, K., J. Hammer, D. Lippoldt. "Food Policy Councils. Past, Present, and Future." In Remaking the North American Food System. Strategies for Sustainability, edited by C. Hinrichs and T. Lyson, Lincoln: University of Nebraska, 2007.

Community Food Security Coalition, North American Food Policy Council. www.foodsecurity.org/FPC/.

Drake University Agricultural Law Center, State and Local Food Policy Councils. www.statefoodpolicy.org.

Schiff, R. "The role of Food Policy Councils in Developing Sustainable Food Systems. Journal of Hunger and Environmental Nutrition 3(2/3) 2008: 206–28.

World Hunger Year, Food Policy Councils. www.worldhungeryear.org/fslc/faqs/ria_090.asp?section=8&click=1.

See Also: Local Food; Policy, Food; Food Security

Angie Tagtow

Food Safety

Food safety is the process of serving safe and edible food to all populations. Food safety is the most important responsibility of a food production operation. If the food operation is for profit, then profit is the second most important management responsibility after food safety. Practicing safe food handling may increase profits by reducing waste, which generate lower food costs. A well-orchestrated food safety system will help protect a company from lawsuits and possible closure, due to reports of food-borne outbreaks.

A food-borne outbreak is when two or more people become ill from eating the same contaminated food. A food-borne illness is an incident when one person gets sick from eating contaminated food.

There are three types of food contaminates: biological, chemical, and physical. Biological contaminants are the most common type. They include bacteria, viruses, parasites, and fungi. Chemical contaminates are those arising from chemicals, pesticides, or metals found within the food service operation or the flow of food. Physical contaminates pertain to any objects found in the food, natural or foreign, that pose a threat to the consumer (e.g., metal, bones, bandages).

The majority of food-borne outbreaks are caused by biological contaminates or microorganisms. There are four types of microorganisms that can contaminate food: bacteria, viruses, parasites, and fungi.

Bacteria is the most common type of pathogen to contaminate food. Bacteria are living, single-celled animals that reproduce exponentially given the right conditions. FATTOM is an acronym for the ideal conditions bacteria needs to reproduce: food, acidity, time, temperature, oxygen, and moisture.

Viruses are another microorganism that require a living cell to reproduce. Viruses do not grow in food, but can be transmitted through food. Viruses can best be prevented by practicing good personal hygiene (washing hands).

Parasites are microorganisms that require a living host to survive. Parasites are most commonly passed on to humans through the meat people eat. Food-borne illnesses caused by parasites can be controlled by purchasing from reliable vendors, cooking foods to proper temperatures, avoiding cross-contamination, and observing proper hand washing.

Fungi may be the smallest microorganisms to cause food-borne illness, although not all fungi negatively impact food. Edible molds found in cheeses, yeast in bread, and nonpoisonous mushrooms are examples of non-contaminate microorganisms. Molds that cause food to spoil are examples of a negative microorganism. Molds can grow under many conditions that would otherwise kill bacteria.

However, the most common cause of food-borne illnesses are personnel failures, such as poor personal hygiene, cross-contamination, and cooking, cooling, and holding foods at the wrong temperature. Such issues, if properly monitored, should be controlled through a trained staff that practices food safety.

Good personal hygiene includes proper hand washing, personal cleanliness, clean uniforms, and sanitary habits. One opportunity for food contamination to occur is when employees transmit pathogens, which are on their hands, bodies, or uniforms, to the food customers eat. Therefore, employees should use proper hand washing techniques immediately upon entering and re-entering a food operation. Likewise, proper hand washing should be performed any time job tasks are changed and anytime a potentially contaminated object is touched. Proper hand washing may seem like an obvious protocol before handling food, yet it is commonly poorly performed and even skipped in many instances.

"Cross-contamination" is the term used to describe the spread of pathogens from one food service or food to another. Cross-contamination can occur if raw foods are mixed with ready-to-eat foods, food surfaces are not cleaned and sanitized, or employees handle raw foods or contaminated objects and then touch ready-to-eat foods. Cross-contamination can be prevented by using clean and sanitized equipment and surfaces, using specific equipment for certain food-related tasks, and preparing raw foods and ready-to-eat foods separately.

One of the most important yet basic components of a food safety system are the proper equipment and cleaning facilities. The dishwashing machine is one of the most important areas within a safe food handling environment; however, it is often the most abused. The dishwashing area needs to be well planned in order to have cleaned and sanitized pots, pans, and dishes in one area, not to be cross-contaminated by dirty dishes and pots from other areas. Dishwashers can either use hot-water temperatures or chemical sanitizers for sanitizing foodware, glassware, flatware, and dishes. Hot-water temperatures need to reach 180°F for sanitation. The dish machine needs to be kept clean and sanitary, along with all equipment and food service areas. Food is only as safe as the environment in which it is prepared.

Time and temperatures conditions are the factors easiest for food operators to control. Time and temperature controls involve what is known as the temperature danger zone (TDZ). The TDZ are temperatures between 41 and 135°F. This is the ideal temperature for microorganisms to grow. By cooking food to 140°F and above or reducing temperatures to 41°F or below, most microorganisms are either killed or impaired to grow. In addition, time is an element that needs to be controlled. Limiting the amount of time food remains in the TDZ will control the growth of microorganisms.

Foods should be checked for correct temperature with a sanitized, calibrated thermometer. A calibrated thermometer is an instrument that has been set correctly using either the ice-bath method or the boiling-water method. If no food thermometers are used, it could increase the chance that food is not being cooked to proper temperatures, thus not destroying most bacterial food pathogens (salmonella).

The storing of food is a very important step to ensure a safe food product. All food products should be labeled and dated. Foods should be rotated on a first-in first-out (FIFO) system, so as to use the oldest food first. Food should be put in clean containers and never mixed, such as combining old product with new product to consolidate and save space. Potentially hazardous foods should be kept out of the TDZ, and storage areas should be kept at the correct temperature.

During the food preparation stage, time and temperature standards should be maintained and cross-contamination conditions should be restricted. Cooking foods to the proper temperature can kill many microorganisms; therefore, employees should be instructed on minimum required temperatures.

The next step in the food path is cooling foods. This is a very important step and one that is frequently abused to expedite production. After cooking, if food is to be stored, it must be cooled in two stages: stage 1 from 135°F to 70°F in two hours, and stage 2 from 70°F to 41°F in four hours. A total of six hours may be used to cool foods to the proper temperature. There are a few methods to cool foods properly; the most common is the ice-bath method. This is when food is redistributed into smaller, clean containers, and these containers are subsequently submerged in an ice bath. Another method of cooling foods is to use a frozen paddle, which contains ice to stir the food product, reducing the internal temperature of the food product.

Holding foods for food service is an important step to continue the process of serving safe food. In this stage, foods are prepared and ready to eat; therefore, preventing time and temperature abuse and cross-contamination is a major priority.

For food service, reheated foods should reach internal temperatures of 165°F for fifteen seconds within two hours to be considered safe. During service, foods should be checked every four hours to ensure temperatures are 135°F or higher. Foods that are not at correct temperatures or have remained in the TDZ for four hours should be discarded. Proper hot and cold storage equipment should be used for service.

Serving foods is the final step in the "flow of food," although this flow can be considered a cyclical process. While serving food, correct utensils with long handles should be used. Disposable gloves should be worn during all service of ready-to-eat food. All glassware, flatware, and dishes should be handled correctly and food contact areas should not be touched. Food service areas should have sneeze guards to protect food, and food should be held at the proper temperatures. Food should be labeled and replenished using FIFO. Also, customers should not reuse dirty plates when getting second helpings.

Food service employees should practice safe food handling at all times. Keeping clean equipment and a clean work environment are the primary steps toward providing safe food. Gloves should be worn for ready-to-eat food. Cross-contamination should be eliminated by using the correct, sanitized utensils and monitoring food surface areas. Last, food service employees should practice good personal hygiene at all times (e.g., wash hands, maintaining clean uniforms, keeping hair restrained).

It is almost impossible to eliminate all microorganisms from contaminating food during processing. However, almost all cases of food-borne illnesses can be prevented if food is properly handled and prepared. The most important steps a food service handler can do to prevent food contamination is to apply safe time and temperature controls, eliminate cross-contamination, and practice good personal hygiene every day.

Further Reading

USDA Food Safety and Inspection Service. "Fact Sheet: Food Safety Handling." February 26, 2009. http://www.fsis.usda.gov/factsheets/Safe_Food_Handling_Fact_Sheets/.

See Also: Agrichemicals; Endocrine Disruptors; Food and Drug Administration; Health Concerns; Organic Foods, Benefits of

Sylvia Smith

Food Security

"Food security" means having dependable access to sufficient quantities of safe and nutritious food to meet basic dietary needs and cultural preferences, by socially acceptable means. True food security cannot exist in the absence of any one of these conditions. People do not have food security if they have access to enough food for today, this week, or even this year, but are not confident they will have enough to eat at some time in the foreseeable future. They must have dependable access to food. People have food security only if they have enough food that is safe to eat and that provides sufficient nutrition for normal physical and mental development and a healthy, active lifestyle. Food security also means having access to foods that are culturally appropriate, in that they are not considered repugnant or repulsive due to eating customs or ethical or religious beliefs. Finally, people lack food security if they have to resort to means such as emergency food aid, begging, or stealing to acquire enough food. Food security requires far more than having access to enough food calories to keep the body alive for another day.

A family, community, or nation has food security only if every member of the family, community, or nation is secure in access to food. Likewise, global food security requires food security for all people, everywhere. It is doubtful that any family, community, or nation is completely food secure in all essential aspects, and global food security most certainly does not exist. Food insecurity is most commonly linked to poverty. People with sufficient income generally are able to buy enough food, although this may not be true for people who are mentally ill or caught up in natural disasters, civil disruptions, or wars. In late 2007, it was estimated that more than 800 million people globally were chronically hungry due to extreme poverty and up to two billion people lacked dependable access to food due to chronic poverty. Various natural disasters and ongoing civil and multinational wars undoubtedly added a few millions to the 2007 total of people facing hunger. Global uncertainties, such as fossil energy supplies, climate change, population growth, water scarcity, and soil degradation have created uncertainties regarding the possibility of ever achieving long-run global food security. In addition, uninformed or insensitive public policies, such as subsidizing the diversion of food crops for fuel, represent ever-present possibilities of food scarcity totally unrelated to the vagaries of nature.

In addition, more than half of the earth's population live in metropolitan areas and are dependent on complex, interdependent systems of food production, storage, processing, and distribution. In economically developed countries, such as the United States, any disruption in the food distribution network could create critical food scarcities within a matter of days, as has been apparent during various natural disasters and urban riots. At any given time, the food distribution pipeline contains only about seven to ten days' supply of food. A significant crop failure anywhere in the world can have major impacts on global hunger, as less than a sixty days' supply of grains may be carried over from one crop year to the next. The time required to increase poultry or livestock production ranges from months to years. Today's complex, interdependent food system may be very economically efficient but it embodies inherent food insecurities.

Today's global food system relies heavily on the global market economy to provide national food security. Few nations today would be able to provide national food security in the absence of food imports from other countries. Virtually all significant agricultural producing nations, including the United States, export surpluses of products for which they have a competitive economic advantage and import foods that can be produced more efficiently elsewhere in the world. Even in the United States, the value of food imports typically averages only slightly less than the value of agricultural exports. Any nation that depends on global markets for food items necessary for its health and survival lacks food security. In times of food scarcity, global markets will ration available supplies in relation to the ability to pay, without regard for needs. The affluent of today's world will hardly notice the difference in their cost of living, whereas the world's poor will go hungry or starve. For this reason, most nations are unwilling to rely solely on markets for national food security. Food security policies in the United States, for example, include agricultural programs that subsidize domestic farmers for the stated purpose of ensuring adequate supplies of food that is safe and affordable to most people. For those who are still unable to afford enough food, income and food assistance programs are made available. Attention to food quality is limited primarily to nutrition education. However, recent U.S. trade and agricultural policies have made U.S. food security increasingly dependent on global markets. With global food markets increasingly dominated by large multinational corporations, with primary responsibility to multinational investors, U.S. food security has become somewhat tenuous.

The only true food security for any family, community, or nation is in having dependable access to enough productive farmland and enough committed farmers to produce enough safe, nutritious food to meet its basic needs, without relying on producers with no long-term personal or cultural commitment to the family, community, or nation. Food security does not mean self-sufficiency. Food security is about meeting basic needs, whereas self-sufficiency is about meeting both wants and needs. Markets are acceptable means of acquiring foods that make eating more interesting, enjoyable, or convenient, or even foods that make it easier to eat a nutritionally balanced diet. These are desirable but not necessary aspects of food; their loss would not threaten basic food security. Food security does not suggest that each person, community, or nation must be capable of producing enough food to meet its basic needs, but it does suggest that food dependencies must be based on dependable relationships of trust, rather than markets of least cost or greatest convenience. Relationships are more dependable whenever they are reinforced by common historical family, cultural, or ethical values or by deeply held, personally shared values, rather than matters of convenience or practicality. In many lesser developed countries the only real food security available may be family or community food sufficiency. In general, however, individuals, communities, and nations need only have dependable, long-term commitments with those who produce their food. For true food security, those commitments must extend to the farmers who produce it, and those farmers must be committed to maintaining the productivity of the land on which their food is produced.

Historically, organic farming represented a commitment to food security. Organic pioneers, such as Rudolf Steiner of Germany, Sir Albert Howard of Great Britain, and J. I. Rodale of the United States wrote about the necessity of maintaining healthy, organic soils as a means of sustaining healthy people, healthy communities, and healthy nations. They wrote of the historic parallels between declining organic matter in soils and declining nations, with specific attention to the Roman Empire. They saw organic farming as a necessity for national security. More recently the focus of organic foods has shifted toward food safety and environmental quality rather than long-run food security. However, the sustainable agriculture movement has emerged to continue the historical traditions of organic farming by linking the ecological soundness and economic viability of agriculture with social responsibility. A sustainable agriculture must be capable of meeting the food needs of the present, meaning all people of the present, without compromising opportunities of those of future generations to meet their food needs as well. Sustainable agriculture is about food security. Long-term food security requires dependable access to foods produced on sustainable farms.

The local food movement is the most tangible expression of growing concerns for national food security in the United States. Those who show a strong preference for foods made with locally grown, seasonally available, minimally processed foods are called "locavores." They are not simply expressing a preference for greater freshness and flavor; they are attempting to develop meaningful relationships with local farmers, and through farmers, with the land. They are concerned not only about the lack of sustainability of industrial agriculture; they are also concerned about trends toward industrialization and globalization of organic foods. Locavores are promoting community food security not only to achieve personal food security, but also in hopes of eventually linking sustainable community-based food systems together, regionally and nationally, to create a sustainable global food network and achieve global food security.

Further Reading

Allen, Patricia. *Together at the Table: Sustainability and Subsistence in the American Agrifood System*. University Park: Penn State University Press, 2004.

Hinrichs, Clare, and Thomas Lyson. *Remaking the American Food System: Strategies for Sustainability*. Lincoln: University of Nebraska Press, 2008.

Shiva, Vandana, and Gitanjali Bedi. *Sustainable Agriculture and Food Security: The Impact of Globalization*. Thousand Oaks, CA: Sage Publications, 2002.

See Also: Consumers; Food Deserts; Food Policy Council; Food Sovereignty; Gardens, Community; Green Revolution; Globalization; Local Food; Political Ecology of Food

John Ikerd

Food Sovereignty

Food sovereignty exists when an area has self-determination and control over food supplies.

Food sovereignty can be maintained even for countries that produce very little food through regulations on imports. However, complete autonomy over food supplies is not possible in today's world where food is a global trade commodity. Demand for food in other countries will influence food supply and prices around the world. For example, economic and population growth in China has already affected the prices and availability of global food supplies. Even the strictest protectionist policies cannot completely insulate a country from exogenous influences on food without endangering its food security.

The terms "food sovereignty" and "food security" are sometimes, incorrectly, used interchangeably. A country may have access to a sufficient amount of food, which is food security, but may not have food sovereignty. Food security without food sovereignty is perhaps best demonstrated through the example of food aid. International humanitarian efforts have increased food security for many countries even during times of extreme drought. Countries that accept food aid usually have little control over what type of food is received or the quality of that food. In 2002 Zimbabwe refused food aid from the United States because the aid may have included genetically modified (GM) food that was banned in Zimbabwe. Zimbabwean leaders were widely criticized for endangering the lives of citizens, but leaders argued that the potential long-term threat of GM foods was more significant. Aid agencies and the government of Zimbabwe were eventually able to solve the food shortage in a way that maintained food sovereignty and avoided severe famine. The case of the 2002 food crisis in Zimbabwe raises serious questions of when food sovereignty should give way to citizens' right to food.

Food sovereignty can also be threatened by neoliberal policies of the global economy. Members of the World Trade Organization (WTO) are not able to exercise complete control over the flow of food into their country. Imposition of tariffs or bans of foods that do not meet national food standards may be illegal under WTO rules for giving an unfair advantage to domestic producers. Under WTO rules, foods treated with chemicals or GM foods not allowed in domestic production may enter into the domestic food supply.

Neoliberal policies that threaten food sovereignty affect the survival of domestic food production and can increase impoverishment of poor farmers. If a country is forced to quickly open its market to inexpensive imports, many domestic producers will not be able to compete with the sudden flood of high energy input conventionally farmed foreign foods and may be forced to abandon their farms. Local food networks that are largely organic and grow a diversity of crops disappear under this scenario.

Further Reading
IPC Food Sovereignty. "Who We Are." www.foodsovereignty.org/new/whoweare.php.

See Also: Consumers; Development, Sustainable; Food Deserts; Food Security; Free Trade; Globalization; Green Revolution; Political Ecology of Food

Julie Weinert

Fossil Fuel Use

Climatologists with the NASA (the National Aeronautics and Space Administration) believe that an 80 percent reduction of current greenhouse gas emissions will be necessary to stabilize and reduce global warming. About 25 percent of greenhouse gas emissions are embedded in the world's food and fiber system, thus a meaningful solution of world greenhouse gas issues is possible by engaging food and agriculture sectors. Most societal energy has focused on electrical and transportation systems. These areas alone cannot resolve the world greenhouse gas problems. Energy and climate solutions will depend on broad-based efforts including a variety of economic sectors.

Reducing agriculture emissions of greenhouse gases, soil sequestration, and renewable energy all play important roles in greenhouse gas roadmaps and global climate change issues. Beyond its ability to impact greenhouse gases and energy, food and agriculture systems are critical to feeding a growing population. Invariably, the food system is grounded on plant photosynthesis, where sunlight, carbon dioxide, and water are converted into carbohydrates. Plant products based on carbohydrates recycle in the soil and are converted into carbon rich organic matter. Increased soil organic matter is available to help resolve greenhouse gas issues. The effect of the agricultural system has been to decrease soil organic matter and to contribute and not counteract global warming. Through better utilization of energy in agriculture and food, greenhouse gas emissions can be reduced. Through plant mediated genesis of soil organic matter, they can be reversed.

ENERGY USE

It is estimated that about one-third of present energy in agriculture production environment is related to fertilizer use, one-third to the production and use of pesticides, and the remaining third mostly to mechanization that reduces labor requirements.

U.S. agriculture consumes approximately 80 percent of all the fresh water used by humans. And an irrigated wheat crop in the United States requires more than three times the energy needed as the same area of rainfed wheat. The fundamental flaw of the conventional, chemically intensive agricultural system is that water and energy are more restricting to world food production than fertilizers and pesticides. Because energy is a major constraint to world food production, the intensified use of inputs, such as chemicals and irrigation, need to be rethought. In addition, the factory production of animals is not only unhealthy but is also more demanding in energy and promotes more contamination of the environment.

Improved agricultural energy efficiency represents a tip of the climate mitigation iceberg. Soil carbon sequestration, energy efficiency, and alternative energy resources are key in global energy and environmental management. Indeed carbon sequestration and alternative energy represent the less seen and underestimated base. Soil sequestration generates crop productivity and quality improvements. As soil organic matter increases, so do the health of environmental resources. Moreover, organic agriculture has extraordinary potential to increase soil organic matter.

Until 1950, land use changes (i.e., cutting forest and degrading soil organic matter reserves by plowing and erosion) were the principal sources of elevated greenhouse gases. Since 1950, burgeoning use of fossil fuel for electrical energy and transportation have been the principal contributors to excess atmospheric greenhouse gases. A continuing trend of elevated carbon dioxide content mirrors increased use of fossil fuels and rising temperatures. Long-term database data show that global warming is already occurring.

Depleted petroleum reserves are increasingly viewed as a key limitation to economic development. In terms of supporting the planet and its multiplying human population base, water, climate resources, and food are just as important for ensuring the continuation of civilization. Scarcity of both energy and natural resources threaten future existence.

PREVENTION

About 25 percent of national greenhouse emissions are encompassed in the national food, fiber, and agriculture system. Production activities are only a small part of this total, with processing and food distribution contributing the vast majority. So, a more holistic accounting of energy embedded in food and agriculture is a better vehicle for deliberating on food and energy policy issues. These analyses will provide a framework for understanding these issues and resolving them. Farmers, consumers, policy makers, businessmen, and others can all be involved in critical efforts to counteract global warming.

BACKGROUND

Just before World War I, German scientists Carl Bosch and Fritz Haber pioneered a process employing fossil fuel to provide high heat and pressure needed to promote the condensation of nitrogen in ambient air to aqueous ammonia. This process formed the background of the twentieth-century commercial fertilizer (nitrogen) industry.

Currently most of the world energy system is devoted to burning petroleum, coal, or natural gas fossil fuels as nonrenewal energy resources. With unprecedented use, fossil fuel reserves are dwindling faster than new exploration can replace them. The skyrocketing use and demand with shrinking supplies is called "peak oil crisis."

As fossil fuels are becoming increasingly scarce and expensive, there is growing recognition of the large environmental, health, and climatic footprints they represent. Development of alternative renewal energy has been more easily said than done. Increasingly, energy efficiency and conservation are seen as prime resources to provide cost-effective energy in the near term. The transformation of agriculture and food systems to incorporate more efficient use of energy is part of a foundation to resolve greenhouse gas issues. The current energy and greenhouse gas quandary is rooted in an abundance of fossil fuels at reasonably low prices that developed from 1920 to 1970. This contributed to overuse and dependency on nonrenewable fossil fuels.

There have been growing concerns that petrochemically based food and agriculture is not sustainable or worthy of the environment and health prices paid. Rodale Institute studies show 33 percent of the energy in corn and soybean production systems can be reduced under organic farming methods compared with chemically based conventional methods. In maize production, the largest energy draw is the consumption of natural gas for producing nitrogen fertilizer using the Borsch and Haber process of fixation. In this process natural gas is burned to provide the energy needed to condense atmospheric nitrogen into liquid ammonia using high temperature and pressure. About 40 percent of all energy in a conventional corn and soybean field crop system is allocated to ammoniated fertilizer. Fugitive nutrients from fertilizers and pesticide toxicities testify to the unplanned side effects of chemically based conventional agriculture.

According to a 2004 report by the United States Geological Survey, at that time almost two-thirds of excess nitrates in the Chesapeake Bay were attributable to nitrogen fertilizer, and one-third to manure. At the Rodale Institute (2006) over 70 percent reduction of fugitive nitrates were shown with the use of compost rather than raw manure or synthetic mineral fertilizer. Although fertilizers, nitrogen in particular, are seen as panaceas, biological nitrogen fixation using legumes can be much more energy efficient. Promoting energy-efficient production agriculture and trapping greenhouse gases in the soil through organic farming methods provides an extremely beneficial ecological service which is needed to address greenhouse gas elevation.

When energy was a cheap input, there was no emphasis in its conservation or efficient use. Economic and environmental pressures have worked to change this view and associated behavior. If the overall potential of agriculture is to positively impact greenhouse gases, both emission reduction and sequestration should be considered. Net greenhouse gas potential must track both fossil fuel requirements and use but also soil organic matter conservation.

SOLUTIONS

Starting in early 1900s, Rudolf Steiner, Sir Albert Howard, and the Rodale family championed new paradigms of biologically based agriculture depending on natural, not artificial, processes.

By the 1960s, the scientific writing of Rachel Carson pointed to the devastating potential of chemically based food production systems. The founders of organic farming are increasingly seen as visionaries for their understanding of the bounty of natural production practices.

Natural, sustainable, low-input agriculture is being confirmed energetically, environmentally, economically, and in terms of health and well being, with scientific measurement and mechanism to back it up. Natural processes better conserve and increase natural resources on which life is dependent. Organic agriculture is not only energetically sound but also economically and environmentally advantageous.

Solutions put forth include the following: Set energy/environment goals and work to meet them. Use organic food to reduce energy use—compared with conventional food, the energy requirement is 25 to 50 percent less. Use fresh, unprocessed, unpackaged, and local foods because a majority of the energy requirement for the food and agriculture system is in processing, packaging, and distribution. Increase soil organic matter. Cover cropping, reduced tillage, and compost addition favor carbon sequestration and the reduction of energy in agriculture while contributing to crop productivity, quality, and the environment.

Further Reading

Denver, J. M., S. W. Ator, L. M. Debrewer, M. J. Ferrari, J. R. Barbaro, T. C. Hancock, M. J. Brayton, and M. R. Nardi. "Water Quality in the Delmarva Peninsula—Delaware, Maryland, and Virginia, 1999–2001: U.S. Geological Survey Circular 1228." 2004.

Hepperly, P., D. Douds Jr., and R. Seidel, R. "The Rodale Farming Systems Trial 1981 to 2005: Long Term Analysis of Organic and Conventional Maize and Soybean Cropping Systems." In *Long-Term Field Experiments in Organic Farming*, edited by J. Raupp, C. Pekrun, M. Oltmanns, and U. Köpke, 15–32. Bonn: International Society of Organic Agriculture Research (ISOFAR), 2006.

Pimentel, D. *Impacts of Organic Farming on the Efficiency of Energy Use in Agriculture.* Foster, RI: The Organic Center, 2006. www.organic-center.org/science.environment.php?action=view&report_id=59.

Pimentel, D., P. Hepperly, J. Hanson, D. Douds, and R. Seidel. "Environmental, Energetic, and Economic Comparisons of Organic and Conventional Farming Systems." *Bioscience* 55, (2005) 573–82.

Pimentel, D., B. Berger, D. Filiberto, M. Newton, B. Wolfe, B. Karabinakis, S. Clark, E. Poon, E. Abbett, and S. Nandagopal. "Water Resources: Agricultural and Environmental Issues. *Bioscience* 54(10) 2004: 909–18.

U.S. Geological Survey. Synthesis of U.S. Geological Survey Science for the Chesapeake Bay Ecosystem and Implications for Environmental Management." Circular 1316. 2008.

See Also: Biodiversity; Climate Change; Environmental Issues; Agriculture, Conventional; Vegetarian Diet

Paul Reed Hepperly

Free-Range Poultry

Consumer concerns about the environment and the humane treatment of animals have driven growth in the organic and specialty poultry market in the United States. One of the labels often used in the specialty poultry market is "free-range," a term used for both poultry meat and eggs. While consumers are typically seeking an alternative to the intensive production techniques used to raise conventional poultry, farmers can use free-range poultry systems to increase income and diversify crop and other livestock systems.

Free-range poultry is also often called "pastured poultry," "free roaming," or a number of other related terms. In general, it describes a production practice in which poultry have access to outdoors through different types of systems, from completely free-ranging birds out on pasture, usually with portable shelters or feeding units, to permanent poultry houses that allow birds to access the outdoors through paddocks. Generally, start-up costs for farmers using these systems are low, and often free-range poultry is raised on a much smaller scale than conventional poultry, although large-scale production does occur. Feeding costs may sometimes be lowered through access to pasture, although this may be countered by higher mortality rates due to predators and higher labor costs overall. In addition, these systems are often seasonal due to the fact that birds need access to the outdoors. Concerns about free-ranging birds catching diseases from wildlife, most recently avian influenza, are also important.

At the retail level, the market for niche poultry products is growing, and farmers can often realize premiums. Annual average growth rates for organic poultry meat alone from 2004 through 2007 were almost 40 percent. The use of the free-range label, however, is controversial, because the USDA currently does not have specific regulatory definitions. Producers labeling poultry meat as free range or free roaming must demonstrate to USDA's Food Safety Inspection Service that the poultry has been allowed access to the outside. No specific amount of time outside or stocking density is required, and this label does not require third-party certification. In addition, egg production is not covered.

The labeling of poultry as organic, however, does provide regulations concerning outdoor access for poultry: the mandatory third-party organic certification states that producers must provide their animals with access to the outdoors, shade, exercise areas, fresh air, and direct sunlight, but minimum levels of access have not been set. Indoor confinement (caging of animals is not allowed) must be temporary and justified due to weather, stage of production, health and safety of animal, and risks to soil or water quality.

Further Reading

Fanatico, Anne. Alternative Poultry Production Systems and Outdoor Access. Fayette, AR: Appropriate Technology Transfer for Rural Areas, National Center for Appropriate Technology. 2006.www.attra.org/attra-pub/PDF/poultryoverview .pdf.

Oberholtzer, Lydia, Catherine Greene, and Enrique Lopez. *Organic Poultry and Eggs Capture High Price Premiums and Growing Share of Specialty Markets.* Washington, D.C.: USDA, Economic Research Service, 2006. www.ers.usda.gov/Publications/ LDP/2006/12Dec/LDPM15001/ldpm15001.pdf.

See Also: Animal Welfare; Factory Farming; Livestock Production, Organic

Lydia Oberholtzer

Free Trade

International economic regulation is based on the premise that international trade is beneficial, that the benefits outweigh the costs to those negatively affected, and that expanding and freeing markets of tariffs, taxes, subsidies, and regulations will increase the standards of living in all societies. This is the position taken up by proponents of neoliberal economic policies that—more than any other philosophy—has influenced international agreements such as the General Agreement on Trade and Tariffs (GATT) since the postwar period, culminating in the establishment of the World Trade Organization (WTO) in 1994. Discussion of free trade in agriculture, the most highly distorted sector in the global economy, has brought significant controversy and debate to WTO trade negotiations.

Free trade is based on the doctrine of comparative advantage, which originally sought to explain why two countries could benefit from international exchange unhindered by government interference. In *On the Principles of Political Economy and Taxation* (1817), David Ricardo famously elaborated on the theory, illustrating how it could play out in terms of trade in cloth and wine between England and Portugal. Although Portugal might have been able to produce both wine and cloth for less work than England (an absolute advantage for Portugal, the early logic of free trade used by Adam Smith), it was in the interests of Portugal to produce wine (the cheaper of the two for Portugal) and import cloth, and for England to produce cloth (the cheaper of the two for England) and import wine, if England had a comparative advantage in cloth. Here, Ricardo assumed Portugal would have to give up more wine to produce cloth than England would (a higher opportunity cost). Assuming the two countries reached an agreement on terms of trade, and assuming that all other costs remained constant, if each country traded for commodities in which they had a comparative advantage, the two countries would make a profit and the outputs of both would increase through the international exchange. The implication was that each country should specialize in producing what they produce best and trade it on the free market. Ricardo's pioneering work remains persuasive to economists in the development of more nuanced international trade models.

Critics argue that, although the theory of comparative advantage rather accurately predicts that global welfare will rise as trade in goods and services expands, it does little to explain how that welfare will be distributed. One of the most serious concerns for policymakers has been the impact of free trade on individual countries, and on specific sectors within those countries. Many less-developed countries hold the view that their agricultural sectors are at a disadvantage vis-à-vis industrialized country agriindustry due to government policies that keep agricultural commodity prices artificially low. Disagreements between the United States and less-developed countries, such as India and Brazil, and the European Union— which called on the United States to commit to reducing its agricultural subsidies—led to a collapse in WTO agricultural trade talks in 2006.

People often mistake free trade with fair trade. But the terms are very different. Although free trade advocates would argue that free trade *is* fair trade, fair trade proponents use the term to explain an attempt to make terms of trade more equitable for low-income producers. Fair trade proponents argue that producers should be provided with a minimum price, despite current market conditions, and that can be gained by more closely linking these producers with consumers. Much focus on fair trade has taken place in the agricultural sector, notably in cocoa, bananas, and coffee.

Further Reading
Diaz-Bonilla, E., S. E. Frandsen, and S. Robinson, eds. *WTO Negotiations and Agricultural Trade Liberalization: The Effect of Developed Countries' Policies on Developing Countries*. Cambridge, Mass.: CABI, 2006.
Lowenfeld, A. F. *International Economic Law*. New York: Oxford, 2003.

See Also: Fair Trade; Food Sovereignty; Globalization; Political Ecology of Food

Brian J. Gareau

Fungicides

Fungicides are pesticides used to control fungi, the major cause of plant diseases. There are three main roles of fungicides: to control diseases during establishment and growth of a crop, to increase crop productivity and reduce blemishes, and to improve postharvest storage life and quality. Fungicides are used because other methods of disease management, such as cultural practices and resistant cultivars, in some situations are inadequate.

Most fungicides are preventative, applied before or at first symptoms of a disease. They generally do not cure plant diseases. Movement in plants, role in protection, activity, the mode of action, or chemical structure can be used to classify fungicides. Contact fungicides stay on the plant surface and can be phytotoxic if absorbed. Systemic fungicides are absorbed by plants and moved in the vascular system. Some fungicides are multisite, affecting a number of pathways in many organisms. Other fungicides are single site, targeting one metabolic pathway in a single fungus species. They have a range of modes of action including damaging cell membranes, inactivating necessary proteins, interfering with important processes (such as energy production or respiration), or impeding specific metabolic pathways. Some fungicides are derived from carbon-based compounds, whereas others are derived from inorganic compounds (e.g., copper or sulfur). Some fungicides, such as sulfur, can be used in organic farming. But the use of synthetic fungicides and synthetic chemical pesticides overall is prohibited in organic methods.

Fungi can develop resistance to fungicides through random mutations. Resistance is most likely in single-site fungicides, which a single gene mutation can overcome. Resistance is often caused by reduced uptake or increased detoxification of the fungicide. Integrated management approaches, such as using resistant crop varieties, crop rotation, sanitation, and tank mixes, can reduce the development of resistance. Farmers can use disease forecasting and action thresholds so fungicides are used only when necessary, reducing development of resistance.

Fungicides can cause unintended environmental problems. Fungicides may alter complex soil ecosystems. Fungicides can be toxic to fish or other animals. Captan, a common fungicide, is toxic to honeybees. In some instances, fungicides can injure crops. Fungicides can also become contaminated with other pesticides. In one instance, Benlate (benomyl) was removed from the market due to high legal costs associated with crop injury from contamination with the herbicide atrazine.

Most fungicides are ineffectively absorbed and generally do not cause acute toxicity in mammals. There is strong evidence that some synthetic fungicides, such as captafol, captan, and folpet, cause cancer in laboratory animals and possibly humans. In 1989, the EPA banned captan on a large range of fruits and vegetables because it posed a risk of cancer.

Further Reading

McGrath, M. T. "What Are Fungicides?" The American Phytopathological Society Education Center Web Site. http://www.apsnet.org/education/top.html.

Pesticide Management Education Program. Extension Toxicity Network. Cornell University. http://pmep.cce.cornell.edu/profiles/extoxnet/index.html.

Schumann, Gail, and Cleora D'Arcy. *Essential Plant Pathology.* St. Paul, MN: The American Phytopathological Society, 2006.

Shabecoff, Philip. "E.P.A. Restricts Fungicide's Use On 42 Products." *New York Times,* February 17, 1989. http://query.nytimes.com/gst/fullpage.html?sec=health&res=950DEEDE163BF934A25751C0A96F948260&scp=1&sq=E.P.A.%20Restricts%20Fungicide's%20Use%20On%2042%20Products&st=cse.

Staub, T. "Fungicide Resistance: Practical Experience with Antiresistance Strategies and Role of Integrated Use." *Annual Review of Phytopathology* 29 (1991): 421–42.

See Also: Agrichemicals; Herbicides; Pesticides

Lauren Flowers and John Masiunas

G

Gardens, Children's

The number of children's gardens has been growing rapidly since the late 1980s. Of the many factors that have stimulated this growth, two of the most fundamental are concerns for the health of the environment and the health of people. The growth of franchises and chain stores has lead to fundamental changes in how food, both plant and animal, is being raised, processed, and distributed; these changes have led to ecological uniformity and the narrowing of food choices. The poor quality of food has resulted in increased obesity among the young as well as loss of knowledge about the growing and preparation of healthy food. Involving children in gardening has proved to be successful in addressing these problems as well as serving an array of other important functions. A description of three children's gardening projects illustrates a variety of strategies.

Established in 1992, the Greater Boston Food Project involves sixty youth each year, ages fourteen to seventeen, drawn from both city and suburbs. They manage a thirty-one-acre rural farm and two and one-half acres of remediated urban land, selling their organic produce through two urban

Children plant spinach and learn about where their food comes from. Courtesy of Leslie Duram.

farmers markets and a Community Supported Agriculture (CSA) program. They also use their produce to prepare and serve lunches in local soup kitchens and homeless shelters. Workshops each week cover diversity awareness, hunger and homelessness, sustainable agriculture, and reflections on their experience in the program.

The Edible Schoolyard, founded by the Chez Panisse Foundation, consists of a one-acre organic garden and kitchen classroom at a public middle school in Berkeley, California. Students participate in garden classes that introduce them to the origins of food, plant life cycles, and community values. Kitchen classes assist them in preparing and eating the produce they have grown in their garden. Since 1996, over three thousand students have graduated from the Edible Schoolyard, which has inspired and encouraged several hundred kitchen and garden programs across the country.

The Youth Enterprise in Food and Ecology Project of the Community Design Center of Minnesota, begun in 1995, is a neighborhood children's garden project that has expanded through partnering with local organizations, schools, and government agencies to sponsor three interrelated programs: the Garden Corps, the Conservation Corps, and classes in cooking and nutrition. The Garden Corps manages seven produce, herb, and flower gardens. They sell their produce in two farmers markets and through a

CSA. In the off-season they use garden materials to make value-added products: vinegars, wreaths, and garden ornaments. The Conservation Corps involves youth in helping develop and maintain a two-acre bird sanctuary and thirty rain gardens, and in designing and planting a neighborhood park. Cooking classes are offered in twenty-four elementary schools.

These examples suggest the extent to which children's gardens, whether city, school, or neighborhood based, provide children and youth with opportunities to work together in meaningful ways, ways that enhance their communities as well as helping them learn through direct hands-on experience about healthy food, principles of ecology, as well as skills for living and for employment.

See Also: Campus Programs; Diet, Children's; Farm to School; Gardens, Community

Cynthia Abbott Cone

Gardens, Community

Community gardens, broadly defined as any garden cultivated by a group of people, are the primary urban agriculture in the United States. Growing both ornamental and edible plants, community gardens have been employed historically as coordinated responses to political economic crises. Community gardens currently enjoy renewed interest in the United States for their potential to serve as an alternative to conventional agriculture, and they provide critical space for new food movements advocating agroecology.

BRIEF HISTORY

Community gardens have a long history in the United States. The first organized community garden program emerged in Detroit in 1893 as a response to urban hunger, with the City of Detroit providing land to local residents to grow supplemental food. In the much-cited overview of community gardening in the United States, Thomas Bassett (1981) explained that the history of community gardening is best understood as seven distinct, but overlapping "movements" of community gardening programs: Potato Patches (1894–1917), School Gardens (1900–20), Garden City Plots (1905–10), Liberty Gardens (1917–20), Relief Gardens (1930–39), Victory Gardens (1941–45), and Community Gardens (1970–present). Notably, these gardening movements emerged as direct responses to specific crises or emergencies, most often war or economic crises and their accompanied social changes.

Community gardens were introduced as a vehicle to address a number of social concerns. During the economic depression of the late nineteenth century, the Potato Patches of Detroit provided food for hungry urbanites through the promotion of self-sufficiency. In the early 1900s, school gardens were used to help socialize urban school children to urban life and

Community garden during World War II. (AP Photo)

work, focusing on self-discipline and a connection to nature. In an effort to address urban issues associated with rapid industrialization, the Garden City Plots of the early twentieth century served to beautify cities and combat urban decay. Liberty Gardens emerged in the wake of World War I, and Victory Gardens dotted the American landscape during World War II. Both provided an outlet for, and promoted the display of, patriotism. Materially, the garden movements emphasized subsistence and the role of gardens in domestic food production, freeing up economic capacity for war-related production. Self-sufficiency became a national duty, a sign of national power. Liberty and Victory Gardens were a means to conserve resources that were desperately needed for the war effort. In between the two world wars, Relief Gardens played a vital role in relieving the social and material devastation of the Great Depression. Finally, the contemporary era of community gardening builds on previous rationales for gardening, including economic necessity and urban disinvestment, but is also augmented by the growth of environmental concern, particularly the environmental degradation wrought by conventional agriculture.

SHIFTING LANDSCAPE

Present-day community gardening is rooted in the community gardening movement of the 1970s, with the rationales for gardening remaining fairly consistent since then. However, it is useful to think of the contemporary landscape of community gardening as decidedly neoliberal, placed within the current political economic context that emphasizes market supremacy above all else.

One of the hallmarks of neoliberalism, the roll back of government spending on social programs, has had significant consequences for the community gardening movement. The end of the federal government's sponsorship of community gardening in 1993, with the folding of the Department of Agriculture's Urban Garden Program, marks an important historical distinction and the beginning of an eighth movement of community gardening. This eighth movement is characterized by shifts in how, and at what scales, community gardens are governed.

A defining moment in the contemporary history of community gardens played out throughout New York City in the late-1990s. In 1999, then-Mayor Rudolf Giuliani placed 114 community gardens on the auction block, prioritizing private property rights over public space, exemplifying the neoliberal insistence that property ownership be secured in the realm of market exchange. The fight over community gardens in New York City was ingeniously framed by the Giuliani administration as focused on the right to affordable housing, which would replace community gardens. In so doing, two common allies—housing advocates and community gardeners—were pitted against each other. Gardeners challenged Giuliani through a variety of tactics, including an insistence on their right to cultivate the city. In the end, land trusts had to purchase the gardens at auction, a move made possible in large part by the effective mobilization of garden advocates to connect their struggle to broader politics as well as the fundraising efforts and public relations support provided by Bette Midler, the famous singer-actress. Many gardens thus became quasi-public property, ushering in a new landscape of gardening with yet unknown consequences. The struggle of New York City gardeners to save their land serves as a catalyst for more effective organizing by community gardeners in the City and beyond.

CONTEMPORARY CONTEXT

Today, community gardens enjoy renewed interest as movements for organic, sustainable, and local foods flourish. Community gardens take a variety of forms—from guerilla gardening on public land to formal not-for-profit organizations with large budgets and professional staff. One such organization, the American Community Gardening Association (ACGA) plays a key role in the contemporary context of community gardening and provides important support to community gardeners, including education, information sharing, networking, and advocacy.

The contemporary landscape of community gardens also mirrors their earlier history in many ways, as vacant lots in the crumbling cities of the Rust Belt are cultivated in response to a myriad of political economic shifts associated with neoliberalism. Community gardens often serve as a means to fulfill various duties (e.g., poverty alleviation) formerly met by the state and provide buffers against the ravages of the market. In so doing, community gardens may actually work to support neoliberalism by allowing the state to shirk its responsibilities.

Community gardens are championed as providing a wide variety of social benefits. Within the United States, hunger assumes a particularly urban form; that is, the geography of hunger in the United States closely coincides with the geography of urban poverty. In this urban landscape of hunger and poverty, community gardens often serve as a vital tool for addressing hunger while providing a number of additional benefits. Gardening is said to improve the quality of life for people and to help to build community. Community gardens are also employed as a tool for urban beautification, for the creation of green space, and for the provision of recreation and education opportunities. Gardens, supporters claim, help reduce crime, boost local property values, and stimulate economic development.

These benefits notwithstanding, some challenges remain to the long-term sustainability of community gardening in the United States and elsewhere, often a result of the lack of state support for urban agriculture. A primary concern for community gardening, as the recent history of New York City illustrates, is land tenure. Neoliberalism prioritizes private property, effectively undermining community garden efforts that are frequently located on public property. Furthermore, urban planners often view community gardens as temporary at best, rather than identifying them as permanent fixtures of the urban landscape. Ecological constraints, such as water shortages, lack of adequate soil for production, and pests and diseases, are some of the most pressing challenges to the viability of community gardening as a widespread solution to urban food requirements. Urban air and water pollution, along with contaminated soil, also pose great problems to urban food production. Lead, in particular, is a dangerous contaminant in cities throughout the world. Many types of produce, especially green, leafy vegetables, easily transport such heavy metals. Finally, community gardening is continually undermined by the need to keep people engaged. All of these challenges, however, have proved surmountable time and time again, indicating that they can be managed and overcome.

The recent growth of community gardening in the United States and across the world provides hope for the future of the movement. In part, the expansion of urban agriculture across the globe is attributed to increasing urbanization throughout much of the world. Accompanied by concerns about the exponential growth in food prices, community gardening is boosted by renewed interest in organic, sustainable, and local foods and an increase in environmental consciousness driven by global climate change. Urban agriculture can be an important space of resistance to the social and economic consequences of neoliberalism. From Baghdad to Havana, New York to Moscow, community gardeners are struggling to make the protest call ring true—another world is indeed possible.

Further Reading

Allen, Patricia. *Together at the Table: Sustainability and Sustenance in the American Agrifood System*. University Park, PA: Pennsylvania State University Press, 2004.

American Community Gardening Association. www.communitygarden.org.

Bassett, Thomas. "Reaping on the Margins: A Century of Community Gardening in America." *Landscape* 25 (1981): 1–8.

Lawson, Laura. *City Bountiful: A Century of Community Gardening in America*. Berkeley: University of California Press, 2005.

Smith, Christopher, and Hilda Kurtz. "Community Gardens and Politics of Scale in New York City." *Geographical Review* 93 (2003): 193–212.

Staeheli, Lynn, Don Mitchell, and Kristina Gibson. "Conflicting Rights to the City in New York's Community Gardens." *GeoJournal* 58 (2002): 197–205.

See Also: Campus Programs; Community-Supported Agriculture; Farmers Markets; Farm to School; Food Security; Gardens, Children's; Local Food; Urban Agriculture

Evan Weissman

Genetically Engineered Crops

Consumer concerns about genetically engineered (GE) crops have been growing. One way consumers act to avoid GE foods it to eat organic foods, because organic certification prohibits the use of GE products in production and processing.

GE crops have their genetic information (genes) modified using molecular biology techniques (also called genetic engineering) to directly alter the deoxyribonucleic acid (DNA) containing gene(s). The DNA is isolated and manipulated in vitro before being inserted into the crop. The most common GE crop has DNA from another species incorporated into their genome. Certified organic crops are prohibited from using genetically modified materials.

How do GE crops differ from crops developed from traditional breeding? Some consider molecular biology as the latest continuum in the thousands of years of human modification of the genetic information in plants. Both methods require variation in the trait (e.g., difference in disease resistance) and use selection. Traditional breeding methods use variation in interbreeding plants, whereas molecular biology techniques use genes found in any organism. Traditional breeding transfers half the genome from each interbreeding parent and uses repeated backcrossing to develop a final variety. The crop's genes are only indirectly manipulated when using traditional breeding; molecular techniques directly manipulate genes.

According to the agribusiness corporations that develop them, GE crops have potential benefits. GE crops have been developed with enhanced yield, taste, and quality, increased nutrient levels, increased stress tolerance, reduced maturity time, greater competitiveness against weeds, and improved resistance to diseases, insect pests, and herbicides. GE crops are being improved for processing into biofuels and industrial products. The use of GE crops may protect the environment, allowing the use of new production techniques that conserve soil, water, and energy. New GE crops may be developed containing environmentally friendly biopesticides or with greater productivity.

Yet, concerns linger, as consumers do not readily trust GE crops. Looking back, a slower decaying tomato was the first crop developed through molecular biology to reach supermarkets and also the first to be withdrawn. In 1994, Calgene Company introduced the "Flavr Savr" tomato. Flavr Savr had a gene inserted slowing fruit decay, allowing them to be harvested more fully ripe and still be shipped long distances. The U.S. Food and Drug Administration did not require labeling because Flavr Savr tomatoes presented no known health risks and were considered essentially similar to nonmodified tomatoes. Calgene's lack of experience with tomato production ultimately led to the demise of Flavr Savr. The variety was never profitable due to lack of flavor, lower yields than conventional varieties, high development costs, and competition with new long-shelf-life varieties.

> "Organic farming has been shown to provide major benefits for wildlife and the wider environment. The best that can be said about genetically engineered crops is that they will now be monitored to see how much damage they cause."
> —Prince Charles, during his speech at the 1998 Soil Association Organic Food Awards, The Savoy Hotel, London.

Monsanto Company has been at the forefront of developing and marketing GE crop seed. The current Monsanto Company is a product of the revolution in genetics occurring in the last three decades. Monsanto aggressively acquired plant genetics and biotechnology companies, including Agracetus, Calgene, Asgrow, DeKalb, Channel Bio, NC+ Hybrids, and Seminis. In 1996, Monsanto introduced Roundup Ready soybeans, resistant to its popular glyphosate herbicide (Roundup), and Bollgard cotton, resistant to cotton bollworm, tobacco budworm, and pink bollworm. An application of glyphosate will control all emerged weeds, making Roundup Ready technology extremely popular among farmers, and extremely profitable for Monsanto. The majority of Monsanto's profits come from sales of genetically modified seed.

In 2006, 10.3 million farmers planted genetically modified crops on 252 million acres in twenty-two countries. The majority of genetically modified crops are herbicide- and insect-resistant soybeans, corn, cotton, canola, and alfalfa. U.S. farmers have widely adopted genetically modified crops. For example, 73 percent of dent corn (also called "field corn") and 91 percent of soybeans are genetically modified. There is no single U.S. government agency that regulates genetically modified crops.

The European Union has generally not accepted GE crops. It refuses most GE crop imports and produces a minimal amount of GE crops. Foods containing GE ingredients must be labeled. The European Food Safety Authority oversees all genetically modified crops. Developing countries may benefit the most from GE crops but they have had mixed adoption of the technology. Latin American countries have adopted GE corn and soybeans. Similar to Europe, African countries have been hesitant to adopt GE crops. China and India are developing their own biotechnology industry and GE crops.

Many products from GE crops are used in processed foods sold in the United States. These ingredients are considered as essentially identical to the nongenetically modified version, so are not labeled. Most vegetable oils, such as soybean, corn, canola, and cotton, are made from GE crop seed. High-fructose sweeteners from GE corn are used in products from sodas to cereals. Even ingredients such as lecithins are made from GE soybeans. In Europe, the use of ingredients from GE crops is more controversial, and labeling indicates GE content. Products labeled "organic" and bearing the USDA organic symbol cannot contain GE ingredients.

Under pressure of consumer demands, some food processing companies and restaurant chains have resisted using ingredients from GE crops. For example, Monsanto introduced NewLeaf, GE potatoes resistant to Colorado potato beetle, a major insect pest. The potatoes contained a bacterial gene from *Bacillus thuringiensis* var. *tenebrionis* that produces a protein killing the Colorado potato beetle. Initial acceptance was limited, because Monsanto created insect resistance in a small number of varieties, required 20 percent of acreage planted with susceptible varieties, and retained ownership of the seed potato, not allowing farmer replanting. In 2000, the biggest purchaser of potatoes, McDonald's, refused to use NewLeaf GE potatoes. Other fast food restaurant chains and potato chip companies soon followed suit. In 2001, Monsanto discontinued selling NewLeaf potatoes.

The introduction of Roundup Ready sugar beets was delayed due to corporate resistance (from Mars and Hershey). But sugar beets represent an ideal crop for genetic modification. Sugar beets make up half of the U.S. sugar production and only 3 percent is exported. There are few herbicides registered for sugar beets. Processed sugar is pure sucrose and contains no protein or DNA, thus the product of genetic modification does not reach the consumer.

No genetically modified varieties of wheat, barley, and rice are being commercially grown. Wheat and rice are directly consumed instead of being processed into industrial products or ingredients, and a large portion of the U.S. crops are exported. Monsanto retired Roundup Ready wheat due to farmer's protests. Farmers feared losing exports (up to 50 percent of U.S. wheat is exported) especially into Asian and EU markets, which both ban GE crops.

GE rice both holds promise and poses a threat. Like wheat, half of the U.S. rice crop is exported to foreign markets concerned about GE crops. GE rice could address problems caused by population growth, changing climate, and nutritional deficiencies (e.g., Vitamin A, iron, and zinc). Syngenta, using molecular techniques, developed "Golden" rice containing higher levels of Vitamin A. However Golden rice has not been widely accepted, because Sygenta did not use rice varieties commonly grown in vitamin-deficient areas, and consumers have been concerned about unknown health and environmental effects. In 2006, there were GE rice scares. The U.S. rice crop was found to be contaminated with GE varieties. The EU put a temporary ban on U.S. rice imports. Also, GE rice unapproved for human consumption was illegally sold and grown in China. Major rice producers, Thailand and Vietnam, banned GE rice production.

GE crops also pose potential problems and controversies. The economic gains from GE technology are not evenly shared. Corporations (e.g., Monsanto, Sygenta, Pioneer) that developed the GE crop varieties profit the greatest. Corporations are able to patent individual crop hybrids (varieties), charge "technology fees" (a premium) for those varieties, and require farmers to purchase new seeds each year. Research indicates that benefits from GE crops depend on farm size with small-sized farms profiting the least from GE crops.

There is also concern that a few corporations will control the genetic resources for important crops such as maize. GE crops raise ethical concerns. Some people are unsettled with the idea of altering a crop's genetic code using molecular approaches. Cross-kingdom (i.e., between animals and plants) recombination of genes is often a religious and moral issue, understood in a plurality of ways and typically not well supported by the public.

There is controversy about whether GE crops reduce use of pesticides. Prior to the widespread introduction of GE crops, pesticide use dropped by 20.6 million pounds (during 1996 to 1998). During the six years (1998 to 2004) after widespread adoption of GE crops, pesticide use *increased* by 14.3 million pounds, mainly due to the shift away from low use rate ALS (acetolactate synthase)-inhibitors to glyphosate (Roundup) herbicide. The use of GE crops can result in pest resistance. For example, horseweed resistance to glyphosate has occurred in GE soybean fields in the Midwest. Cross-pollination can transfer resistant genes from GE crops to related weeds (sometimes called "superweeds"), making them difficult to manage. The gene for herbicide resistance in GE canola has been found in related wild radish. GE crops may affect nontarget organisms (e.g., beneficial insects, soil microbes) and biodiversity.

Further Reading

Center for Food Safety. *Genetically Engineered Crops*. 2008. www.centerforfood
  safety.org/geneticall2.cfm.

Ching, Lim Li. "GM Crops Increase Pesticide Use." Institute of Science in Society.
  www.i-sis.org.uk/GMCIPU.php.

Dill, Gerald. "Glyphosate Resistant Crops: History, Status and Future." *Pest Management Science* 61 (2005): 219–24.

Fernandez-Cornejo, J. *Adoption of Genetically Engineered Crops in the U.S.* USDA Economic Research Service. May 15, 2008. www.ers.usda.gov/Data/BiotechCrops/.

Gerdes, L. I. *Genetic Engineering.* Opposing Viewpoints Series. Farmington Hills, MI: Greenhaven Press, 2004.

Human Genome Project. *Genetically Modified Foods and Organisms.* 2008. www.ornl.gov/sci/techresources/Human_Genome/elsi/gmfood.shtml.

Peterson, G., S. Cunningham, L. Deutsch, J. Erickson, A. Quinlan, E. Raez-Luna, R. Tinch, M. Troell, P. Woodbury, and S. Zens. "The Risk and Benefits of Genetically Modified Crops: A Multidisciplinary Perspective." *Conservation Ecology* 4 (2000):13–20.

Pew Initiative on Food and Biotechnology. "Public Sentiment about Genetically Modified Foods." 2006. http://pewagbiotech.org/research/2006update/.

See Also: Agribusiness; *Bacillus thuringiensis* (Bt) Corn; *Diamond v. Chakrabarty*; Precautionary Principle; rBGH; Roundup Ready Soybeans; Substantial Equivalence; Terminator Gene

John Masiunas and Stephen Bossu

Geographic Information Systems and Global Positioning Systems

Geographic information systems (GIS) and global positioning systems (GPS) currently are providing technologies and information aimed to create integrated, environmentally, and economically sustainable agricultural production systems. The promotion of organic farming and sustainable agriculture is being achieved through data flow and site specific farming in a GIS environment. At a time when human population growth is wearing the earth's abundance as never before, agricultural production experts are discovering the power of GIS and GPS to help make the crucial decisions for food production and security. Once an expensive technology favored by research scientists and technocrats, GIS and GPS have emerged recently as the tool of choice in local, state, and national agencies around the world. These technologies make it possible to focus attention on larger problems, identify small problems, and predict other obstacles that might be waiting in the wings. GIS and GPS technologies are helping the agriculture and conservation communities find common ground by providing a framework for the analysis and discussion of crop management issues.

GIS has been defined in many different ways; the definition chosen depends on what one is looking for. A GIS can be seen as a set of tools involving software packages, computer peripherals, and information. A GIS can also be considered as an information system where data can be stored and queried with results providing answers and possible scenarios. Other GIS communities have approached and defined GIS as a science. In some circles, GIS has been viewed as a multibillion-dollar industry providing services for the private businesses, governmental organizations, and educational institutions. GIS in its generic sense can be defined as a computer system for capturing, managing, integrating, manipulating, analyzing, and displaying data which is spatially referenced to the earth. Implicit in the definition is "spatially referenced" information, in other words, geographically referenced data. A GIS stores information about the world as layers of spatial features. These separate layers are interrelated on the basis of shared geography. The spatially referenced data in agriculture involves gathering data on the pattern of variation in crops, soils and other environmental parameters across field.

The technology and functions of GIS have undergone considerable changes since its commencement in the early 1960s. The formative years started with computer mapping at the University of Edinburgh, the Harvard Laboratory for Computer Graphics and the Experimental Cartographic Unit, and the Canadian Land inventory leading to the development of the Canada Geographic Information System. The introduction of the urban street network in the United States Census Bureau's Dual Independent Map Encoding (DIME) system also played a role in GIS development in the 1960s and 1970s. The formative years of GIS coincided with the mainframe computing era. For many years, GIS was viewed and considered to be too difficult, expensive, and proprietary. The introduction of affordable and portable personal computers with improved graphical user interface, and public digital data have broadened the range of GIS applications. The advancements of computer technology and availability of digital data ushered in mainstream use of the technology in the 1990s. The beginning of the World Wide Web and the civilian use of the global positioning system paved the way for the increased use of the technology. Currently, the full potential of the GIS technology is being met at multifaceted fields including agriculture.

In the late 1970s, the GPS was established based on a constellation of radio-emitting satellites deployed by the United States Department of Defense used to determine location on the earth's surface. The system provided the potential to determine position (latitude, longitude, and altitude) anywhere on the earth's surface, twenty-four hours a day, to an accuracy of a few centimeters. Using a GPS receiver device, it is possible to accurately determine one's location at or near the surface of the earth without the need for any reference or land marks. With such information available to farm field machines (e.g., tractors and combines), the treatment applied during field operations could be related to very localized requirement within the field. GPS receivers are now attached to tractors and combines linked to computers to collect location information on planner seed density settings, yield flow rates, and sprayer and manure application rates.

Farming operations of all sizes use GIS and GPS technologies to select the right growing regions for their crops and to design fields and farmlands in ways that improve crop yield while saving money on farm inputs and promoting sustainable agriculture. GIS and GPS technologies have been widely used to find locations with the right soil and climate. The application of GIS, GPS, and other related technologies in agriculture is often referred to as precision agriculture (PA) or site-specific agriculture. The goal of site-specific agriculture is to handle plots of land uniquely to realize the profit-yielding potential and environmental friendliness based on each plot's combination of nutrient, soil, topography, and other characteristics. PA allows precise tracking and tuning of farm production. The technology provides farmers with opportunities of changing the distribution and timing of farming inputs based on spatial and temporal variability in a field. Precision agriculture technologies try to answer farm-specific questions. For example, a farmer could verify that for 75 percent of the time, 80 percent of the corn grown in a field would yield 4.2 metric tons. Knowing the cost of inputs, farmers can also calculate the cash return over the costs for each hectare. The advantage of GIS in answering these questions is to visualize what is really happening at the farm level. GIS provides the agriculture community with a very rich data-visualization tool.

GIS as a technology has found its way to everyday life. Its capability of converting data from tables with location information into maps has provided users and operators a tool for varied analysis. As large-scale agribusiness has proliferated, so too has the role of GIS in food production and agriculture. GIS and GPS technologies allow the farmer to respond to and analyze local conditions in the field with pinpoint accuracy. For example, a section of a corn field might have stain. The area with the stain is inventoried using global positioning system, and then combined with other layers of data such as soil type, soil chemistry, corn variety, pesticide load, and irrigation information to determine why the particular area of the farm is in duress. Organic crop management includes planning to protect vulnerable crops and achieving optimal yield without compromising the environment. GIS therefore provides organic farming a tool to improve and enhance crop management for optimal production and environmental friendliness leading to sustainable agriculture.

GIS and GPS technologies have the potential for enhancing and accelerating organic food production. It is an indispensable and logical tool that organic farmers can add to their toolbox to help them do what they do best. In organic farming, the concept of the farm is thought of as an organism in which the component parts such as soil minerals, organic matter, insects, plants, animals, and humans interact. GIS technology can help analyze the interaction by providing information as layers, and visualizing the relationship for better decision-making. GIS and GPS technologies can be considered as technologies that will provide information to combat the gap between desire and theory and implementation of organic agriculture. The technology and its related data-flow could be a passageway for providing information needed for verifying organic product cycle at any site at a particular point in time.

GIS and GPS technologies have been used widely in the field of agriculture for effective and efficient crop management. The use of GIS and

GPS technologies in organic farming is fairly new, and the adoption will require challenging efforts and training to achieve all the advantages that come with the technology. Notwithstanding the challenges of the technologies, GIS and GPS will provide the organic farming community the toolbox to support decisions on farm inputs to achieve food sufficiency and sustainability at the local, state, and national levels.

Undoubtedly, there is an increasing optimism in many circles that the application of GIS and related technologies to agriculture will help feed the projected 10 billion people by the year 2020 without jeopardizing the environment. GIS technology by all account offers the hope for understanding and correcting the myriad damages being done to our planet.

Further Reading

Clarke, Keith, C. *Getting Started with Geographic Information Systems*. 4th ed. Upper Saddle River, NJ: Prentice Hall/Pearson Education, 2003.

Environmental System Research Institute (ESRI). "GIS for Agriculture." www.esri.com/industries/agriculture/index.html.

Lang, Laura. *Managing Natural Resources with GIS*. California: ESRI Press, 2003.

Longley, P. A., F. M. Goodchild, D. J. Maguire, and D. W. Rhind. *Geographic Information System Science*. 2nd ed. Hoboken, NJ: John Wiley & Sons, 2005.

Runquist, S., N. Zhang, and R. Taylor. "Development of a Field-Level Geographic Information System. *Computers and Electronics in Agriculture* 31 (2001), 201–9.

Westervelt, J. D., and H. F. Reetz Jr. *GIS in Site-Specific Agriculture*. Crete, IL: Interstate Publishers, 2000.

See Also: Environmental Issues

Samuel Adu-Prah

Globalization

Globalization is a process that intensifies relationships between people and places around the world. The intensification of global relationships affects economies, cultures, political systems, and environments. There are many debates that have arisen about globalization that explore the nature and end results of the process (e.g., the driving forces of globalization, the effects of globalization, whether globalization is inevitable). Even though globalization has the potential to bring greater uniformity to global practice, the effects of globalization on the production and distribution of food are varied and uncertain because global linkages between diverse people and places are formed in many different ways.

Technological development is often heralded as a driving force of globalization as improvements in transportation, data processing, and communications have clearly made the functional integration of distant places possible. However, organizational connections across the world are even more critical driving forces of globalization. International governmental organizations (IGOs), like the United Nations and its affiliated institutions that include the World Trade Organization (WTO), the International Monetary Fund (IMF), and the World Bank, and thousands of multinational corporations (MNCs) are influential forces that drive decision-making on a global scale. IGOs and MNCs are the driving forces of globalization "from above" because they consist of powerful people and organizations that have the ability to enact policy and guide investment flows around the world. Globalization can also occur "from below." Technological development used for global coordination of IGOs and MNCs is also available to thousands of nongovernmental grassroots activist organizations (NGOs). People and groups from around the world, who alone would not be very powerful beyond their localities, are able to harness technology to make their needs and goals known around the world and to find other like-minded people and organizations from other localities to build solidarity for various causes, including the promotion of organic, local, and sustainable food.

The neoliberal ideology of free trade dominates globalization discourse. As technological innovation reduces the friction of distance, farmers must compete not only with other local farmers but, in some cases, with farmers on the other side of the world. Strategies of regional specialization, economies of scale, and principles of comparative advantage are more important than ever in agribusiness. The global scope of food production and markets means that rulings of the WTO have played an important role in setting minimum acceptable standards for agricultural practice.

In a globalizing world, grassroots NGOs and IGOs have pushed for standardization of food safety laws to protect consumers and laws governing chemical use in farming. For example, if food is to be traded internationally, some chemicals may not be used or their use must be clearly identified. Standardization can have positive effects in expanding organic production to more places around the world so more farmers would be able to work in conditions free of dangerous chemicals and more consumers would be able to buy organic food. Unfortunately standardization does not benefit all farmers or all countries equally. Wealthy corporate agribusinesses are able to lobby global institutions that determine the rules for international trade to suit their capabilities. International standards for agricultural production can limit the ability of individual countries to set higher environmental standards. The WTO has ruled that a country does not always have the right to ban imports that do not meet domestic environmental or safety standards, thus undermining the ability of domestic agriculture to compete with imports. This globalized system of trade can hurt local producers that would ordinarily have an advantage in market access.

NGOs promoted organic agriculture in less developed countries as a development strategy for poorer smallholder operations. Poor farmers could specialize in organic food production as a means to profit in a niche market. In many cases such systems proved to be successful as demand for organic fair trade goods grew. As demand for organic food grew, large agribusinesses expanded their organic production. Now smallholders find themselves unable to compete with powerful agribusinesses that have more resources and are already well positioned to make connections to wealthy consumers willing to buy their goods.

Modern technological advances have also affected food production and distribution. Technological advances allow for food that is grown nearly anywhere in the world to reach a consumer nearly anywhere in the world. Preserved food has been traded over long distances for thousands of years, and refrigerated shipping of fresh produce began in the 1800s. But, exotic and exogenous foods that were once luxury goods are now standard fare for millions of families around the world. Consumption of counter-seasonal produce is the norm in countries like the United States. Unlike improvements in food preservation and transportation, technological advancements in an era of increasing globalization focus on genetic modifications of plants that will produce fruits and vegetables that are durable, uniform in size and appearance, and will not spoil quickly.

Cultural and ideological effects of globalization also affect food production, distribution, and consumption. It is now possible for the production of food to be disembedded from any particular locale. Food can now be associated purely with shopping in a grocery store instead of as part of a process linked to the natural environment. If one does not connect the production of food with the consumption of food it is much easier to ignore the serious environmental consequences of unsustainable farming. Problems of soil fertility loss, water depletion, and water pollution are experienced by a disembodied "other" far away from the consumer's lived experience.

It is now possible and increasingly common to see a similar selection of foods available in markets all around the world. But, technological changes are not solely responsible for the homogenization of diets. Economic power is more consolidated under MNCs in the current era of globalization. That there are relatively few global food distributors means that the variety of foods will be reduced to increase efficiency through economies of scale to maximize corporate profits. As MNCs grow, their ability to market their particular product around the world grows. One example of an MNC with global reach that has significantly affected diets is McDonald's.

There are several problems associated with homogenization of diets. One is the loss of local cultural traditions. Local cuisine, something that may have been an important source of cultural pride and identity, may disappear. The loss of uniqueness of local foods may undermine the ability of local food growers to remain distinct and viable. Another major problem with the homogenization of diets is that the new globalized diet may be nutritionally inferior to traditional local diets. The nutritional value of some produce deteriorates in the time between harvest and consumption. The foods that bring the greatest profit to an MNC often are not the foods with the highest nutritional value. A more recent global nutritional crisis deals not with famine, but with increasing obesity rates, especially among children.

Finally, homogenization of diets can have serious environmental consequences. For example, increased global demand for beef has led to both intensification and extensification of agricultural cultivation. However, globalization does not always lead to homogenization. In some cases trends towards globalization can be perceived as aggressive and can lead to a resurgence of local cultural pride. Local foods and local methods of agricultural production can be linked to cultural revitalization.

Arguments in favor of more localized food networks say that such a system would by default make people more aware of the environmental effects of food production and more caring about the fates of people that grow their food. Proponents of globalization argue that a globalized food network is not necessarily worse than a more localized system. One effect of globalization is increased knowledge of faraway places, which makes global fair trade networks and promotion of sustainable organic farming more likely to be widely adopted with significant positive impacts. Those in favor of global food networks argue further that it may in fact be easier for inequality and environmentally unsustainable practice to continue when food production and consumption are more localized. Local context-specific food systems are heterogeneous so any agencies that regulate such a system would need to be flexible to accommodate difference. Increased complexity in regulation makes oversight more difficult, and corruption in the system may be more difficult to detect.

Further Reading

Born, Brandon, and Mark Purcell. "Avoiding the Local Trap: Scale and Food Systems in Planning Research." *Journal of Planning Education and Research* 26 (2006): 195–207.

Raynolds, Laura T. "The Globalization of Organic Agro-Food Networks." *World Development* 32(5) 2004: 725–43.

Shiva, Vandana. *Sustainable Agriculture and Food Security: The Impact of Globalisation.* Thousand Oaks, CA: Sage Publications, 2002.

See Also: Development, Sustainable; Fair Trade; Farm Workers' Rights; Food Security; Food Sovereignty; Free Trade; Green Revolution; Political Ecology of Food; Walmart

Julie Weinert

Grassroots Organic Movement

The term "grassroots organic movement" describes a social phenomenon in which farmers and consumers work together at the ground level to reshape agricultural and food systems to be less toxic and more environmentally, socially, and economically sustainable.

The growth of the grassroots organic movement has been a fittingly evolutionary process. The western organic movement began with the work

Proud organic farmers work hard on their land in upstate New York. Courtesy of Leslie Duram.

of Englishman Sir Albert Howard, who developed "organic growing methods" in India in the early 1900s. During the same period, the German Rudolf Steiner and student, Ehrenfried Pfeiffer, were developing "biodynamic" methods, based on the concept that the earth is a living organism that needs to be replenished and revitalized through organic methods. During and after World War II, Jerome I. Rodale adopted these concepts on an experimental "organic" farm in the United States and began publishing *Organic Farming and Gardening* magazine.

Concerns over unhealthy diets, and news of the environmental and human health dangers resulting from chemically intensive agriculture, as reported in Rachel Carson's 1962 book, *Silent Spring*, raised public consciousness and increased the demand for organic products. The interest in organic foods, along with farmers' interests in seeking safer, ecologically sound farming methods, accelerated as more has been learned about the risks associated with conventional production.

The grassroots organic movement emerged in the United States from the ground up, with primary activity in the Northeast, upper Midwest, and West Coast, during the late-1960s and through the 1970s. Concerned farmers and consumers joined together to form various networks, such as buying clubs, farmers markets, food co-ops, marketing co-ops, and farmer/buyer associations. Members of these emerging networks grew, processed, and distributed organic products; established certification standards; organized organic conferences; held farmer-to-farmer field days; and wrote publications, all with little or no government or university support.

Persons active in the grassroots organic movement come from different backgrounds, with a variety of motivations and belief systems. Many began their journey into organic agriculture as part of the "back to the land" movement of the 1970s, rejecting consumerism and seeking harmony with nature. Others, who might be described as "salt of the earth" Christians, reject chemical agriculture and are active in the organic movement as a way to properly care for God's creation.

Some join the movement as an act of rebellion against corporate and government control of agriculture, operating from libertarian, and occasionally, anarchistic, motivations. Others come in crisis mode, choosing organic methods as a last chance to cut input costs and save their farms. Many farmers and consumers have become active in response to high rates of cancer, birth defects, and asthma in rural areas. For others, the choice to go organic is rooted in a strong conservation ethic, commitment to protect biodiversity, and general love of nature.

The grassroots organic movement has evolved from a patchwork of disconnected entities to an effective, and increasingly interconnected, collaboration of farmers, processors, retailers, consumers, chefs, researchers, educators, and activists who impact government policies and market developments at local, state, regional, national, and international levels.

Further Reading

Growing for Market. www.growingformarket.com.

Organic Farming Research Foundation. http://ofrf.org/index.html.

National Sustainable Agriculture Information Service (ATTRA). Organic Certification Resources. February 26, 2009. www.attra.org/organic.html.

Rodale Institute. "New Farm." www.newfarm.org.

See Also: Agriculture, Organic; Family Farms; Transition to Organic; APPENDIX 1: Report and Recommendations on Organic Farming, 1980

Jim Riddle

Green Revolution

The term "Green Revolution" is usually used in reference to the development and widespread adoption of high-yielding varieties of food crops in the late 1960s and early 1970s, but many Green Revolution institutions and goals still exist today. Recent reports from the United Nations note that organic agriculture provides an opportunity for food security in developing nations, but this viewpoint was not always present.

Dire predictions in the 1940s and 1950s of devastating famine in developing countries led to a concerted and coordinated international effort to improve crop yields largely through technological innovation. Green Revolution research is best known for the development of high-yield crops, but research centers also investigated ways to increase cattle productivity, soil health, and management of pests, fisheries, forests, and water.

The first formal research institute devoted to the development of more productive food crops in the developing world was the Mexico-Rockefeller Foundation International Agriculture Program, a joint program between the United States and Mexico. The success of this early foundation led to the development of other affiliated research centers. CIAT (Centro Internacional de Agricultura Tropical [International Center for Tropical Agriculture]), established in 1967 in Colombia, and IITA (International Institute of Tropical Agriculture), established in 1967 in Nigeria, specialized in tropical agriculture; CIMMYT (Centro Internacional de Mejoramiento de Maíz y Trigo (Maize and Wheat for the Developing World), established in Mexico in 1966, specialized in maize and wheat; and IRRI (International Rice Research Institute), established in 1960 in the Philippines, specialized in rice. Each research center was originally funded by private foundations. By the mid-1960s, the Food and Agricultural Organization (FAO), the United Nations Development Program, and the World Bank were called on to help raise money to support the research centers. Representatives from each of the international development groups recognized a need for coordinated research on agriculture. As a result, the Consultative Group on International Agricultural Research (CGIAR) was established in 1971. CGIAR added research center locations in India, Peru, Kenya, the United States, Ethiopia, Italy, the Ivory Coast, Syria, the Netherlands, Sri Lanka, France, and Indonesia. CGIAR is currently the leader in coordination of international agricultural research, but many countries also operate affiliated or independent agricultural research programs.

Green Revolution goals have changed over time. The early emphasis in agricultural research centers was to increase food production. In the 1980s, a need for sustainable food production was recognized. By the 1990s, growing concerns about inequality in Green Revolution technology distribution and benefits led to a greater effort to link agricultural research goals to specific national development programs specifically designed to benefit those most in need. Today the official priorities of CGIAR are "reducing hunger and malnutrition by producing more and better food through genetic improvement, sustaining agriculture biodiversity both *in situ* and *ex situ*, promoting opportunities for economic development and through agricultural diversification and high-value commodities and products, ensuring sustainable management and conservation of water, land and forests, and improving policies and facilitating institutional innovation."

The first intensively studied Green Revolution crops were wheat and rice; in large part because the more developed countries in the 1940s and 1950s had experience in growing these crops. Research on several other tropical food crops like maize, cassava, chickpea, sorghum, potato, and millet soon followed, but progress was slower because the crops had not been scientifically studied as intensively as had wheat and rice. By the year 2000, CGIAR had developed over eight thousand different varieties of eleven different crops. Higher-yielding crops are most commonly created by breeding dwarf varieties. In a dwarf plant more energy is devoted to producing edible material and less energy is expended on growing inedible materials like stalks and husks.

In practice, adoption of Green Revolution high-yielding seeds was accompanied by increased inputs in synthetic fertilizers, pesticides, tractors, and irrigation systems. The first adoption of a Green Revolution crop was wheat in the 1950s in Mexico. Within a few years Mexico was self-sufficient in wheat and the program was deemed a success. Lessons and crops from Mexico were quickly transferred to India, a country that many at the time thought was on the brink of severe famine. Green Revolution technology was given much of the credit for averting the famine. The transfer of Green Revolution technology has had varied rates of success. Green Revolution success was initially limited in Sub-Saharan Africa in the 1960s and 1970s. High-yielding varieties of seeds, proven successful in Latin America and Asia, were quickly introduced to Africa, but were not adapted to the local ecological or social context until the end of the 1980s.

Promotion of organic agriculture was clearly not a part of the Green Revolution technology package of the 1960s and 1970s. The Green Revolution is largely responsible for ending organic agricultural practice in the developing world with the introduction of synthetic fertilizers and pesticides. However, the recent movement towards long-term sustainability of environmental health has been termed the "second green revolution." Some of the same research centers that were at the forefront of the first Green Revolution are now leaders in developing agricultural systems that are environmentally, socially, and economically sustainable.

The impact of the Green Revolution for promotion of local foods is more mixed than for organic production. One of the goals of the Green Revolution was to increase local self-sufficiency in food production. If higher-yielding agricultural systems could be adapted to many different ecosystems, in particular to tropical areas, those places would be less reliant on food aid. Nonetheless, the overall goal was to increase the total stock of food. Green Revolution research focused on places and conditions in particular countries that would maximize total gains. Most crop varieties have been specifically designed to grow best in places with reliable rainfall or irrigation systems, not for areas that have only marginal agricultural potential. Therefore, some of the poorest farmers in the world have not seen much benefit from Green Revolution technology, and improvement in local food production is not equally distributed to all agricultural areas. Additionally, exogenous inputs are compulsory for farmers that adopt Green Revolution agricultural systems. Seeds, fertilizers, pesticides, and tractors are all developed and sold by outsiders. If a Green Revolution variety is adopted, then cultivation of other crops that had previously been grown in the area must be abandoned. Local agricultural systems that had been in place before the Green Revolution were at least transformed and sometimes completely dismantled. Higher yields were accompanied by higher costs, so some farmers had to export foods to wealthier markets to maximize profits.

In some ways the Green Revolution was very much in line with the concept of sustainability. The key goal of the Green Revolution was to sustain food supplies for the entire population of the world. But, the alluring promise of improved yields and greater profits led in some cases to expanded cultivation into areas that were not well suited to Green Revolution technology, and more intensive cultivation can have negative consequences for the natural environment.

Although Green Revolution agricultural research centers have developed thousands of seed varieties, the biodiversity of agricultural crops has declined since the 1940s. The adoption of Green Revolution crops meant the abandonment of other crops, and in some cases local varieties of plants ceased to be cultivated and are now extinct. A lack of both species diversity and genetic diversity reduces long-term sustainability through an increase in vulnerability to crop failure due to plant disease and climate extremes.

Investment in Green Revolution technology was expensive but had potential for significant profit. Only the wealthiest landowners were able to afford Green Revolution inputs. Farmland was consolidated when land from poorer farmers with small land holdings was purchased by wealthier farmers. Wealthy farmers were able to increase their wealth. Poorer farmers either moved to more marginal land to try to farm, which can lead to increased degradation in those marginal areas, worked as farm laborers, or moved away, often to urban areas. Landless farm laborers benefited from wage increases when Green Revolution technology raised profits, but laborers generally did not benefit as much as landowners.

The long-term impacts of the Green Revolution for sustainability are varied. The intensive, high-energy inputs associated with the Green Revolution can clearly degrade soil quality, decrease biodiversity, and increase social inequality. Yet, proponents of the Green Revolution argue that overall the goal of preventing widespread famine has been achieved. In truth, the Green Revolution has never ended and currently greater attention is now paid to enhancing many different facets of sustainability beyond tons of grain produced.

Further Reading

Consultative Group on International Agricultural Research (CGIAR). "Who We Are." 2008. www.cgiar.org/who/index.html.

Freed, Stanley A., and Ruth S. Freed. *Green Revolution: Agricultural and Social Change in a North Indian Village.* New York: American Museum of Natural History, 2002.

Shiva, Vandana. *The Violence of the Green Revolution: Third World Agriculture, Ecology and Politics.* Atlantic Highlands, NJ: Zed Books, 1991.

United Nations Environment Programme. "Organic Agriculture and Food Security in Africa." United Nations Conference on Trade and Development. UNEP-UNCTAD Capacity-building Task Force on Trade, Environment and Development. New York: United Nations, 2008.

See Also: Development, Sustainable; Food Security; Food Sovereignty; Globalization; Shiva, Vandana (1952–)

Julie Weinert

Growth Hormones and Cattle

Many consumers specifically seek out organic meat to avoid synthetic growth hormones. The U.S. Food and Drug Administration (FDA) currently approves the use of six hormones to promote growth in cattle. Three occur naturally in cattle—progesterone, testosterone, and estradiol-17ß. The other three are synthetic hormones—trenbolone acetate (TBA), zeranol, and melengestrol acetate (MGA). Estradiol and progesterone are female sex hormones; testosterone is a male sex hormone; and zeranol, TBA, and MGA are synthetic growth promoters, hormonelike chemicals that can make animals grow faster. Currently hormonal growth promoters are approved in the United States for use only in sheep and cattle. Recombinant Bovine Growth Hormone (rBGH) is also approved for use in dairy cows to stimulate higher levels of milk production.

The cattle growth hormones, except MGA, are put into pellets implanted in the cow's ear, and the hormones are released over time. MGA is put into the cattle feed. The purpose of the hormones is to promote faster weight gain. There are no U.S. legal requirements that beef from hormone-treated cattle be labeled.

Canada has also approved the use of the six growth hormones in cattle. But in 1985, the European Union (EU) prohibited their use within EU countries. Since 1989, the EU has also banned the import of U.S. and Canadian beef from cattle treated with the growth hormones. This ban was challenged by the United States and Canada under the auspices of the World Trade Organization (WTO), the international body that governs trade among the world's nations. This dispute over the safety of beef growth hormones has been one of the longest and most intractable international trade disputes.

In the WTO trade dispute, in 1998, the EU's ban was found not to be in compliance with the WTO Agreement on Sanitary and Phytosanitary (SPS) Measures, the international food safety measure that governs the dispute. This was the first case to apply the SPS measures. The United States and Canada could not compel the EU to accept beef from hormone-treated cattle. Instead, the winners in WTO disputes are permitted to put trade sanctions on imports from the countries of the losers. The United States has imposed $116.8 million per year in tariffs on goods imported from the EU. Canada has imposed trade sanctions on EU imports worth $11.3 million (Canadian dollars) each year.

In 2003, the EU issued a directive on the use of growth hormones that retained the ban, based on a new EU assessment of the risk to human health from eating beef with growth hormone residues. The EU contended that the directive brought the EU into full compliance with the WTO agreements. But in March 31, 2008, a WTO dispute panel ruled that the EU was still not in compliance with the SPS Measures. The EU has appealed that decision to the WTO Appellate Body, arguing that the dispute panel exceeded its powers in the way it judged the science and made an error when it required the EU to show that its new directive was in compliance rather than requiring the United States and Canada to show that it is out of compliance. The EU also argues that the WTO dispute panel "applied inappropriate standards regarding the required scientific basis for health protection measures."

At the same time that the United States is challenging the EU decision to ban imported beef from cattle treated with growth hormones, there is a growing market in the United States for consumers who want beef from cattle raised without the use of the growth hormones. The Agricultural Marketing Service (AMS) of the U.S. Department of Agriculture, through its program of process-verification for livestock, has approved consumer labels for beef that indicate the beef came from cattle that were never treated with growth hormones. In 2007, AMS proposed a "Naturally Raised" label which could be applied to meat from livestock raised entirely without growth promotants or antibiotics and never fed mammalian or avian by-products. The USDA has designated this label the "Never Ever 3" label. A number of organizations, including those raising grass-fed livestock on pasture, oppose the label. They claim that just stating "naturally raised" on the meat product label is not a clear signal to consumers about the underlying standards and implies more about how the animals were treated than the "naturally raised" standard requires.

In addition, a 2007 scientific study calls into question whether the U.S. FDA testing of the safety of treating cattle with growth hormones is sufficient. The FDA has established "acceptable daily intake" for the residues of the synthetic hormones in beef but has never called for studies of the effects of these growth hormones on human populations who eat meat from hormone-treated cattle. In the 2007 study, researchers from the United States and Denmark found that men whose mothers in the United States consumed higher beef levels while pregnant had lower sperm concentration than men whose mothers reported eating less beef. Sperm concentration was not related to the mother's consumption of other meat or the men's consumption of any meat. These results are not surprising, because the growth hormones approved for use in U.S. cattle are anabolic steroids, relatively potent sex hormones, and it is well documented that growth hormone residues may remain in the beef products from treated cattle.

Further Reading

American Grassfed Association. "Comments on USDA 'Naturally Raised' Meat Label." www.americangrassfed.org/pdf/articles/naturally-raised.pdf.

Food Safety Network. www.foodsafetynetwork.ca.

Gandhi, Renu, and Suzanne M. Snedeker. "Consumer Concerns about Hormones in Food. Fact Sheet #37. Program on Breast Cancer and Environmental Risk Factors in New York State." Updated 2003. http://envirocancer.cornell.edu/Factsheet/Diet/fs37.hormones.cfm.

Kastner, J. J., and R. K. Pawsey, "Harmonising Sanitary Measures and Resolving Trade Disputes through the WTO-SPS Framework. Part I: A Case Study of the US-EU Hormone-Treated Beef Dispute." *Food Control* 13(2002): 49–55.

Swan, S. H., F. Liu, J. W. Overstreet, C. Brazil, and N. E. Skakkebaek. "Semen Quality of Fertile US Males in Relation to their Mothers' Beef Consumption during Pregnancy." *Human Reproduction* 22(2007): 1497–1502.

USDA, Agricultural Marketing Service. "United States Standards for Livestock and Meat Marketing Claims, Naturally Raised Claim for Livestock and the Meat and Meat Products Derived from Such Livestock." *Federal Register* 72(2007): 67266–8.

World Trade Organization. United States—Continued Suspension of Obligations in the EC—Hormones Dispute. Dispute Settlement: Dispute DS320. www.wto.org/english/tratop_e/dispu_e/cases_e/ds320_e.htm.

See Also: Agrichemicals; Antibiotics and Livestock Production; Livestock Production, Organic

Martha L. Noble

Gulf Hypoxia

Gulf hypoxia is a condition in which increased nitrogen contributions lead to a reduction of dissolved oxygen in the Gulf of Mexico. The mechanisms that lead to Gulf hypoxia and the feedback loops that in part sustain this condition are complex. Conditions that contribute to Gulf hypoxia are frequently attributed to high levels of nitrogen fertilizer use on agricultural lands in the Mississippi River Basin. Organic agricultural methods do not apply these chemicals, and thus do not contribute to the runoff of these nitrogen compounds in streams and rivers.

Hypoxia is a condition wherein oxygen is not present, or is present in inadequate levels to sustain life. Hypoxia in water bodies is the result of several factors, including high water temperatures and inadequate mixing of water. For many aquatic species, hypoxia exists when oxygen is present in water at a level less than 1 mg/liter. In the hypoxic zone of the Gulf of Mexico, measurements have recorded oxygen levels as low as .002 mg/liter.

The hypoxic zone in the Gulf of Mexico is frequently referred to as the "dead zone" for its inability to sustain aquatic life. Fishermen first noticed a decline in the productivity of the hypoxic area. Previously, this zone of the Gulf of Mexico was home to a productive shrimp fishery. The damage to

Gulf of Mexico fisheries has national implications, as a recent EPA study found, with "approximately 40 percent of the U.S. fisheries landings, including a substantial part of the Nation's most valuable fishery (shrimp), comes from this productive area."

Hypoxia is caused by an excess of nutrients, also referred to as nitrification, in water, particularly the presence of an abundance of nitrogen and phosphorus. These nutrients are essential in small quantities, but in higher concentrations, lead to an excess in growth of certain types of algae. With algae growth comes a proliferation in zooplankton. The excess growth of algae is problematic for native ecosystems because it blocks the penetration of sunlight below the surface of the water. Lack of sunlight prohibits the growth of plant species under the water surface, reducing available habitat and food for native species. In addition, the lack of sunlight for native underwater plant species also prevents the process of photosynthesis from occurring, a process that delivers dissolved oxygen into the nearby water, allowing the local fish species to uptake oxygen. Additionally, nitrification reduces available oxygen by increasing organic decomposition at the bottom of the sea. When the increase in zooplankton that accompanies an increase in algae growth leads to more decay, available oxygen supplies are consumed by the decomposition process.

The hypoxic zone in the Gulf of Mexico is not an isolated occurrence; numerous other estuaries and coastal areas are experiencing similar affects. The Chesapeake Bay off the coasts of Delaware, Virginia, and Maryland is experiencing a similar problem that has impacted crab populations. The San Joaquin-Sacramento Delta in California also has experienced a decline in native fisheries and has seen several species of fish listed on either the California State or Federal Endangered Species Act Threatened or Endangered Lists.

The hypoxic zone in the Gulf of Mexico is a seasonal event, occurring during summer. The oxygen capacity of water is inversely correlated with temperature, meaning that as water temperature increases, the amount of dissolved oxygen the water can hold is lowered. The summer months are also the period during which algae grows most effectively. The U.S. Environmental Protection Agency estimates the size of the "dead zone" on a rolling five year average, from 2000 to 2004, to be approximately 14,000 square kilometers. A study conducted by the Universities Marine Consortium measuring the hypoxic zone found the size of the "dead zone" frequently exceeds 20,000 square kilometers. The Universities Marine Consortium study also found that the size of the hypoxic zone has not reduced in recent years despite efforts to reduce some of the believed causes of Gulf hypoxia. The EPA has set a goal of reducing the size of the hypoxic zone to 5,000 square kilometers by 2015.

Myriad factors contribute to excessive nutrient levels in the water entering the Gulf of Mexico. These factors include natural and anthropogenic factors. Human-induced causes include wastewater treatment releases, runoff from agricultural fields, runoff from Concentrated Animal

Feedlot Operations (CAFOs), and products of combustion of carbon-based fuels. The modification of the Mississippi River from a naturally meandering stream to a river that has been channelized also leads to increased contributions of pollutants into the Gulf of Mexico, because the link between the Gulf and runoff in upstream states such as Missouri and Illinois are now more direct. Historic wetlands in the bayous of Louisiana naturally filtered high concentrations of nitrogen.

The most significant contributor of nitrogen into the Gulf of Mexico is believed to be nitrogen runoff as the result of excessive use of inorganic fertilizers throughout the Mississippi River Basin. The Mississippi drains approximately one-third of the contiguous United States landmass, and contains many highly productive agricultural lands, particularly those in Iowa, Missouri, Illinois, and Ohio. Fertilizers applied to lands in these areas can seep into groundwater or runoff during periods of rainfall, working their way into tributaries of the Mississippi, then into the Mississippi itself, and then into the Gulf of Mexico.

Reduction of nitrogen in the Gulf of Mexico is problematic because it requires a basin-wide management approach; getting farmers in Pennsylvania to change operations because of effects off the coast of Louisiana is difficult. Techniques that allow more efficient methods of delivering nitrogen to the plant, and not to the field as a whole are likely the key to reductions of this form of pollution.

Further Reading

Goolsby, D. A. "Mississippi Basin Nitrogen Flux Believed to Cause Gulf Hypoxia." *Eos, Transactions, American Geophysical Union* 81(29) 2000: 321.

National Research Council. "Nutrient Control Actions for Improving Water Quality in the Mississippi River Basin and Northern Gulf of Mexico." National Academies Press, 2008.

U.S. Environmental Protection Agency. "Mississippi River Gulf of Mexico Watershed Nutrient Task Force." 2008. www.epa.gov/msbasin/.

See Also: Environmental Issues; Pesticides; Water Quality

Michael Pease

H

Hain Celestial Group

The Hain Celestial Group, Inc., is a publicly traded company that provides organic and natural foods and personal care products. It has annual sales exceeding $1 billion. Of this total, approximately 50 percent is from certified organic products. Rather than specializing in one or two types of products, like many of its competitors do, Hain Celestial considers itself a leader in thirteen of the fifteen most popular natural food categories.

Hain Celestial has utilized acquisitions and mergers as a key strategy in its growth, under the direction of CEO Irwin Simon. Simon acquired Hain Pure Food Co. in 1994, building on the natural foods brand name that was originally established in 1926. Dozens more acquisitions followed, and the company's brands now include: Alba, Arrowhead Mills, Bearitos, Boston's, Breadshop, Casbah, Celestial Seasonings, DeBoles, Earth's Best, Estee, Ethnic Gourmet, Freebird, Garden of Eatin', Grains Noirs, Hain Pure Foods, Harry's, Health Valley, Heather's Naturals, Hollywood, Imagine Foods, JASON, Lima, Linda McCartney, Little Bear, MaraNatha, Mountain Sun, Natumi, Nile Spice, Orjene Organics, Rice Dream, Rosetto, Shaman Organics, Soy Dream, Spectrum, SunSpire, Terra Chips, Walnut Acres, Westbrae Natural, Westsoy, Yves and Zia Natural Skincare.

In 2000, Hain Pure Foods merged with Celestial Seasonings, a natural tea company, to form Hain Celestial Group, Inc. Celestial Seasonings was founded by Mo Siegel and Wyck Hay, who sold the company to Kraft in 1984. Kraft later proposed to sell the company to Lipton, which would have resulted in a combined U.S. tea market share of 81 percent. Due to antitrust concerns, the sale to Lipton was blocked, resulting in a venture capitalist–leveraged buyout in 1988. Siegel eventually returned to the company and oversaw an initial public offering in 1993, but retired two years after the merger with Hain.

This merger resulted in the largest natural foods company in the United States. Because of its rapid growth, concurrent with the growth in organic and natural food sales, Hain Celestial's key competitors are now larger, mostly conventional food processors such as Kraft and General Mills. *Food Processing Magazine* ranked Hain Celestial as the eighty-second largest food and beverage processor in North America by sales in 2008. The company also has operations in Europe.

Hain Celestial will typically attempt to sell production facilities after acquiring a brand, and to use the proceeds to pay down the purchase costs. The company may then contract with these same manufacturing plants to supply Hain Celestial's customers. More than half of their production is outsourced through co-packing agreements with nearly one hundred different manufacturers. Company executives have stated that they would outsource all of their production if they were able to find enough reliable suppliers, rather than manufacturing products in-house.

Further Reading
Hain Celestial Group. www.hain-celestial.com.

See Also: Brands; Market Concentration; Trader Joe's; Walmart; Whole Foods

Philip H. Howard

Health Concerns

For decades, consumer interest in organic agriculture has been driven by concerns about the pesticide residues on and in conventionally produced food. Farmworkers are at high risk for pesticide illnesses, both acute and long-term, in the form of cancer, neurological, and reproductive problems. Farming is also the most consistent occupational risk factor for cancer. Children, however, have been the focus of most concern because of their special sensitivities and exposure per pound of body weight. Infants are especially vulnerable because they don't have mature levels of the enzymes that metabolize pesticides. Organic foods contain much lower levels of pesticide residues. An analysis conducted in the United States of a large number of conventional and organic fruits and vegetables found that 75 percent of conventional and 25 percent of organic products contained at least one pesticide. The organic foods also had a much lower level of residues; a fact corroborated in 2006 when researchers reported that substituting most of the conventional diets of three- to eleven-year-old children with organic foods reduced the urinary concentrations of two organophosphate pesticides, malathion and chlorpyrifos, to nondetectable levels. Since this pesticide class increases risk of neurological damage, the authors assume that children eating organic foods would be at a decreased risk.

Pesticide residues are not the only health hazard from conventional agriculture. Another is antibiotic resistance; a problem that has increased because of the overuse of antibiotics in the treatment of human illness, and the use at low doses of antibiotics that are of critical importance to human medicine to enhance growth in food animals. The latter fosters the development of antibiotic resistant bacteria, and it is now known that antibiotic resistant diseases in human populations can occur through exposure to food. Organic regulations forbid the use of antibiotics in meat and poultry products sold with the organic label, so the risk of exposure to antibiotic resistant bacteria is substantially decreased.

The bacterial pathogens that can become resistant to antibiotics pose their own problems. The U.S. literature is sparse, but shows that salmonellae and campylobacters are as prevalent on organic as conventional poultry. However bacteria isolated from organic poultry are much less likely to be antibiotic resistant. The appearance of pathogenic bacteria on fresh fruits and vegetables because of the use of manure as a fertilizer became much more problematic in the late 1980s. The largest study done so far reported that approximately 10 percent of organic and about 2 percent of conventional produce were contaminated with *Eschericia coli*. Produce grown on certified

Environmental Working Group Top Fifteen Foods to Buy Organic

The Environmental Working Group (www.ewg.org) Food News carefully assesses the pesticide exposures associated with popular fruits and vegetables in the United States. Their analysis is based on actual food samples from grocery stores nationwide and takes into account consumers' typical methods of preparation (washed, peeled, etc.). Here are the top fifteen fresh foods that shoppers may want to eat organic to reduce their exposure to pesticides:

1 (worst)	Peaches	100 (highest pesticide load)
2	Apples	96
3	Sweet bell peppers	86
4	Celery	85
5	Nectarines	84
6	Strawberries	83
7	Cherries	75
8	Lettuce	69
9	Grapes (imported)	68
10	Pears	65
11	Spinach	60
12	Potatoes	58
13	Carrots	57
14	Green beans	55
15	Hot peppers	53

organic farms did not have a significantly different prevalence of *E. coli* from that produced on conventional farms. The use of manure or compost that had been aged for less than twelve months caused a nineteen-fold increase in the prevalence of *E. coli* compared to farms using older materials.

Heavy metals at high levels can be toxic to plants, animals, and humans. Since most elements (e.g., cadmium, lead) are not degraded in soil, they may be present for a long time. Therefore it is not surprising that researchers have found, in general, that there are few differences in the heavy-metal content of foods produced organically or conventionally.

Further Reading

Baker, B., C. Benbrook, E. Groth, and K. Lutz Benbrook. "Pesticide Residues in Conventional, Integrated Pest Management (IPM) Grown and Organic Foods: Insights from Three U.S. Data Sets." *Food Additives and Contaminants* 19(5) 2002: 427–46.

Lu, C., K., Toepel, R. Irish, R. Fenske, D. Barr, and R. Bravo. "Organic Diets Significantly Lower Children's Dietary Exposure to Organophosphate Pesticides." *Environmental Health Perspectives* 114(2) 2006: 260–63.

Luangtongkum, T., T. Morishita, A. Ison, S. Huang, P. McDermott, and Q. Zhang. "Effect of Conventional and Organic Production Practices on the Prevalence and Antimicrobial Resistance of Campylobacter Spp. in Poultry." *Applied and Environmental Microbiology* 72(5) 2006: 3600–3607.

Mukherjee, A., D. Speh, E. Dyck, and F. Diez-Gonzalez. "Preharvest Evaluation of Coliforms, Escherichia coli, Salmonella, and Escherichia coli 0157: H7 in Organic and Conventional Produce Grown by Minnesota Farmers. *Journal of Food Protection* 67(5) 2004: 894–900.

U.S. Environmental Protection Agency. "Children's Environmental Health Centers. Children's Exposure to Pesticides and Related Health Outcomes" http://www.epa.gov/ncer/childrenscenters/pesticides.html.

See Also: Agrichemicals; Antioxidants; Corn Syrup; Diet, Children's; Endocrine Disruptors; Organic Foods, Benefits of; U.S. Food and Drug Administration

Kate Clancy

Herbicides

Herbicides are pesticides that kill or inhibit growth of undesirable plants called weeds. Herbicides have been used since ancient times. Before synthetic herbicides, farmers used ashes, salts, copper sulfate, and oils to control weeds. These materials killed desirable plants as well as weeds, required many applications, persisted in the soil, and were environmental hazards. Organic farming uses natural methods of weed control; synthetic herbicides are prohibited.

The first synthetic herbicide, 2,4-dichlorophenoxyacetic acid (2,4-D), was developed during World War II. 2,4-D gained widespread use in cereal crops such as corn and wheat. Numerous herbicides have been developed in the past six decades. They caused profound changes in crop production, increased yields, and substantially reduced agricultural labor. By the 1980s, herbicides were the most commonly used pesticides. Recently, crops such as cotton, corn, and soybeans have been genetically modified to tolerate glyphosate (Roundup) and other herbicides.

Herbicides can be classified according to application timing, type of application, selectivity, and how they kill plants. Herbicides kill plants through inhibition of cell division, destroying pigments (i.e. chlorophyll), inhibiting photosynthesis, acting as synthetic auxins (plant hormones), inhibiting synthesis of essential amino acids, blocking lipid formation, and disrupting plant membranes. Some herbicides are selective, killing only certain types of plants, whereas other herbicides are nonselective, killing all plants they contact. Nonselective herbicides are used for total vegetation control or are spot-sprayed directly onto a weed. Herbicides can be applied before the crop is planted and weeds have emerged (preplant) and are often incorporated into the soil to improve their activity against emerging weeds. Preemergence herbicides are applied shortly after planting but before weeds emerge and rain or irrigation move the herbicide into the soil. Postemergence herbicides are applied after emergence of weeds and the crop.

Herbicides can have unintended environmental and health effects. One of the most notorious herbicides was 2,4,5-triclorophenoxyacetic acid (2,4,5-T), a component of Agent Orange used to defoliate jungles in the Vietnam War. The 2,4,5-T was contaminated with dioxin and has been implicated in long-term health problems of Vietnamese and veterans. Herbicides can drift from where they are applied and injure desirable plants. Weeds can evolve resistance to herbicides because of frequent use and high selection pressures. Herbicides have been detected in surface and ground waters. Herbicides may cause acute or chronic health effects, including cancer and endocrine system impacts. Recent research in the United Kingdom has found that herbicides impact biodiversity of songbirds through reduction in habitat and foods such as insects and weed seed.

Further Reading

Ching, Lim Li. "GM Crops Increase Pesticide Use." Institute of Science in Society. 2003. www.i-sis.org.uk/GMCIPU.php.

Dill, Gerald. "Glyphosate Resistant Crops: History, Status and Future." *Pest Management Science* 61(2005): 219–24.

Pike, David, and Aaron Hager. "How Herbicides Work." University of Illinois College of Agricultural, Consumer, and Environmental Sciences. PIAP 95–4. http://web.aces.uiuc.edu/vista/pdf_pubs/HERBWORK.PDF.

Zimdahl, Robert. *Fundamentals of Weed Science*. 3rd ed. Burlington, MA: Academic Press, 2007.

See Also: Agrichemicals; Fungicides; Pesticides

Lauren Flowers and John Masiunas

Howard, Sir Albert (1873–1947)

Sir Albert Howard laid the foundation of organic agriculture in his works on humus and soil health. Howard grew up at an English country home in Shropshire, England, and was educated at London and Cambridge Universities, where Professor Marshall Ward at Cambridge shaped his broad enthusiasm for botany and plant diseases. After he graduated, Howard was appointed as a lecturer emphasizing plant diseases at the agricultural college on Barbados. In 1902, he returned to England as a botanist at the agricultural college at Wye in Kent, where he learned about plant breeding.

In 1905, Howard left to become Imperial Economic Botanist to the Government of India. Howard spent the next twenty-five years conducting agricultural research in India, where he became disenchanted with reductionist scientific approaches and the emphasis on chemical farming. The highlight of Howard's career was as Director of the Institute of Plant Industry at Indore between 1924 and 1931. During this time, Howard developed the "Indore Method" of composting, focusing on returning nutrients to the soil and creating humus. Howard concluded that the secret of plant and human health lay in a fertile soil containing a high proportion of humus. He took a decentralized approach, focusing on the whole of India, not individual provinces, and on the whole of soil health, not individual components. Howard's crowning scientific achievement was the book *The Waste Products of Agriculture* (with Y. D. Yad) published in 1931 and describing his humus methods. After twenty-five years in India, Howard retired to England where he was knighted in 1934.

In England, Howard began a campaign to convey the importance of humus in the soil. In 1940, he published *An Agricultural Testament*, which emphasized the importance of recycling organic waste material and introduced the "The Law of Return." Howard described the scientific method of composting that he developed in India. He deemed chemical fertilizers unnecessary if the farmer recycles nutrients back into the humus. His second book *Farming and Gardening for Health or Disease* (later renamed *The Soil and Health*), published in 1945, presented a holistic approach to preventing disease of both crops and humans. Howard believed we must understand the whole process of plant growth, including solar power and soil health, to understand disease. One of Howard's greatest legacies was his influence on a generation of organic agriculture pioneers, such as J. I. Rodale and Lady Eve Balfour.

Further Reading

Addison, Keith. "Albert Howard." Journey to Forever, Small Farms Library. http://journeytoforever.org/farm_library/howard.html.

Barton, Gregory. "Sir Albert Howard and the Forestry Roots of the Organic Farming Movement." *Agricultural History* 75(2001): 168–87.

Conford, Philip. *Origins of the Organic Movement.* Edinburgh, UK: Floris Books, 2001.

Howard, Albert. *An Agricultural Testament.* London: Oxford University Press, 1940.

Howard, Albert. *Farming and Gardening for Health or Disease (Soil and Health).* London: Faber and Faber, 1945.

Stephen Bossu and John Masiunas

Hydroponics

Hydroponics is the practice of growing plants without soil, using water and a fertilizer solution. The word is derived from the Greek words, "hydro" (water) and "ponics" (work). The advantages of hydroponics are: higher quality plants are produced faster and with greater yield; growing conditions can be controlled; plants can grow in areas where it is otherwise impossible, such as Antarctica; and soilborne pests are eliminated. The most common commercial hydroponic crops are tomato, lettuce, and cucumber. Canada, Netherlands, and Israel lead the world in hydroponics production.

Hydroponic systems provide plants with nutrients, water, and aeration. There are four types of hydroponic systems. A liquid system has no support medium for roots. In aggregate systems the roots are supported by solid medium such as gravel, sand, rockwool, or vermiculite. Open systems do not reuse the nutrient solution after it passes through the roots, whereas closed systems recycle the nutrient solution.

Several types of hydroponic systems have been developed. Drip systems are the most popular hydroponic method. These systems drip nutrient solution on the plant base from small tubes. In ebb and flow systems, nutrient solution from a reservoir floods the roots in a grow tray on timed intervals. The solution then drains to the reservoir and is recycled. Nutrient film systems use a thin layer of nutrient solution that constantly flows along the bottom of a grow tray containing plant roots. The nutrient solution is recycled through a reservoir. Passive systems have plant roots hanging in a bath of nutrient solution. Air is pumped into the nutrient solution, bathing the plant roots. In aeroponics, no media is used to support the plant roots. Instead, nutrient solution is sprayed directly on roots suspended in a chamber.

There are both pros and cons in hydroponics. On the positive side, in hydroponics, crops can be produced on nonarable land, with no soil insects or diseases. There is no need for weeding or tillage. Hydroponic systems control the root system environment (easy adjustment of nutrient concentration and pH). High planting densities and yields are possible. Hydroponic systems can efficiently use water and nutrients. Vegetables grown using hydroponics can be more flavorful (e.g., tomatoes are picked fully ripe). On the negative side, however, hydroponic systems are expensive. For example, a commercial hydroponics operation can cost $600,000 per acre, not including land costs. Hydroponic systems have high-energy costs for cooling and heating. They have high maintenance costs and the crops can quickly die during equipment failures. Diseases and insects may be more difficult to control in hydroponic systems. The technology is usually limited to crops with high economic value.

Further Reading

Bridgewood, Les. *Hydroponics: Soilless Gardening Explained.* Marlborough, UK: Crowood Press, 2003.

Roberto, Keith. *How-to Hydroponics.* 4th ed. Farmingdale, NY: Futuregarden, 2003.

Rorabaugh, Patricia A. "Introduction to Controlled Environment Agriculture and Hydroponics." University of Arizona Controlled Environment Agriculture Center. 2005. http://ag.arizona.edu/ceac/CEACeducation/PLS217.htm.

Wilkerson, Don. "Hydroponics." Texas A&M University Department of Horticulture. http://aggie-horticulture.tamu.edu/greenhouse/hydroponics/index.html.

Lauren Flowers and John Masiunas

I

Insects, Beneficial

Beneficial insects refer to any insect group, both wild and domesticated, that provides valuable ecological services to society. Beneficial insects are particularly important in agriculture, where they pollinate crops, cycle soil nutrients, and regulate pest populations. However, the intensive tillage and pesticide applications typical of conventional farming have dramatically reduced the diversity and abundance of beneficial insects in agricultural landscapes. Consequently, conventional farmers must replace or supplement ecological services with inputs such as pesticides, synthetic fertilizers, and imported honeybee colonies. Many of the practices that are fundamental to organic agriculture (cover cropping, reduced pesticide applications, reduced tillage, and reduced weed control) have helped to restore native insect communities and the services they provide.

There are approximately four thousand native bee species in the United States. Many of these species contribute to the pollination of high value crops such as melons, strawberries, tomatoes, peppers, and orchard crops. Farms adjacent to wild land, where solitary bees nest and forage, may receive the majority of their pollination services from native bees. Even in simple landscapes, native bee populations can be restored in farms by protecting nesting sites, creating undisturbed habitat where bees may nest and forage, and reducing pesticide applications or applying pesticides at night when bees are in their nests. Soil nesting bees require areas of undisturbed soil, while stem-nesting bees find refuge in piles of branch clippings, hedgerows, or weedy field margins.

Soil dwelling insects (e.g. ground beetles, springtails, ants, sow bugs, and nematodes) play an important role in the decomposition of organic matter. Insect detritivores physically break down plant residue and animal waste and convert the organic matter through their frass, or excrement, into available nutrients for plant uptake and growth. Thus, detritivores not only are important in the management of crop residue and animal waste, but their activity recycles nutrients that improve or sustain soil fertility and crop production. A long-standing principle of organic farming has been to manage the soil as a living component of the agroecosystem to sustain soil fertility. Organic growers use practices such as cover cropping, compost additions, and reduced tillage to add organic matter to the soil and thus support a diverse community of soil dwelling organisms, including insect detritivores.

Insect predators and parasitoids are another important group of beneficial insects, a group that regulates insect and weed pest populations. Predators eat their prey, whereas parasitoids use and ultimately kill their prey by laying their eggs near or on them, which then develop and feed on the host. Lady beetles, syrphid flies, lacewings, damsel bugs, and minute pirate bugs are some of the important predators of aphids and other soft-bodied insects. Many ground beetle species are predacious on insect eggs, larvae, and weed seeds. Insect parasitoids, which include several wasp and fly families, are important regulators of lepidopteran (moth), dipteran (fly), and hompoteran (aphid, leaf hopper, etc.) pest populations. Native predators and parasitoids can be conserved and restored in farm systems by establishing seminatural habitat, such as hedgerows or flower strips, in undisturbed areas of the farm. Optimally, added habitat provides food resources (pollen and nectar), alternate hosts when host prey is scarce, and shelter and nesting sites.

Further Reading

Losey, John E., and Mace Vaughan. "The Economic Value of Ecological Services Provided by Insects." *Bioscience* 56 (2006): 311–23.

Vaughan, Mace, Matthew Shepherd, Claire Kremen, and Scott Hoffman Black. *Farming for Bees: Guidelines for Providing Native Bee Habitat on Farms.* 2nd ed. Portland, OR: Xerces Society for Invertebrate Conservation, 2007.

See Also: Agroecology; Bees; Integrated Pest Management; Pest Control, Biological; Pesticides

Tara Pisani Gareau

Integrated Pest Management

Integrated Pest Management (IPM) refers to a particular approach in crop protection that is characterized by the employment of a combination of multiple compatible methods for reducing pest populations and maintaining them at levels that do not damage the crop, without causing harm to nonpests or beneficial organisms, humans, and the environment. Sustainable farmers rely on IPM techniques to reduce or avoid agrichemical usage. Organic farmers use IPM and other techniques to completely avoid synthetic chemical applications, which are prohibited on organic farms.

An IPM program is based on a thorough knowledge of pest biology, its life cycle, and its interaction with the biotic and abiotic factors in the environment. Devising an IPM program also requires knowledge of the level of pest population that would cause significant damage to the crop and thus would result in reduced quality and quantity of yield and farm profits. The pest population level that is injurious to the crop and would likely cause yield loss is called an "action threshold." IPM practitioners use the action threshold to determine the optimum timing of employing pest control methods.

Because an IPM program aims to prevent or reduce pest problems before crop loss occurs, various techniques are employed before, during, and after the growing season. Some pest control measures that can be employed before planting a new crop include removing residues left by the previous crop because it might be harboring pest propagules, disinfecting farm tools and equipment before using them to plant the new crop, removing weeds that harbor and support pest populations, using disease-free seeds and other planting materials, taking soil samples prior to planting to test for the presence of soilborne pests and diseases, avoiding planting in fields with a history of pest problems, and selecting planting sites that are not prone to conditions that favor pest outbreaks.

Some pest control techniques that are employed at planting or during the cropping season include timing planting dates to coincide with periods when pest populations are low or when the pests are known to be inactive; using mulches on beds to suppress weed emergence or growth; using reflective mulches to repel insects that feed on the crop, especially those that may also vector diseases while they feed; rotating crops to disrupt pest life cycles; planting disease-resistant crop varieties, when available; removing and destroying infected plants or plant parts; and growing plants (hedge rows) that attract beneficial insects.

Some of the commonly used pest monitoring techniques include installing sticky traps to detect the presence of insects that fly; regularly inspecting plants for the presence of insects or insect egg masses; and inspecting plants for symptoms of disease (e.g., wilting, leaf spots, galls, mosaic, and cankers). Some pest control measure that are employed in an IPM program, usually when preventive measures are not enough to suppress pest problems, include the application of biologically based pesticides, botanical extracts, or soft chemicals to control insects, fungal diseases, or certain species of weeds.

Further Reading

Mississippi State University Extension Service. "Organic Vegetable IPM Guide." Publication 2036. 2008. http://msucares.com/pubs/publications/p2036.pdf.

U.S. Environmental Protection Agency. "Integrated Pest Management (IPM) Principles." 2008. www.epa.gov/opp00001/factsheets/ipm.htm.

See Also: Agroecology; *Bacillus thuringiensis* (Bt); Crop Rotation; Insects, Beneficial; Pest Control, Biological; Pesticides; Wine, Organic

Camilla Yandoc Ables

Intercropping

Intercropping is the practice of growing two or more different crop species with similar cultural requirements at the same time and in the same field. Intercropping is a common practice in Latin America, Asia, and Africa, primarily as a way for increasing farm productivity, because more crops are grown per unit area within one growing season, and for ensuring harvestable yield from the field in the event that the main crop fails. Other benefits of intercropping are weed suppression and reduction of insect pest problems. Insect damage reduction possibly results from planting certain species of plants that attract and support populations of beneficial insects to the field or planting diverse crops to impede the movement and feeding of insect pests. Farmers who employ sustainable or organic methods often rely on intercropping to increase their crop yields and reduce pest problems.

Intercropping systems differ in terms of the economic value of the crops that are grown in the same field. In some systems, two high-value crops are grown together; in other systems, one high-value crop is grown in rows and a lower-value crop is planted in between the rows, serving as a "smother crop" (i.e., a crop that suppresses the emergence and growth of weeds) or as a ground cover for preventing soil erosion. Intercropping systems also differ in the way the crops are planted on the field (arrangement and timing); row intercropping refers to the culture of two or more crop species at the same time, with at least one of the crops sown in rows; strip intercropping refers to the culture of two or more crop species in swaths or broad strips that allow for separate mechanized harvesting of each of the cultivated crops; mixed intercropping refers to the culture of two or more crop species on the same field with no defined rows; and relay intercropping refers to the culture of two crop species on the same field but with one crop sown ahead of the other and reaching maturity before the second crop is planted. Planting one crop ahead of another is done to minimize crop

Cows grazing on an organic farm in western Washington. (AP Photo/John Froschauer)

herds that newly transition to organic production allows the transitioning cows to be fed crops and forage from land included in the farm's Organic System Plan during the third year of transition. The crops and forage must be grown on land that has been free of prohibited substances for at least twenty-four months prior to harvest of the feed. These "third-year transitional" crops and forage may be consumed by the dairy animals on the farm during the twelve-month period immediately prior to the sale of organic milk and milk products. Under this provision, an existing dairy farm can be converted to organic production in 3 years, with the land and animals simultaneously eligible for certification.

Once a herd is converted, all dairy animals must originate from animals that were managed organically during the last third of gestation. The animals must be fed and managed organically at all times to produce organic milk.

Livestock used as breeder stock may be moved from a nonorganic operation onto an organic operation at any time. However, if the livestock are gestating and the offspring are to be raised as organic slaughter stock, the breeder stock must be brought onto the organic farm and managed organically no later than the last third of gestation. Bulls used for breeding purposes only do not need to be managed organically.

Unless breeder or dairy animals have been fed and managed organically their entire lives, beginning the last third of gestation prior to their birth, they cannot be sold as organic slaughter stock. They can produce

L

Livestock Production, Organic

In order to sell meat, milk, live animals, or livestock products as "organic" in the United States, the animals must be managed in accordance with the Organic Foods Production Act (OFPA) of 1990, as implemented in 2002 under the National Organic Program (NOP) Final Rule.

According to the NOP regulation, "livestock" are defined as "any cattle, sheep, goat, swine, poultry, or equine animals used for the production of food, fiber, feed, or other agricultural-based consumer products; wild or domesticated game; or other non plant life, except such term shall not include aquatic animals and bees for the production of food, fiber, feed, or other agricultural-based consumer products."

In order to be sold as "organic," all animals, except for poultry, dairy, and breeding stock, must be under continuous organic management from the last third of gestation prior to the animal's birth. Poultry or edible poultry products must be from poultry that has been under continuous organic management beginning no later than the second day of life.

Organic milk or milk products must be from animals that have been under continuous organic management at least one year prior to the production of the milk or milk products. A one-time feed exemption for whole interference or the competition of the crops for water, nutrients, and sunlight, which can result in reduced yields. Crop interference can also be reduced by carefully considering the planting density (seeding rate) and the architecture of the crops that will be planted together in the field (tall crops planted adjacent to short crops). Some examples of crops that have been grown together as intercrops include barley and pea; lettuce, fava bean, and pea; field beans and wheat; leeks and celery; wheat and flax; canola and flax; sunflower and black lentils; corn and peanuts; and sorghum and pigeon peas.

Further Reading

Sullivan, Preston. "Intercropping Principles and Production Practices." http://attra.ncat.org/attra-pub/intercrop.html.

Yandoc, Camilla B., Erin N. Rosskopf, and Carolee Bull. "Weed Management in Organic Production Systems." In *Emerging Concepts in Plant Health Management*, Robert T. Lartey and Anthony J. Caesar, eds., 213–54. Kerala, India: Research Signpost, 2004.

See Also: Agriculture, Organic; Agroecology

Camilla Yandoc Ables

organic offspring or organic milk, but they cannot be slaughtered for organic meat.

Organic dairy and slaughter stock lose their organic status if they are removed from the organic farm and managed on a nonorganic operation. They cannot be rotated back into organic production. For example, a dairy calf born from an organic cow cannot be raised nonorganically for one year and then transitioned to organic for the second year to produce organic milk the third year.

The NOP requires that all organic operations, including livestock and dairy producers, must maintain records that disclose all activities and transactions, are auditable, demonstrate compliance with all applicable requirements, are maintained for at least five years, and are made available to organic inspectors and certification agencies.

In addition, organic livestock operations must maintain records to preserve the identity of all organically managed animals and edible and nonedible animal products produced by the operation. This means that all organic animals must be tagged, named, grouped in flocks, or otherwise identified, with corresponding records maintained of all health events and medications or activities and all feeds and feed supplements purchased and consumed. Records must also be maintained of all products produced, including meat, milk, eggs, wool, etc.

All organic livestock producers must complete Organic System Plans that describe their production practices, list and describe all substances used and planned for use, describe the monitoring practices used to ensure that the operation follows the requirements, describe their recordkeeping system, describe steps taken to prevent contamination or commingling, and provide other information requested by the certification agency.

One hundred percent organic feed is required for all organic livestock. "Feed" is defined as "edible materials, which are consumed by livestock for their nutritional value. Feed may be concentrates (grains) or roughages (hay, silage, fodder). The term "feed" encompasses all agricultural commodities, including pasture ingested by livestock for nutritional purposes." In other words, all agricultural components of the ration must be certified organic. Fields, including pastures, used to grow feed for organic livestock, must be certified. Records must be kept of all farm-raised and purchased feed and feed additives. Crop producers who grow livestock feed for sale to organic livestock producers must be certified.

Natural substances, such as oyster shells or calcium carbonate, and synthetic substances that appear on the National List may be used as feed additives and supplements. Trace minerals and vitamins, approved by the Food and Drug Administration (FDA), may be used as feed supplements. No synthetic colorings, flavorings, dust suppressants, or flowing agents are allowed.

Livestock feed used for organic production must not contain the following items, some of which are included in conventional production: animal drugs, including hormones, to promote growth; feed supplements or additives in amounts above those needed for adequate nutrition and health maintenance; plastic feed pellets; urea or manure; mammalian or poultry slaughter by-products fed to mammals or poultry; or feed, additives, or supplements in violation of FDA rules.

Organic livestock producers must establish and maintain livestock living conditions, which accommodate the health and natural behavior of animals, including access to the outdoors, shade, shelter, exercise areas, fresh air, and direct sunlight suitable to the species, its stage of production, the climate, and the environment; pasture for ruminants during the growing season; appropriate clean, dry bedding (which must be organic, if consumed by the species); and shelter designed to allow for natural maintenance, comfort behaviors, and opportunity to exercise; temperature levels, ventilation, and air circulation suitable to the species; and reduction of potential for livestock injury. "Pasture" is defined as "land used for livestock grazing that is managed to provide feed value and maintain or improve soil, water, and vegetative resources."

Organic livestock producers may provide temporary confinement for an animal because of inclement weather; the animal's stage of production; conditions under which the health, safety, or well-being of the animal could be jeopardized; or risk to soil or water quality.

Organic livestock producers must take steps to prevent the contamination of water and minimize soil erosion. Livestock producers must ensure that their animals do not cause stream bank erosion or contaminate water resources.

Organic livestock producers are not allowed to use lumber treated with arsenate or other prohibited materials for new installations or replacement purposes in contact with soil or livestock. The prohibition applies to lumber used in direct contact with organic livestock, and does not include uses, such as lumber used for fence posts or building materials, if the animals are isolated from the lumber by use of electric fences, or other methods approved by the certification agent.

Organic livestock producers must establish preventative livestock health care practices, including the selection of species and types of livestock that are suitable for site-specific conditions and resistant to prevalent diseases and parasites. They must also provide a feed ration sufficient to meet nutritional requirements, including vitamins, minerals, protein or amino acids, fatty acids, energy sources, and fiber (for ruminants). Appropriate housing must be established with pasture conditions and sanitation practices to minimize the occurrence and spread of diseases and parasites. Conditions that allow for exercise, freedom of movement, and reduction of stress appropriate to the species (no caged laying hens, for example) must be maintained. Physical alterations needed to promote the animal's welfare must be accomplished in a manner that minimizes pain and stress.

When preventive practices and veterinary biologics are inadequate to prevent sickness, organic livestock producers are allowed to administer synthetic medications, if the medications appear on the National List. These medications may have specific withdrawal or "veterinarian prescription only" restrictions, and use of these medications without following the specific restrictions can result in loss of certification.

Organic livestock operations must not sell as "organic" any animal or edible product derived from any animal treated with antibiotics or any synthetic substance that is not on the National List. The NOP also prohibits the use of animal drugs, other than vaccines and other biologics, in the absence of illness; hormones for growth promotion; synthetic parasiticides on a routine basis; parasiticides for slaughter stock; or animal drugs in violation of the Food, Drug, and Cosmetic Act.

Organic livestock producers must not withhold medical treatment from a sick animal in an effort to preserve its organic status. When methods acceptable to organic production fail, all appropriate medications must be used to restore an animal to health. Livestock treated with a prohibited substance must be clearly identified and not be sold, labeled, or represented as organically produced.

Organic livestock operations must manage manure in a manner that does not contribute to contamination of crops, soil, or water by plant nutrients, heavy metals, or pathogenic organisms, and which optimizes the recycling of nutrients.

In order to be labeled "organic," meat, milk, and egg handling and processing operations must also be certified. Organic products must not be commingled with nonorganic products or come in contact with prohibited substances during slaughter, handling or processing. All ingredients and other substances used in or on organic products during processing must appear on the National List. Records must be maintained of all processing activities.

Further Reading

Cornucopia Institute. "USDA's Proposed Organic Pasture/Livestock Rule." www.cornucopia.org.

USDA Agricultural Research Service. "National Organic Program." 2008. www.usda.ars.

See Also: Agriculture, Organic; Free-Range Poultry; Growth Hormones and Cattle; Organic Valley; APPENDIX 2: Organic Foods Production Act of 1990

Jim Riddle

Local Food

"Local food" commonly describes food that has been produced, processed, and distributed within a particular geographic boundary or is associated with a particular geographic region. Local food has been gaining the attention of academics, policymakers, food system advocates, and producers as an alternative to the conventional mode of food production and marketing. The term "local" has been contested and debated, but is largely defined by

Local food is popular because consumers want to know the farmers who grow their food. Courtesy of Leslie Duram.

the particular context in which it is being used, often defined and redefined to meet the needs of the retailer, producer, or organization in each situation. For the most part, definitions of "local" tend to be associated with issues such as sustainability, quality, authenticity, and community. Consumers often define local food in a geographic sense, but don't necessarily assign a minimum distance. They are more likely to buy from the closest possible source for each food item.

Local food systems and alternative food supply chains have developed in response to issues on both the production and consumption sides of the food system. Farmers are dealing with locally driven demands arising from diverse social and economic pressures in the countryside, which create new opportunities as well as new constraints on agricultural production. Many farmers try to generate more income and respond to declining commodity prices by creating local food systems. These food systems incorporate a variety of direct marketing techniques, such as farmers markets, pick-your-own produce farms, road-side farm stands, community-supported agriculture, food box schemes, local restaurants, bakeries, and specialty processors. For the most part, local food is produced on small to midscale farms and sold directly to the consumer. Value-added processing at the local level is often done on the farm and at a small-scale basis, using raw products from the farm along with products purchased from neighboring farms.

Do you know where to buy food that has been produced locally? Click on the map at the Local Harvest Web site (www.localharvest.org). Here you'll find farmers markets, farms, community-supported agriculture, restaurants, grocery stores, co-ops, and even online stores where you can buy sustainably grown foods in your area.

Local food systems often developed out of the sustainable and organic farming movements and community food security endeavors, or as part of a rural development program. The effort to relocalize food systems in North America has been centered more around community food security and opposition to conventional agriculture; in Europe, however, local food has more ties to rural economic development and food safety issues. In either case, there are strong links to the political desire to create an alternative system of food production and consumption organized around reducing negative externalities in the food system and fostering a sense of community. Others approach local food systems without incorporating notions of community or sustainability and instead include notions of health, rural economic development, and food security. Nevertheless, there is usually a connection to the idea of an alternative food system, which encompasses a variety of direct marketing schemes, where the consumers are aware of the local nature of the product, such as farmers markets, food-box schemes, community-supported agriculture, and specialty shops.

In some instances local food is associated with a particular region that is well known for growing a particular product. Products may be tied to a place by association rather than geographically produced and consumed in one region. The emphasis is on identifying a particular geographic place to distinguish the product through labeling and certification, and associating the product with values attributed to the region. Its association with a particular region becomes the value-added component that makes the product marketable. This categorization emphasizes the economic aspects of local food; it is still assumed to be distinct and separate from the conventional food system, and therefore part of an alternative food system. Local food in this sense is assumed to be part of alternative food networks on a local scale, which are connected to a broader, more conventional system of production. The economic reality is that most producers and consumers are participating in both local and global food systems.

CONSUMERS

Consumers have a growing interest in high-quality food, not only in taste and appearance, but additionally in the way it is produced. These attributes are often associated with locally produced food. Consumers have identified a number of reasons for demanding and supporting a local food economy. One reason indicates a desire to protect local farmers and local businesses and to maintain a sense of agricultural tradition in the local area.

Many people are reacting to mass-produced food by demanding flavorful, nutritious, and safe food from local independent sources. Other consumers are interested in local food for the environmental aspects. Desire for a connection to the origins of food production is another area of growing interest in local food. People want to travel to the country and meet the farmer that grew or raised their food, see the land where it was produced, and feel a connection with that process. Some models of local food, such as pick-your-own produce farms and community-supported agriculture, allow customers to participate in some aspect of food production. Consumers feel invested in the process and the product. This benefits producers by providing interested and supportive customers and by reducing their labor needs.

BENEFITS OF LOCAL FOOD

The trend toward local foods is seen by some as a strategy for helping farmers and regions to capture value by reducing the number of intermediaries through which food passes from producer to consumer. This value can be economic, in the form of better returns on labor and capital, or social, in the form of reattaching a human face to food production and the development of codependence between the farm and community sectors. In a variety of ways, the relinking of food with producer identity, place specificity, preferred and vouched-safe production practices, and human health outcomes has become associated with notions of quality in food products. From a rural development perspective, local food systems provide opportunities for food enterprises to retain added value in the community or region, improve employment benefits, strategically build a regional identity, and support other local activities, such as tourism. The rural development benefits of a local food system are juxtaposed against the growing interest in a local food system's ability to build access to healthy food and promote community food security. However, these potential benefits are frustrated in some cases by the fact that the current agro-food system is structured in a way that largely discourages farmers from selling products locally and impedes consumers from accessing products from their own region.

Environmental benefits of local food include using fewer chemicals, less energy, and more sustainable production practices than food shipped from large conventional farms. Less packaging and reduced emissions from transportation is another benefit. Increasing local food production will reduce the number of large, monocropped farms and will increase the number of diverse and sustainable farms. There are many who believe that locally produced food tastes better and is more nutritious than conventionally produced food. Local food encourages social interaction, transfer of information, and a sense of community that is not developed through the conventional food system.

Food security and rural economies are related issues that can be addressed with local food systems. Local food systems create sound local economies, which create more jobs and allow more money to circulate through the local economy. Although local food can be more expensive, most advocates of food security believe that the cost of food is not the major issue. People need to have better job security, income levels need to be higher, and housing costs need to be lower, so that they can afford to pay for better food.

CHALLENGES

On a small scale, direct marketing gives producers and consumers the social interaction and information-laden products that many people desire from local food systems. That scale is an important aspect of local food, but to meet demand local food will need to scale up. There are logistical barriers to bringing local food up in scale. To make local food more widely available, producers will need to find ways to get their product processed and distributed to large purchasers, such as institutional buyers, large retailers, and wholesalers. This puts more links in the supply chain and begins to disconnect the consumer from the origin of production. One way this has been addressed is through labeling and advertising that allows the producer to attach their story or production information to the product. Consumers are able to receive assurances about the product without having to connect with the producers by traveling to the farm or attending a farmers market. Some organizations have worked to provide some standards and labeling schemes to help assure customers of the local value of the product.

Some of the barriers in scaling up the local food system are infrastructure, financial support, market access, consistent supply, and a lack of understanding and education on the part of the buyer. Connecting producers with purchasers can be challenging. Farmers are not necessarily trained or skilled marketers, and they often don't have time to devote to developing new markets. Working with local producers takes time and coordination. To meet the needs of large institutional or retail buyers, farmers need plenty of advance notice of the products desired and the quantities needed. Local level production has to work around variables such as weather, pests, seasonality, and other growing conditions that cannot be controlled by farmers. Organizing transportation and distribution is another area that needs additional time and finances to coordinate. The biggest barrier to selling local products to institutional buyers or large retailers is the purchasing system they have implemented, which requires all food purchases to go through a central warehouse and distributor.

ADVOCACY

To increase and support the development of local food systems, many steps could be taken to address barriers and challenges. Local food promotion has become a popular project initiative for many levels of government and nongovernmental organizations in recent years. Local food initiatives, with the assistance of local food animators, can offer viable opportunities to improve access to healthy food and new marketing strategies for producers. The capacity to provide opportunities and support for producers marketing to direct local markets depends on technical, financial, institutional, political, and social factors within each region. Local food advocacy has the ability to shape the way people think about and define local food. Much of the advocacy work has an underlying theme associated with health, food security, or agricultural protectionism. Food policy councils are one segment of organizations that have been developed to encourage the local food systems around urban centers.

Policies that encourage and support a regional food system will also promote food security. Policies to protect farmland and small to midsized farms are the foundation of a secure food system. These policies need to address the loss of farmland, challenges to new farmers, and the viability of small and midsized farms. There is also a need for more local processing facilities, distributors, and retailers. Increasing local infrastructure, including distribution networks, processing facilities, and market outlets, are key aspects to target in local food policy. Food policy should include aspects of housing and employment to make food more accessible for everyone. Marketing coordination and advertising campaigns to promote local food should target all segments of the population. Policies to support small to midscale farmers, encourage new farmers, and protect farmland around urban centers will also support local food systems.

Further Reading

Food Routes Network. www.foodroutes.org.

Halweil, Brian. *Home Grown: The Case for Local Food in a Global Market.* Washington, D.C.: Worldwatch Institute, 2002.

Hinrichs, Clare. "The Practice and Politics of Food System Localization." *Journal of Rural Studies* 19(1) 2003: 33–45.

Kingsolver, Barbara, with Steven L. Hopp and Camille Kingsolver. *Animal, Vegetable, Miracle.* New York: HarperCollins, 2007.

Local Harvest. www.localharvest.org.

Morris, Carol, and Henry Buller. "The Local Food Sector: A Preliminary Assessment of Its Form and Impact in Gloucestershire." *British Food Journal* 105(8) 2003: 559–66.

O'Hara, Sabine, and Sigrid Stagl. "Global Food Markets and Their Local Alternatives: A Socio-Ecological Economic Perspective." *Population and Environment: A Journal of Interdisciplinary Studies* 22(6) 2001: 533–54.

Selfa, Theresa, and Joan Qazi. "Place, Taste, or Face to Face? Understanding Producer-consumer Networks in 'Local' Food Systems in Washington State." *Agriculture and Human Values* 22(4) 2005: 451–63.

Starr, Amory, Adrian Card, Carolyn Benepe, Garry Auld, Dennis Lamm, Ken Smith, and Karen Wilken. "Sustaining Local Agriculture: Barriers and Opportunities to Direct Marketing between Farms and Restaurants in Colorado." *Agriculture and Human Values* 20 (2003): 301–21.

See Also: Agriculture, Alternative; Community-Supported Agriculture; Consumers; Family Farms; Farmers Markets; Food Miles; Food Policy Council; Gardens, Community; Processors, Local Independent; Sustainability, Rural ; Slow Food; Urban Agriculture

Shauna M. Bloom

M

Market Concentration

Market concentration describes the extent to which a small number of firms dominate a given industry. Concentration is increasing in many industries due to consolidation. This tendency for competitors to merge or acquire other firms, and for firms to leave a market (such as through bankruptcy), facilitates reduced competition. This means that the largest firms are able to more easily signal their willingness to cooperate in certain areas, including maintaining higher prices or restricting output. Such cooperation can occur without any formal communication, simply by observing and responding to the behavior of other firms. Increasing concentration can occur through two processes, horizontal integration or vertical integration.

HORIZONTAL INTEGRATION

Horizontal integration involves mergers and acquisitions between competitors at the same stage of production. An example of a horizontally integrated firm is Tyson Foods. Tyson Foods started as a poultry processor and has acquired or merged with other poultry processors in recent decades, including Holly Farms and Hudson Foods. The company has also made acquisitions in the larger protein sector, such as the beef and pork processor IBP (formerly Iowa Beef Processors, Inc.) and several seafood processors. As a result, by 2001 Tyson processed nearly one-fourth of all meat sold in the United States and tallies $26.86 billion in annual sales.

In the organic industry, Hain Celestial is an example of a horizontal integrator. It is the result of merger between Hain Foods and Celestial Seasonings in 2000, after Hain Foods had already acquired more than a dozen organic and natural food processors. The combined firm has made more than a dozen acquisitions since then, and is currently the leader in sales ($1.11 billion) of organic and natural foods and personal care products in the United States.

VERTICAL INTEGRATION

Vertical integration involves the joining together of firms that are at different stages of production, such as suppliers of inputs with manufacturers, or distributors with retailers of finished products. An example of a highly vertically integrated firm is ConAgra. ConAgra promotes the fact that it is involved in food production "from the farm gate to the dinner plate." The company sells fertilizers, pesticides, and seeds; buys, transports, and stores grain; produces animal feed; processes seafood and meat; and manufactures processed and frozen foods.

In the organic industry, the Canadian corporation SunOpta is an example of a vertical integrator. It is involved in sourcing, processing, manufacturing, and distributing organic foods. The company has integrated vertically by making over thirty-five acquisitions in the past decade, including firms in the United States, Mexico, and the Netherlands. Some of their products include soymilk and frozen fruit. It is also the world's largest producer of oat fiber.

MEASURING CONCENTRATION

Two measures are commonly used to assess concentration in a given market, the concentration ratio and the Herfindahl index. They both assess horizontal integration, as vertical integration is more difficult to measure.

The concentration ratio is the simplest measure. It refers to the percentage of total sales made by the largest firms, typically the four largest. If the four largest firms each have a 10 percent share of the market, for example, the concentration ratio is 40 percent. Economic theory suggests that when four firms control more than 40 percent of the market, it is no longer competitive.

The highest levels of concentration are typically found in a monopoly, which is defined as a noncompetitive market with more than 90 percent of sales from one firm. An oligopoly is a noncompetitive market dominated by two or more firms. A four-firm concentration ratio of more than 60 percent is considered a tight oligopoly, and 40 to 60 percent is considered a loose oligopoly.

The more complicated measure of market concentration is the Herfindahl Index, sometimes called the Herfindahl-Hirschman Index (HHI). This measure is obtained by summing the squares of the market share for every firm in an industry. The maximum index is 10,000, which occurs when one firm has a 100 percent market share ($100^2 = 10,000$). This measure is more sensitive to differences in the size of the largest firms than a concentration ratio. The U.S. government currently defines markets with an HHI of at least 1,000 as moderately concentrated, and 1,800 as concentrated. For comparison, a concentration ratio of 40 percent could result in an HHI as low as 400. Mergers or acquisitions that would raise the HHI by 100 points raise concerns with the Department of Justice, which shares its antitrust enforcement powers with the Federal Trade Commission.

DEFINING MARKETS

A key challenge in measuring concentration is defining the market. If a market is defined too narrowly or too broadly, then measures of concentration may be misleading in terms of their effects on competitive behaviors. The recent acquisition of Wild Oats by its competitor Whole Foods, for example, led to contested definitions of the market location of these organic/natural foods retailers. The Federal Trade Commission attempted to block the acquisition, using the reasoning that these were the two largest firms in the premium natural and organic supermarkets category. Whole Foods responded that they were competing in a much larger market, which includes natural food discount chain Trader Joe's, natural food stores recently introduced by traditional supermarket chains (such as Supervalu's Sunflower and Publix's Greenwise), as well as Walmart (the leading retailer of organic milk in the United States, and the largest corporation in the world).

IMPACTS OF CONCENTRATION

Increasing concentration may have positive impacts, negative impacts, or both. One argument in favor of consolidating industries is that it may result in economies of scale. This means that as a firm becomes larger the cost of producing a good or providing a service decreases. Economies of scale are most common in highly capital-intensive industries, such as large aircraft manufacturing. If these savings result in lower retail prices, then consumers are better off than they were when the market was less concentrated. A similar argument for increasing concentration is that it may result in economies of scope, or synergies when marketing and/or distributing a wider range of goods and services. Pepsi-Cola's merger with Frito-Lay in 1965, for example, allowed the combined company to be more efficient when distributing beverages and snacks, as both products are often sold by the same retailers.

Arguments against concentrated markets include inefficiency and inequality. Diseconomies of scale or scope may occur when larger firms become more inefficient, due to factors such as the increased difficulty of managing a large and extensive organization. Inequality may result when a small number of firms have disproportionate power to determine prices, volume, quality, and other aspects of goods and services, as well as to maintain barriers that prevent new competitors from entering their industry. High levels of economic power may also translate into political influence, and the capacity to shape regulations in ways that benefit the largest firms in an industry.

One of the most concentrated industries in food and agriculture is the vertically integrated poultry system in the United States. Although there are thousands of chicken farmers, they typically enter contracts with one of a few dozen firms who provide nearly all of their inputs (e.g., chicks, feed, medicine), and buy their chickens at a prearranged price. These firms also control the processing and marketing stages of production. As a result, consumers now benefit from historically low prices for processed chicken. Farmers, however, have little decision-making power and the majority earn less than poverty wages for their efforts. For workers, poultry processing is one of the lowest paid and most dangerous occupations in the United States, and benefits are rare. This contrasts with high levels of compensation for executives in these vertically integrated firms, such as the $24 million paid to the CEO of Tyson in 2007.

Further Reading

Agribusiness Accountability Initiative. www.agribusinessaccountability.org.

Hannaford, Stephen G. *Market Domination! The Impact of Industry Consolidation on Competition, Innovation, and Consumer Choice.* Westport, CT: Praeger, 2007.

Heffernan, William, Mary Hendrickson, and Robert Gronski. "Consolidation in the Food and Agriculture System." Report to the National Farmers Union, Department of Rural Sociology, University of Missouri, 1999.

Vorley, Bill. *Food, Inc.: Corporate Concentration from Farm to Consumer.* London: UK Food Group, 2003.

See Also: Hain Celestial Group; United Natural Foods

Philip H. Howard

Marketing

Historically, farmers have excelled at raising crops and livestock. Marketing their products, however, has been a longtime challenge, as farmers have struggled with finding markets, selling at reasonable prices, dealing with buyers, defining product quality, and maintaining quality during the journey from the farm to the consumer. Yet marketing is the key to success. Organic farmers are no different from "regular" farmers in this regard, as they face these same issues plus some issues specific to the organic sector: proving that their products are organic and maintaining the organic integrity of products as they move through the supply chain.

Figure 1. Organic handlers move products through the supply chain

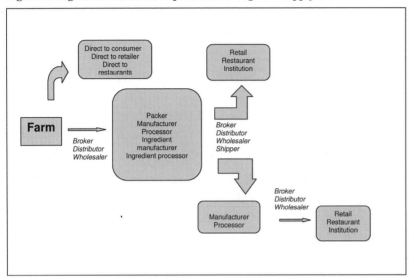

Courtesy of USDA ERS, 2008.

Food passes through many hands as it moves from farm to consumer. Some foods are fresh when delivered (e.g., apples and eggs) whereas others are processed before delivery (e.g., pasta and bread). Regardless of whether they are fresh or processed, higher quality products generally have a higher selling price. As a result, farmers have a strong incentive to produce and sell commodities of the highest possible quality. Yet, since most foods pass through a number of intermediaries as they move from the farm to the consumer, maintaining quality along the marketing chain is a challenge. To do so, each agent along the marketing chain must begin by moving the product to the next agent quickly. Farmers need to sell their perishable commodities immediately after harvesting, and middlemen (mainly distributors and wholesalers) need to get fresh products to retailers as quickly as possible. Because fresh foods rapidly deteriorate, they must be delivered to the market quickly. Processed foods, on the other hand, have a longer shelf life, but the products that go into them must be harvested at the right time, delivered at the right time, and satisfy the processor's quality requirements.

IDENTIFYING ORGANIC PRODUCTS

In the United States, organic products are identified in the marketplace by the label "organic" as specified by the national organic standards. The labeling requirements under the standards apply to raw, fresh products and processed products that contain organic ingredients and are based on the percentage of organic ingredients in a product. Agricultural products labeled "100% organic" must contain (excluding water and salt) only organically produced ingredients. Products labeled "organic" must consist of at least 95 percent organically produced ingredients. Products labeled "made with organic ingredients" must contain at least 70 percent organic ingredients. Products with less than 70 percent organic ingredients cannot use the term "organic" anywhere on the principal display panel but may identify the specific ingredients that are organically produced on the ingredients statement on the information panel. The USDA organic seal—the words "USDA organic" inside a circle—may be used on agricultural products that are "100% organic" or "organic."

The organic standards apply to the methods, practices, and substances used in producing and handling crops, livestock, and processed agricultural products. Although specific practices and materials used by organic firms may vary, the standards require every aspect of organic production and handling to comply with the provisions of the Organic Foods Production Act (OFPA). All firms that produce, process, manufacture, broker, distribute, pack, or otherwise handle organic products, with sales of $5,000 or more per year, must comply with the regulation. The standards require that the organic integrity of a product be maintained as it moves along the supply chain, which means that organic products cannot commingle with conventional products at any time.

ORGANIC PRODUCTS FROM FARM TO MARKET

Each commodity, depending in large part on whether it is fresh or processed, follows an individualized path from farm to market. The supply chains for three types of agricultural products—dairy products, fresh fruits and vegetables, and processed foods (e.g., breakfast cereals, pasta, and frozen and canned foods)—are described here. These three products are the best-selling organic products in the United States. Organic fresh produce accounted for 34 percent of the $18.9 billion of organic retail sales in 2007; organic processed foods (excluding beverages), 33 percent; and organic dairy, 16 percent.

Organic Dairy Products

The first step in marketing organic dairy products is simply producing the milk. To produce organic milk, cows must be fed organically grown grain and hay, must be grazed on organically managed pastureland, and must be given access to the outdoors. Because procuring organically grown feed is often challenging, many dairy farmers produce at least a portion of the feed they use. Organic milk produced by small farms is usually sold locally. However, some farmers may resort to selling their organically grown milk in conventional markets when they have access to no other markets. There are a few large producers of organic dairy products, including Horizon, Organic Valley, and Aurora Dairy (which produces store branded organic milk). In addition to bottling milk from their own cows, these companies pasteurize and bottle the milk of other dairy farmers, either through a marketing cooperative or through direct contracting with small and midsized family farms.

The milk and milk products are distributed nationwide to a wide variety of retail venues in the natural product and conventional market channels. Natural product channels include natural product supermarkets such as Whole Foods and independent natural product stores such as My Organic Market (in Maryland). Conventional channels include supermarkets (such as Safeway), big box stores (such as Target) and club stores (such as Costco). In 2007, approximately 80 percent of organic milk and cream sales took place in conventional market channels.

Organic Fresh Fruits and Vegetables

For fresh fruits and vegetables, the first stage in the marketing chain—the production and preparation of produce for shipment—involves growers, packers, and shippers, working together in a number of possible combinations. In some cases, one firm grows, packs, and ships the produce, whereas in others, one firm grows and another packs and ships. After it is shipped, produce can either be sold to retailers directly or by a broker or specialty wholesaler or delivered to a terminal market, where it is sold to retailers by wholesalers. In practice, most organic produce is sold through specialty wholesalers or brokers rather than in a terminal market. In some instances, when a specific variety, quality, or quantity is desired, larger retailers may buy fresh fruits and vegetables directly from the shipping point.

Organic fresh produce is sold to consumers in a wide variety of venues, including natural products stores, conventional supermarkets, and directly to consumers through farmers markets, community-supported agriculture, and farm stands. Direct sales to restaurants are increasing, as local, organic food is growing in popularity in restaurants. Data are not available indicating the value or percent of fresh organic produce sold direct to consumers, direct to restaurants, and in the natural products and conventional channels. Some information, however, is available for packaged fresh produce, such as bagged salads and baby carrots. Seventy-three percent of packaged organic produce was sold in conventional channels in 2007, with the other 27 percent sold in natural products channels.

Processed Organic Foods

Market channels for organic processed foods, which include frozen vegetables, pasta, canned vegetables, and sauces in jars, are complex. Specific products must be used to manufacture these foods. For example, pasta processors need to use a particular variety and grade of wheat, and frozen fruit and vegetable processors need produce that is a specific size and quality. All processors want uniform quality of the ingredients used, so they can offer products that consistently taste the same. Consequently, the biggest challenge facing organic manufacturers is how to secure a steady supply of organic ingredients of a consistent quality. The next biggest challenges are how to transport their processed goods to the supermarket and how to secure shelf space.

There are two basic supply chains for organic processed foods. In the first, the farmer produces raw commodities such as grains or fresh vegetables. These commodities are then sent to the manufacturer, who converts them into a processed product, such as pasta. A distributor acts as a middleman, moving processed products from manufacturers to retailers. In the second scenario, a middleman (e.g., an ingredient wholesaler or broker) facilitates the transaction: the wholesaler procures raw commodities from farmers and delivers them to manufacturers, while the broker arranges the sale. Both the wholesaler and broker secure the quantities needed, and ensure that the commodities are high quality and meet the manufacturer's organic standards.

One problem periodically faced by the organic sector is producing enough organic commodities. For example, at different points in time milk and feed grains have been scarce. Given tight supplies, organic processors and manufacturers often arrange to buy their needed supplies in advance, through contracts or closely aligned transactions between buyers and sellers. The organic sector uses contracts to procure needed ingredients at a much higher rate than the conventional sector. Nearly half of the volume (46 percent) of organic products bought by organic handlers is obtained with written, negotiated contracts. Another 24 percent is procured through verbal agreements or ongoing relationships between suppliers and handlers. The remaining 27 percent of ingredient volume is acquired through spot markets, or anonymous transactions. In comparison, the conventional sector uses contracts or vertically coordinated relationships for 40 percent of their purchases, and the remaining 60 percent is procured in the spot market.

Similar to other products, organic processed foods are also sold in the conventional and natural products channels. Most baby food (76 percent) is sold in the conventional channels, as is cold cereal (65 percent) and canned soup (60 percent). Organic breads and grains are sold more often in the natural products channels (56 percent).

THE ORGANIC CONSUMER

Ultimately, the consumer is the force driving growth in the organic market, in which retail sales have increased from $3.5 billion in 1997 to $18.9 billion in 2007. Underlying this tremendous growth is the fact that organic products have shifted from being a lifestyle choice for a small share of consumers to being consumed at least occasionally by two-thirds of U.S. consumers. Organic foods were once purchased mainly for environmental or social reasons, but more recently, consumers have developed broader reasons for purchasing organic food. Health and nutrition, taste, and food safety have emerged as the most important reasons for buying organic food, and as a result, retail sales of organic food products have soared. This growth in consumer demand provides ample marketing opportunities for organic farmers and other firms in the organic sector.

Further Reading

Dimitri, Carolyn, and Catherine Greene. *Recent Growth Patterns in U.S. Organic Foods Market.* Agricultural Information Bulletin 777. U.S. Department of Agriculture, Economic Research Service, 2002.

Dimitri, Carolyn, and Nessa J. Richman. *Organic Food Markets in Transition.* Policy Studies Report No. 14. Henry A. Wallace Center for Agriculture and Environmental Policy, 2000.

"NFM's Market Overview." *Natural Foods Merchandiser* (June 2008). 1.

U.S. Department of Agriculture, Economic Research Service. Organic Agriculture Briefing Room. 2008. http://ers.usda.gov/Briefing/Organic/.

See Also: Marketing, Direct; Markets, Ethnic; Prices; Appendix 4: Recent Growth Patterns in the U.S. Organic Foods Market, 2002

Carolyn Dimitri

Marketing, Direct

Direct marketing of agricultural products in the United States has increased over the last decade due to a growing consumer interest in having high-quality, local foods and supporting local farms. For farmers, much of the attractiveness of direct marketing is the potential to increase profitability. Agricultural direct marketing is the marketing of agricultural products directly to the end consumer, although it can also encompass sales to institutions (such as schools) and restaurants. Direct marketing of agricultural products includes, among others, sales through farmers markets, farm stands, pick-your-own produce operations, community-supported agriculture (CSA) or subscription sales, and sales through the Internet.

The Census of Agriculture shows a marked increase from 1997 to 2002 in the value of agricultural products sold directly to individuals for human consumption; in 2002, 25 percent more farms took part in direct marketing (116,733 or 5.5 percent of all farms in 2002) than in 1997, although the total value of direct sales represented less than 1 percent of all agricultural sales by farms. The average direct marketing sales per farm was also up 18 percent to almost $7,000 per farm from the 1997 census. This is mirrored by the growth in direct marketing venues. Over the last decade, farmers markets have increased over 70 percent to number 4,685 in 2008. Virtually unheard of fifteen years ago, CSA farms have grown to over 2,000 farms today.

Many direct marketing farmers are marketing products that need little processing, such as produce and meat, and specialty or niche products, such as organic foods, of special interest to urban consumers. Value-added products, or agricultural products that are processed or modified to increase their value, such as cheese or jellies and jams, can also be successfully direct marketed. Because it cuts out the middleman, direct marketing has the potential to provide a larger share of the food dollar directly to the farmer.

Many also see direct marketing as a good fit for smaller farms, which cannot compete as effectively in wholesale markets.

On the part of farmers, direct marketing commands different marketing skills than those used for wholesale markets, with an increased focus on individual consumers and needed knowledge about quickly changing markets and products. In addition, location is important for direct market opportunities. Often farmers with access to urban consumers are less concerned with attracting consumers because they have a large pool of potential clients, but find problems with the cost of business on the urban fringe. Those in rural areas often have more difficulties reaching consumers, including problems with accessing urban markets, and transportation issues.

Further Reading

Adam, Katherine. Direct Marketing. Fayette, AR: Appropriate Technology Transfer for Rural Areas, National Center for Appropriate Technology. 1999. http://attra.ncat.org/attra-pub/directmkt.html.

Alternative Farming Systems Information Center, National Agricultural Library. 2009. http://afsic.nal.usda.gov.

See Also: Agritourism; Consumers; Family Farms; Marketing; Markets, Ethnic; Processors, Local Independent

Lydia Oberholtzer

Markets, Ethnic

Ethnic marketing is defined as selling goods and services across various ethnic consumers. Ethnic produce is defined as vegetables, fruits, and herbs that are not traditionally sold in the mainstream supermarkets, but are imported or currently grown on a limited scale in the local geographical region. Ethnic and specialty vegetables are also referred to as exotic, unusual, world vegetables, or high-value crops. Specialty crops are non-commodity crops, and have unique characteristics. These specialty crops are usually targeted toward a specific ethnic population with strong preference for ethnic cuisines. Most food decisions are made by consumers based on cultural, psychological, and lifestyle factors, and food trends. In many communities, food is an integral part of the culture, and there is an established linkage between food and culture. The type of food cooked, patterns of purchasing produce, and amount spent on food are very centric to and dependent on cultural trends. These underlying sociocultural and economic factors tend to drive consumers to buy specialty produce or product items.

Many Americans are of European ancestry, mostly from the United Kingdom, Germany, and Ireland, and many Americans report multiple ancestries. Other majority and minority ethnic groups include African Americans, Chinese, Asian Indians, Filipinos, Koreans, Japanese, Mexicans,

Ethnic markets contain unique items. Courtesy of Leslie Duram.

Puerto Ricans, and other Spanish-speaking people. Economic opportunities have arisen in the last decade for specialty crop agriculture catering to the ethnically diverse consumers in the United States. In addition, the rapid expansion of the ethnic populations presents significant opportunities for fruit and vegetable producers in the United States to take advantage of their close proximity to densely populated areas. Increasingly, these producers adapt new crops or create new value-added products to remain economically viable. Growing demand for ethnic crops presents opportunities for producers to exploit existing comparative advantages associated with serving densely populated local ethnic markets to increase profitability and sustain farming operations. The coordination of production and marketing are critical to avoid the threats of rapid overproduction and to overcome inadequate marketing infrastructure to move product into community markets. When farmers of small- and medium-sized farms are not able to compete with larger produce growers in the regular produce market, they might want to focus on the rapidly growing ethnic produce market. Establishing or extending existing cooperative marketing associations with small and medium-sized farms in the United States can create an improved market system that provides appropriate year-round supplies of ethnic produce to the densely populated regions.

Further Reading
Bhugra, D. "Cultural Identity and Its Measurement: A Questionnaire for Asians." *International Review of Psychiatry*, 11(2/3) 1999: 244–50.
Govindasamy, R., R. VanVranken, W. Sciarappa, A. Ayeni, V. S. Puduri, K. Pappas, J. E. Simon, F. Mangan, M. Lamberts, and G. McAvoy. "Demographics and the Marketing of Asian and Hispanic Produce in the Eastern Coastal U.S.A." New Jersey Agricultural Experiment Station. 2007. www.dafre.rutgers.edu/documents/ramu/nri_ethnic_crops_results_nov2007.pdf.

See Also: Marketing; Marketing, Direct

Ramu Govindasamy

Media

Media as technology for information transmission has fertile connections to organic growing. Two specific ways of thinking about media and organics together involve considerations of how organic farming and commodities are represented in the media and in what ways some media are organic. Both considerations demonstrate similar trends in how the concept "organic" circulates discursively.

Representations of organic growing in the media are shaped more generally by representations of agriculture. It has been noted that Hollywood has participated in the construction of the "agrarian myth." The agrarian myth is an important construction of U.S. national identity, finding some of its earliest articulations in the writings of Thomas Jefferson. Jefferson's idealization of the U.S. citizen as farm owners and workers celebrated such work as honest and virtuous—the core of sound American values. Today, the view of farmers as "salt of the earth" remains resonant in large part due to how farmers are represented in popular culture, a representation that tends not to include industrial agricultural operations. In Jefferson's time, 90 percent of U.S. citizens were farmers; in contrast, today only about 1.5 percent of the U.S. population engages in farm labor. Speaking to an increasingly urbanized population devoid of much rural experience, cinematic narratives tend to romanticize "farm life." Although they depict such a life as difficult and a site of often Herculean struggle, it is also largely represented as a scene of family and individual perseverance. The agrarian myth, as represented in most forms of popular culture, typically does not address industrial farm operations and corporate farming models that dominate U.S. agricultural production today.

The contemporary discourse of organics circulates within this myth, often located within the ideal of small, local operations over and against large-scale practices of industrial agriculture. "Organic" remains an ambiguous signifier; what exactly constitutes organic production is subject to shifting and vague regulations. For some, it refers to a lack of chemical pesticides and fertilizers. For others, it means locally grown on small, often family-owned operations. For many, it privileges traditional growing means rather than modern, industrial methods. Whatever the nuance of the definition, it is in many ways food that is produced by means that fit with the idealizations of the agrarian myth.

In media that is concerned with advertising and selling products (which, it could be argued, includes all media), organic becomes a purr word whose positive connotations promote products, thus making it a key component of "green consumerism." To buy organic is to form one's green identity, creating that identity through ritualized acts of purchasing. As organic is linked to other environmental issues and as environmental issues take center stage in the contemporary culture, "organic" becomes an attractive marketing tool. An *Advertising Age* marketing poll recently found that 75 percent of respondents believe that organics are not a trend about to end, although the same report acknowledges general confusion over what constitutes organic. The success of magazines like *Organic Gardening* and *Organic Style* demonstrate the popularity of organic foods as a lifestyle choice and the viability of the markets they reach.

The typically higher cost of organic products and the ambiguity about what makes a product organic has also led to some significant popular criticism of organics. Some of this criticism comes directly from agri-industry, but there has been a fair amount of it in popular media as well. A representative example comes from the premium cable TV show, *Penn and Teller Bullshit!* An episode from the first season titled "Eat This!" debunks organic foods. At one point, the hosts interview shoppers coming out of a Whole Foods Market, asking them to define what organic and GMO (genetically modified organisms) mean, demonstrating a wide variety of uninformed responses. The hosts then claim that moves to have non-GMO and organic-only agriculture cause famine in Sub-Saharan Africa. In addition to linking organics to anxieties over GMOs, the show also criticizes organic foods as being in the same category as fad diets. The message is clear: organic food is as much a sham as "get thin quick" quackery. Moreover, in addition to duping privileged U.S. consumers, this supposed sham is impacting the availability of food in Third World countries.

Although the celebration of organic as a positive lifestyle choice or the denigration of it as a scam mark the ends of a contentious spectrum, what seems lost in this popular media discussion is a meaningful exploration of what actually constitutes organic production. Although news coverage occasionally addresses these concerns, the bulk of media outlets focus either on a carte blanche celebration of organics or derisive criticism of such. If there is insufficient media coverage to help educate consumers about how their food is produced and the consequences of that production, that may be due in part to the imbrications of media within markets of exchange. This is the point where the representation of organic agriculture in media meets the concept of organic media and a general concern over media consolidation.

If the later part of the twentieth century saw an increase in industrialized, large-operation farm-works, that has been paralleled by the loss of independent media outlets and the formation of large media consortiums. Media consolidation results in centralized production of content and, increasingly, the Federal Communication Commission (FCC) has relaxed regulations preventing a single company from owning a functionally exclusive amount of media outlets in a given market. The reduction of diversified and competing information outlets results in a reduction of information about key issues of concern to local publics. Although the media may not solely shape the political economy of particular regions, they nonetheless significantly influence perceptions. The media may not tell the public what to think, but they do tell the public what to think about. Media outlets are controlled by an ever decreasing number of large, multinational corporations that influence and benefit from governmental policies of deregulation. Much as large-scale agriculture tends to deplete the genetic diversity of foodstuffs and focus on appealing appearance over healthy substance, so too does media consolidation form its own monocultures of slick packaging.

Whereas this observation about media consumption is suggestive of a similar relationship between practices of production and more basic consumption of food, environmental author Bill McKibben goes further to link local food and local media. In his description of Barre, Vermont, he describes a town struggling simultaneously with producing its own local media and its own local, organic food. A town, in other words, resisting the trends of globalization that encourage the transport of industrial farmed foods from half a globe away or the centrally produced global culture of consolidated media. He tracks with several chilling examples the harms that centralized, corporate food and media production has on smaller communities.

Organic often (but not ubiquitously) designates local, "home-grown," independent media. It is explicitly a response to globalization, the homogenization of cultures into a monoculture by globalizing economic forces. Several cultural theorists note this celebration of the local as a return to popular culture's origins in folk traditions—a reaction of resistance to the flatness of globalized monoculture. There are certainly differences between such a turn to the local in the contemporary media landscape and actual folkloric traditions, but the connection of independent media to folklore resonates with the agrarian myth—a desire for simpler times and communities, a yearning for more organic means of cultural production.

Where this trend toward designating certain media practices as organic has seen the most proliferation is in advertising agencies, especially those working on the Internet. *Organic Inc.*, founded in 1993, provides an interactive advertising agency that offers a holistic and experiential approach to creating advertising campaigns. *Organic* president and CEO, Jonathan Nelson, stated: "We believe the power has shifted from the producers of goods, services and media to the customer, and the Net has effected the change." The company utilizes the more open media outlets of the Internet to facilitate consumer- rather than producer-oriented advertising campaigns. The client-focus along with the open-access Internet venues are key components that make this company's self-avowed approach organic.

A related example is *Organic Media*, a company operating in Puerto Rico that uses the term "organic" to designate two components of its business plan. First, it identifies itself as a 100 percent local company dealing specifically with Puerto Ricans advertising to Puerto Ricans. Second, it promotes advertising in unexpected venues, taking advantage of underutilized surfaces like sailboat sails and popcorn bags. Similarly, *Organic Entertainment* is a New York independent music booking agency that provides a label for hire for independent musicians. Again, the company provides the resources of a big record label to local, independent musicians. It uses "organic" to stress its emphasis on local, not (yet) mainstream music.

Finally, *EcoMedia Design*, which also markets itself under the designation "Organic Media Design," is an advertising agency that specializes in LOHAS (Lifestyle of Health and Sustainability) advertising. The company proudly asserts that "green" doesn't have to be bland and boring and offers advertising campaigns that are fresh and organic. For this company, organic is more about a green lifestyle than a commitment to independent, bottom-up advertising or a focus on local clientele. However, *Organic Media Design* does emphasize a niche focus on green consumerism and making green products appealing to consumers. The company appears to have a commitment to promoting only sustainable practices and products, but also strays at times into greenwashing tactics that help companies with less environmentally sound practices and products market them as green.

What these organizations demonstrate with their self-identified use of the term "organic" is that the term remains a fairly ambiguous concept. For some, it stresses a commitment to local needs and concerns, offering an alternative to large, multinational corporations. For some, it stresses bottom-up marketing as opposed to top-down message construction, or at the very least more of a generative partnership between company and clientele and consumers. For some, "organic" is isomorphic with "environmentally sustainable," from products to practices to lifestyle.

The prevalence of the term "organic" beyond agricultural practices is telling. If federal agencies, companies, and consumers struggle over what exactly the term means, part of this struggle comes from the wide variety of ways the term circulates in public discourse. There is ample evidence that "organic" is an example of what has been referred to as an ideograph.

An ideograph carries a certain degree of ambiguity and ideological content—examples include terms such as "progress" and "justice." Such terms play key roles in public debates, often being taken up by different positions in a dispute and variously inflected by different rhetors. One system of ideographic analysis suggests identifying such terms and tracking their different meanings across groups and over time, using such analysis to identify the key ideological struggles the terms encapsulate.

When associated with media and public discourse, the term "organic" oscillates between associations with green consumerism, fad scams, sustainable lifestyles, emphasis on the local, and emphasis on smaller, noncorporate businesses. Organic also links to the perseverance of the agrarian myth in U.S. culture, a celebration of farm work and lifestyle without acknowledging the actualities of modern agricultural practices. Looking across representations and arguments, it is clear that organic is a concept that continues to develop and shift its meaning and significance in public discourse, with popular media providing a fertile growing material for such meanings.

Further Reading
Deliso, M. "Organics Here to Stay, But Category Could Use a Cleanup. *Advertising Age* 77(43) 2006: 4.
Killingsworth, M. J., and J. S. Palmer. "Liberal and Pragmatic Trends in the Discourse of Green Consumerism." In *The Symbolic Earth: Discourse and Our Creation of the Environment*, J. G. Cantrill and C. L. Oravec, eds., 219–40. Lexington: University Press of Kentucky, 1996.
Laufer, W. S. "Social Accountability and Corporate Greenwashing." *Journal of Business Ethics* 43 (2003): 253–61.
McChesney, R. W. *The Political Economy of Media: Enduring Issues, Emerging Dilemmas.* New York: Monthly Review, 2008.
McGee, M. C. "The 'Ideograph': A Link between Rhetoric and Ideology." *The Quarterly Journal of Speech* 66 (1980): 1–16.
McKibben, B. "Small World: Why One Town Stays Unplugged. *Harper's Magazine* (December, 2003): 46–54.
Retzinger, J. "Cultivating the Agrarian Myth in Hollywood Films." In *Enviropop: Studies in Environmental Rhetoric and Popular Culture*, M. Meister and P. M. Japp, eds., 45–62. Westport, CT: Praeger, 2002.

Jonathan M. Gray

Methyl Bromide

There is much controversy about the use of the chemical methyl bromide in agricultural production. It is prohibited in certified organic agriculture, but its widespread use had broad environmental impacts. Methyl bromide (chemical formula Ch_3Br), also known as Bromomethane, is a halogen compound. It is colorless, nonflammable, and odorless. It is naturally emitted into the atmosphere, mainly from oceans. On land, smaller quantities are emitted into the air in wildfires. Methyl bromide also has anthropogenic sources; it has been manufactured from bromine salts that occur naturally in sources like the Dead Sea, by reacting methanol with hydrobromic acid.

Since its registration in 1961, the primary use of methyl bromide in the United States has been as a fumigant, mainly for preplant soil treatment, but also for commodity quarantine and preshipment treatment. In the United States, it is used most often as a preplant fumigant for strawberry and tomato production in California and Florida. Methyl bromide played a key role in the development of the lucrative strawberry and tomato industries in the United States due to its ability to control *Verticillium* wilt and other soilborne diseases, nematodes, and weeds, which allows growers to plant multiple harvests on the same soil and thereby increase their yields.

Scientific evidence shows that both methyl bromide and its more famous "cousins," chlorofluorocarbons (CFCs) are photolyzed (i.e., molecules broken down by light) by the sun's ultraviolet radiation in the stratosphere, thus releasing chlorine and bromine that in turn destroy ozone molecules. These halogenated compounds are responsible for global ozone loss as well as for the "ozone hole" in the Arctic and Antarctic regions. Scientific evidence shows that the anthropogenic addition of methyl bromide to the atmosphere disrupts the delicate balance of stratospheric ozone. Ozone loss is tied to increased cases of skin cancer and eye cataracts and negative impacts on agricultural crops and sea life. Therefore, in 2005 methyl bromide was phased out by the U.S. Environmental Protection Agency (EPA) as mandated by the Clean Air Act, and in the entire industrialized world as agreed on by the Montreal Protocol on Substances that Deplete the Ozone Layer. Less-developed countries will complete a phaseout in 2015.

Methyl bromide is so effective as a pesticide that its use continues in the industrialized world through "critical use exemptions" to the Montreal Protocol phaseout. Although conventional alternatives to methyl bromide show comparable results, the United States has convinced countries that the complete reliance on alternatives would have a negative market impact on its strawberry growers. Therefore, the United States and several other industrialized countries (though to a lesser degree) have successfully put off a complete phaseout of this ozone-depleting substance.

California-based research shows that organic growing techniques (e.g., host resistance, organic amendments, high nitrogen organic fertilizers, fungicidal crop residues, and crop rotations) often can control soilborne diseases and pests within economic thresholds, thus making methyl bromide an unnecessary input. However, the expansion of organic practices may be limited in the present sociopolitical context because a large shift to organic production would likely result in a price drop that would not cover increased production costs resulting from reduced yields and greater labor requirements.

Further Reading

Gareau, Brian J. "Dangerous Holes in Global Environmental Governance: The Roles of Neoliberal Discourse, Science, and California Agriculture in the Montreal Protocol." Ph.D. dissertation, University of California, 2008. http://repositories .cdlib.org/cgirs/diss/CGIRS-DISS-1-2008.

Martin, F. N., and C. T. Bull. "Biological Approaches for Control of Root Pathogens of Strawberry." *Phytopathology* 2002: 1956–62.

World Meteorological Organization. "Scientific Assessment of Ozone Depletion: 2006, Global Ozone Research and Monitoring Project." Report No. 47. Geneva: WMO, 2007.

See Also: Agrichemicals

Brian J. Gareau

Milk, Organic

Organic milk is an excellent calcium source for humans and also provides high levels of conjugated linoleic acids, which is thought to fight cancer. This is a direct result of organic cows consuming fresh green grass and clover and lesser amounts of cereal grains than nonorganic cows. A mounting body of evidence indicates that organic milk is more nutritious than nonorganic milk. A study by the University of Liverpool found 68 percent more omega-3 essential fatty acids in organic milk. Several studies have reported higher levels of omega-3 fatty acids as well as beta carotene and vitamin E.

Organic milk is produced according to the USDA organic standard, the strictest organic food labeling standard in the world. Farms are inspected and audited annually by an independent third party to ensure compliance to the standards. Every truckload of milk is tested for antibiotics, bacteria, and temperature. Organic milk processing plants are inspected to verify that organic milk is never mixed with conventional milk, disapproved cleaning materials, or disapproved ingredients.

To qualify for organic designation and to have their milk sold as organic, dairy cows must live on pasture and crop land that has been managed as organic for at least three years. That means no applications of synthetic pesticides, herbicides, or fertilizers. Seeds planted in organic or transitioning crop land should be certified organic if at all possible. If the variety is not available in a certified organic form, an untreated seed may be used after a thorough search for organic seed. The use of genetically modified (GM) products is strictly prohibited. All inputs to the land must be on the national list of allowed substances for crop production and justified through soil tests.

Dairy animals must be managed organically for twelve months prior to organic certification in order for the milk to be sold as certified organic milk. The feed provided to cows must be 100 percent organic. Antibiotics and hormones are not allowed. Overuse of antibiotics in farm animals has been linked with increasing bacterial resistance in animals and humans. Since hormones are very powerful, minuscule amounts can result in dramatic changes. All inputs must be on the national list of allowed substances for livestock production. Health care treatments are primarily herbal or homeopathic. Mineral supplements are carefully reviewed for compliance to the standard, and veterinary treatments may only be used as indicated on the national list. Vaccinations are encouraged based on regional needs. Necessary medical treatment may not be withheld to maintain the organic status of the animal. If antibiotics or other disapproved substances are required, the animal must be treated and marketed as conventional.

Organic cooperatives further require that all dairy cows be fed primarily certified organic pasture during the grazing season. This ensures a higher quality of life for the cows, superior health by nutrition and outdoor exercise, and greater nutritive value in the food provided by the animal. Organic farming is good for the cows and for the environment. It provides a better environment for wildlife; keeps the air, soil, and water free of contaminants; and provides flavorful, nutritious food.

Further Reading

Ellis, K. A., G. Innocent, D. Grove-White, P. Cripps, W. G. McLean, C. V. Howard, and M. Mihm. "Comparing the Fatty Acid Composition of Organic and Conventional Milk." *Journal of Dairy Science, American Dairy Science Association* 89 (2006): 1938–50.

See Also: Organic Valley; rBGH

Wendy Fulwider

N

National Organic Standards

The National Organic Standards (NOS) are the complete set of rules that govern organic crop and livestock production and processing. The NOS determine what practices and substances are legally required and prohibited to obtain and maintain organic certification for a farm or processing facility. The standards also set out the parameters for labeling organic products and legal use of the organic label. Creation of the NOS was authorized by the 1990 Farm Bill, along with the National Organics Program (NOP) and the National Organic Standards Board (NOSB). In that legislation, the NOSB was designated as the body that would be charged with developing the NOS.

The modern organic movement began during the 1960s in the midst of a cultural revolution that had many people questioning traditional institutions, including how food was produced. Rachel Carson's 1962 book *Silent Spring* brought to many the realization that the chemicals being used to control weeds and insects over large areas of the landscape were having unintended negative consequences on nontarget species, calling into question the safety of agrichemicals for humans. This prompted a small number of U.S. farmers to seek nonchemical alternatives to managing pests and maintaining soil fertility, and to grow food using more natural methods.

During this time most organic production in the United States occurred on small farms, with the produce being sold locally.

As the demand for organic food grew, the number of organic farms increased. Soon, enterprising entrepreneurs were processing and selling organic products beyond local markets. The term "organic" was freely used and became synonymous with more natural, healthy, and chemical-free food, but there was no official standard by which consumers could be assured that "organic" was truly organic. Sometime in the early 1970s it became clear that a system of verification was needed to define the term "organic" and protect the legitimacy of the organic brand.

During the late 1970s and early 1980s, third-party certification emerged as a means to provide that legitimacy. By the late 1980s, there were numerous organizations that had developed rules for organic certification. There was little coordination between certifying organizations, so the rules and processes differed, leading to more confusion in the industry and for consumers wondering what the term "organic" really meant. In an unprecedented move, the organic industry approached the U.S. government and requested that the USDA draft one set of organic certification rules for all to follow.

The Organic Foods Production Act of 1990, part of the 1990 Farm Bill, authorized the establishment of the NOP, the NOSB, and the NOS for organic production, processing, and trade. The stated purpose was to "1) establish national standards governing the marketing of certain agricultural products as organically produced products; 2) assure consumers that organically produced products meet a consistent standard; and 3) to facilitate interstate commerce in fresh and processed food that is organically produced."

The NOSB is a fifteen-member board, appointed by the USDA secretary of agriculture, made up of farmers and growers, handlers and processors, a retailer, a scientist, consumer and public interest advocates, environmentalists, and certifying agents. The mission of the NOSB was to assist the secretary in developing the NOS. Though the NOS have been written, the NOSB remains a viable, working board, annually considering modifications to the NOS.

In 1995, the NOSB passed the official definition of organic. It is an important part of the NOS.

NOSB DEFINITION OF ORGANIC

Organic agriculture is an ecological production management system that promotes and enhances biodiversity, biological cycles and soil biological activity. It is based on minimal use of off-farm inputs and on management practices that restore, maintain and enhance ecological harmony.

"Organic" is a labeling term that denotes products produced under the authority of the Organic Foods Production Act. The principal guidelines for organic production are to use materials and practices that enhance the ecological balance of natural systems and that integrate the parts of the farming system into an ecological whole.

Organic agriculture practices cannot ensure that products are completely free of residues; however, methods are used to minimize pollution from air, soil and water.

Organic food handlers, processors and retailers adhere to standards that maintain the integrity of organic agricultural products. The primary goal of organic agriculture is to optimize the health and productivity of interdependent communities of soil life, plants, animals, and people.

In 2001, more than ten years after the Organic Foods Production Act was passed, the NOS were complete and ready to be implemented. Complete compliance was required starting October 21, 2002. For the first time, all organic certifiers had to be accredited by the USDA and all had to meet the same organic standards for USDA Organic Certification.

The complete set of the NOS can be found at the National Organic Program's (NOP) Web site, www.ams.usda.gov/nop/.

The NOS is arcane in its formatting. The complete documentation is hundreds of pages long, including the preambles and comments for each subpart, but only a small portion is directly applicable to producers and processors looking for rules on how to produce organic products. The actual rules pertinent to organic production, processing, labeling, certification, and accreditation are contained in a section called "Part 205" (§205). Part 205 is divided into "subparts" denoted by capital letters—A through G. Subparts are further divided into smaller increments of the number 205. For example, §205.206 under Subpart C contains the "crop pest, weed, and disease management practice standard." Subparts can be further subdivided by lowercase letters—(a) through (z), which can be further divided by numbers—(1), (2), (3), etc., which can be further divided by lowercase roman numerals—(i), (ii), (iii), (iv), etc. Though confusing at first, the formatting system makes it easier to find sections and text pertaining to particular topics. For example, §205.206 (c) (6) states that weed problems in organic systems may be controlled through "plastic or other synthetic mulches: *Provided*, that, they are removed from the field at the end of the growing or harvest season." Each subpart includes empty sections with number designations to allow future additions to the rules.

Part 205 is divided into seven subparts: Definitions (A); Applicability (B); Organic Production and Handling Requirements (C); Labels, Labeling, and Market Information (D); Certification (E); Accreditation of Certifying Agents (F); and Administration (G). What follows is a brief summary of, and highlights from, each section.

SUBPART A: DEFINITIONS

Subpart A is a glossary of words and terms used in the NOS. According to the preamble for Subpart A, the definitions generated many comments and suggestions. As a result, some definitions were changed and some terms were removed altogether from this section. Words have meaning, and meanings can change depending on the perceptions of those using the word. The term "compost" provides a good example. There are many recipes for compost, some that may include the addition of substances prohibited in organic production. The term "compost" is listed in Subpart A and defined as, "The product of a managed process through which microorganisms break down plant and animal materials into more available forms suitable for application to the soil..." The definition also includes specifics for how compost should be made, which mirror §205.203 (c) (2) (i–iii), in Subpart C, Organic Production and Handling Requirements, the section that establishes what is allowed in the production of compost that is to be used in organic systems. It includes requirements for initial C:N ratios, temperatures that must be maintained for specific lengths of time, and the number of times the "cooking" compost must be turned. The definition provides a clear idea for what is referred to as "compost" in the NOS. Growers and certifiers both understand what will be accepted and not accepted as compost in the context of organic production.

SUBPART B: APPLICABILITY

Subpart B (§205.100 through §205.106) outlines what has to be certified, along with legal exemptions and exclusions to certification. It also spells out the proper use of the term "organic" and recordkeeping requirements of certified operations. In addition, Subpart B lays down the basic requirements for allowed and prohibited substances, methods, and ingredients in organic production and handling operations. The last item demonstrates something that is common in the NOS. Basic principles are established in §205.105 for what can and cannot be used; other sections (§205.600-606) are referenced for the specifics on what those allowed and prohibited substances are. For instance, §205.105 (b) states, "To be sold or labeled as '100% organic,' 'organic,' or 'made with organic . . .,' the product must be produced and handled without the use of: . . . (b) Nonsynthetic substances prohibited in §205.602 or §205.604." The sections indicated give a list of natural substances that are not allowed for organic production or handling. The list includes substances such as ash from manure burning, arsenic, strychnine, and sodium nitrate.

SUBPART C: ORGANIC PRODUCTION
AND HANDLING REQUIREMENTS

Subpart C contains §205.200 through §205.299. It establishes the requirement of an organic production and handling system plan for each organic operation. The plan is submitted to the organic certifier along with the application for certification. Other regulations in this section cover land requirements, soil fertility, seeds and planting stock, crop rotation, pest management in the field, livestock origin, feed, health and living conditions, and matters of handling the crop after harvest.

Subpart C, §205.202 (b), establishes the three-year time period during which a farm must be free from the application of any prohibited substances before the harvest of a certified crop. This subpart also provides the standards for how and when to apply manure and what constitutes acceptable compost. The subpart contains much of the information that dictates organic farming practice and establishes what is allowed and prohibited on the farm.

SUBPART D: LABELS, LABELING, AND MARKET INFORMATION

Subpart D contains §205.300 through §205.311, and covers legal use of the term "organic" and the USDA seal. "Organic" is now a legal term with a legal definition. There are legal consequences for using the term in a way that is not in accordance with the law. The rules laid down in this section are very specific. The term "organic" may only be used on labels and in labeling of raw or processed agricultural products, including ingredients that have been produced and handled in accordance with the NOS. This applies to organic livestock feed products and products being shipped to and from countries outside the United States.

There are provisions in this section for processed foods containing less than 100 percent organic ingredients. The term "100% organic" can only be used to label processed food products the contents of which are all certified organic. To legally use the term "organic" alone, 95 percent of a product's ingredients must be certified organic. The label designation "made with organic" is for processed foods containing at least 70 percent organic ingredients. If a product contains less than 70 percent organic ingredients, the word "organic" cannot legally be used on the label. Subpart D also explains how to calculate the percentage of organically produced ingredients, and provides rules for packaging organic foods, the size and placement of the organic label on packaging, and other information that must appear on retail and nonretail packaging.

SUBPART E: CERTIFICATION

Certification is the process by which a farmer or processor gains and maintains the legal status of "organic" for their product and is, at the same time, granted the right to use the term in the labeling of their product. Subpart E (§205.400 through §205.406) establishes what a producer or processor seeking certification must do: (a) comply with all the applicable regulations in the NOS, (b) maintain a current production or handling plan, (c) permit annual onsite inspections, (d) maintain all applicable records for at least five years, (e) pay applicable fees to the certifying agency, and (f) immediately notify the certifying agent in the case of any contamination, commingling, or change in the operation.

This subpart also establishes the processes and responses required by the certifying agency, and how and when onsite inspections should take place. It also specifies how certification will be granted and under what circumstances certification will be denied. Certification is an annual process. Each individual crop must be certified separately, thus requiring annual application and onsite inspection.

SUBPART F: ACCREDITATION OF CERTIFYING AGENTS

Organic certifiers are now required to be accredited by the USDA. Subpart F (§205.500 through §205.510) establishes the requirements for certifiers seeking accreditation. It also establishes the process by which certifying agencies apply for and maintain accreditation. Accreditation lasts five years and certifiers are required to demonstrate sufficient expertise in organic production and handling, employ a sufficient number of qualified personnel, and prevent conflicts of interest.

SUBPART G: ADMINISTRATIVE

The last subpart (§205.600 through §205.690) is a lengthy section with many parts related to the administration of the organic program. One of the most important sections is the National List of Allowed and Prohibited Substances (§205.600 through §205.607). Interestingly, the list is divided into two main categories, a) synthetic substances that are allowed in crop and livestock production, and b) nonsynthetic ("natural") substances that are prohibited for use in organic crops and livestock production. There are separate lists for nonagriculturally (nonorganic) produced substances that are allowed as ingredients in processed organic products and nonorganically produced agricultural products that are allowed as ingredients in processed organic products. The lists change as new items are added. Section 205.607 contains provision for amending the lists.

Subpart G also contains provisions for the establishment of state organic programs. Such a program would allow a state to act as a certifying agency for organic farmers and processors in that state. Other items covered in Subpart G include regulations on fees charged for certification and accreditation (§205.640 through §205.642), procedures for dealing with noncompliance for organic farmers, certifying agencies and state organic programs, and mediation for those appealing noncompliance (§205. 660 through §205.668). Additional topics covered in this subpart deal with the authority of the USDA to test organic products for contamination of prohibited substances, the ability to exclude certain products for sale as organic, and policies dealing with emergency treatments for pests or disease (§205.672).

Further Reading

Heckman, Joseph. "A History of Organic Farming—Transitions from Sir Albert Howard's War in the Soil to the USDA National Organic Program." Weston A. Price Foundation. 2006. www.westonaprice.org/farming/history-organic-farming .html.

National Sustainable Agriculture Information Service. "Organic Farming." Updated February 2009. http://attra.ncat.org/organic.html.

USDA Agricultural Research Service. "National Organic Program." www.ars.usda .gov/nop/.

See Also: USDA National Organic Program; USDA National Organic Standards Board; APPENDIX 2: Organic Foods Production Act of 1990

Dan Anderson

Natural Food

Anything that is found in nature can be considered natural but it is debatable whether "natural" necessarily means better. "Natural food" has been defined as food that has undergone minimal processing and that contains no preservatives or artificial additives. Since the late 1980s, the Food and Drug Administration (FDA) has declined to formalize a legal definition of "natural" on more than one occasion "because of resource limitations and other agency priorities." Because of this, "natural food" has become one of the more controversial issues in the history of food. People tend to confuse "natural" with "organic," because many foods are labeled "natural-organic." Organic foods are sustainable, and produced without any chemical fertilizers, pesticides, hormones, or antibiotics.

Natural and organic foods are becoming very popular for their health and wellness benefits. In spite of some controversy, organic and natural foods are perceived by consumers to be healthier, safer, and more sustainable than conventionally grown foods. Natural food products are typically sweetened with honey, maple syrup, sucanat, or agave syrup, and sea salt is used instead of table salt. With the increasing demand of natural and organic products,

Natural food store in Marquette, Michigan. Courtesy of Zita Riesterer.

big retailers like Whole Foods Market are growing along with food co-ops and health-food stores.

Labels concerning natural foods have the potential to be misleading. For example, salt and water are both natural ingredients, but when a chicken is injected with saline solution, it is questionable whether that should that be considered an "all-natural ingredient." When buying meats, consumers are given no real information by the label "naturally raised" because it has no legal definition. It does not necessarily mean what many consumers assume—that the animals were grass-fed or free-range. A fruit juice labeled "all-natural ingredients" is not necessarily free of added chemicals. The label could mean that the added colors or flavors came from "natural" sources. A specific example would be the added color in pink lemonade ("all natural"), which comes from beets. Many processed foods, even those containing growth hormone, claim to be "all natural." The soft drink 7UP is labeled "all natural" but contains high-fructose corn syrup. Ironically, although table sugar is made from sugar cane, it is highly processed and refined and thus not considered natural.

The United States Department of Agriculture (USDA) has weakened the definition of "organic" to essentially mean no pesticides, hormones, or genetically modified organisms. Sustainability, ethics, and source of food are not addressed in the USDA definition. Companies producing items such as organic cereals or organic yogurts can replace chemicals and pesticides when producing them, but then pollute the environment while shipping the products nationwide. Shopping at local farmers markets or a true family-run organic farm within a hundred miles from home are really the only way to ensure purchasing locally grown, "natural organic" foods. Since the term "natural foods" is somewhat vague, by choosing "natural organic" foods, consumers can have foods that are grown within the most healthful environment possible.

Further Reading

Fortin, Neal D. "Naturally Confused." Food Product Design. February 12, 2008. www.foodproductdesign.com/hotnews/naturally-confused.html.

Morford, Mark. "The Sad Death of 'Organic': How Weird and Depressing Is It Now That Kellogg's and Wal-Mart Are Hawking 'Natural' Foods?" *SFGate, San Francisco Chronicle*, October 13, 2006. http://sfgate.com/cgi-bin/article.cgi?file=/gate/archive/2006/10/13/notes101306.DTL.

See Also: Trader Joe's; Walmart; Whole Foods

Sharon Peterson

Neem

The Neem tree (*Azadirachta indica*, in the family Meliaceae) is native to the Indian subcontinent, but is grown in many dry, tropical, and subtropical regions of the world, including Africa and parts of Asia, Australia, and the Americas. It is a fast-growing evergreen with alternate pinnate leaves and produces drupes, a fruit similar to olives, with a fleshy covering over a hardened seed-containing endocarp. In its native India, all parts of the neem tree are widely used for medicinal properties, as well as an ingredient in soaps and cosmetics. In horticultural applications, neem oil, which is extracted from the fruits and seeds, is used as a biopesticide and fungicide and is approved for use in organic systems.

Neem oil is often used in conjunction with a surfactant, such as in insecticidal soaps, for improved efficacy. Several volatile compounds have been identified in neem seeds, mostly derivatives of di-n-propyl- and n-propyl-1-propenyl di-, tri-, and tetrasulfides which are believed to be responsible for some of neem's repellent and insecticidal properties. Neem oil or neem-containing insecticidal soaps have been recommended for use against aphids, thrips, mites, white fly, Colorado potato beetle, downey mildew, and powdery mildew, for example. There is concern that neem may affect some beneficial insects, such as honey bees and lady bugs, and judicious use of neem products along with complementary control practices is encouraged. Neem cake is made from the byproducts of neem oil production, and can be mixed into soil as an additional fertilizer source with purported plant growth promoting, nemacidal, and antifungal properties. The Organic Materials Review Institute (OMRI) provides a list of neem products that are approved for use in certified organic systems (www.omri.org).

Further Reading

Balandrin, M. F., S. M. Lee, and J. A. Klocke. "Biologically Active Volatile Organosulfur Compounds from Seeds of the Neem Tree, *Azadirachta indica* (Meliaceae)." *Journal of Agricultural and Food Chemistry* 36 (1988): 1048–54.

See Also: Pesticides

Mark A. Williams and Audrey Law

Newman's Own

The late Paul Newman, a U.S. actor and celebrity, whose face graces the label of all the Newman's Own products, and his partner, A. E. "Hotch" Hotchner, started Newman's Own company in 1982 as a charitable organization. The company started with the idea of making all-natural salad dressings. After Newman was persuaded to put his caricature on the label, he demanded the proceeds go to charity. After all, to make money on a face would be tacky, Newman was heard to say. Although Newman's Own is a for-profit company, all after-tax profits go to charities. The Newman's Own Foundation has given over $250 million to charities and grants within the United States and abroad.

Beyond philanthropic programs and salad dressing, Newman's Own brand is also associated with a certain level of quality. This may have something to do with the uniqueness of the ingredients incorporated into the products: sesame ginger, balsamic, pineapple salsa, etc, or it may be because of the price point.

One of the specific charities Newman's Own sponsors is Hole in the Wall Camp. Hole in the Wall Camp is a nonprofit organization that offers children with serious illnesses and life-threatening conditions a fun and free camp experience. The camp was founded in 1988, and since the beginning, Hole in the Wall Camp has provided children from all over the world a summer camp experience missing from their burdened childhood. The camp is staffed with 24-hour trained medical staff and volunteers to provide children a safe and fun environment to achieve things they never thought possible.

In addition to their traditional brand of salad dressings, Newman's Own has an organic line of products. The organic product line started with pretzels and includes other organic products, such as chocolate, cookies, pet food, and coffee. Newman's Own Organics was started in 1993 by his daughter Nell Newman and became a separate company in 2001. All of the Newman's Own Organics products are certified organic by Oregon Tilth. Newman's Own Foundation has given over $250 million to thousands of charities worldwide since 1982 and is a strong supporter of Hole in the Wall Camps, which bring seriously ill children to the outdoors.

Further Reading

Gergen, Christopher, and Gregg Vanourek, *Life Entrepreneurs: Ordinary People Creating Extraordinary Lives.* San Francisco: Jossey-Bass, 2008.
Newman's Own. www.newmansown.com.

See Also: Brands

Sylvia Smith

O

Organic Consumers Association

The Organic Consumers Association (OCA) is a nonprofit 501 (c) (3) organization founded in 1998 amid intense public debate over the drafting of U.S. organic agriculture standards. The organization originated out of the "Pure Food" campaign, a product of activist Jeremy Rifkin's Foundation on Economic Trends. Ronnie Cummins, a longtime political, environmental, and social justice activist remains its National Director.

A twenty-member Policy and Advisory Board made up of farmers, activists, and representatives from a range of public service organizations helps to guide decisions about the OCA's issues and focus. This body makes recommendations to a smaller Board of Directors, which then organizes their implementation and coordinates OCA's actions with its eleven staff members throughout the country.

OCA claims over 850,000 members, subscribers, and volunteers, including thousands of natural and organic foods businesses. Its stated mission is "the promotion of food safety, organic farming, and sustainable agricultural practices in the U.S. and internationally."

OCA is both an educational and activist organization which provides information for individuals, media, and public interest activists through its popular Web site, and its well-developed worldwide communication network. It also organizes letter-writing campaigns, boycotts, and protests, and participates in lobbying and litigation efforts. OCA publishes two newsletters: the quarterly *Organic View* and the weekly electronic newsletter *Organic Bytes*.

OCA was instrumental in defining the content of the U.S. organic standards that now guide the United States Department of Agriculture's (USDA's) National Organic Program (NOP). Specifically, it played a crucial role in generating over 275,000 comments to the USDA in opposition to the inclusion of irradiation, genetically modified organisms (GMOs), and sewage sludge in the "First Draft Standard" issued by the Agency in 1997. OCA continues to promote the integrity of organic standards through its "Save our Standards" (SOS) campaign.

Recent actions include a boycott of milk products from companies that purchase from organic dairy farms that fail to provide adequate access to pasture, which is required under NOP rules. Within the past year, OCA has also organized successful protests against a major coffee chain over the inclusion of genetically engineered bovine growth hormone (rBGH) in its milk products. It has also led a campaign against poor working conditions in meat processing facilities that sell their products through the largest "natural foods" retailer in the United States.

OCA's overall political program is the Organic Agenda 2005-15, a six-point platform calling for the conversion of U.S. agriculture to at least 30 percent organic by 2015; the promotion of fair trade and economic justice; a global moratorium on GMOs in agriculture and food; a "phaseout of the most dangerous industrial agriculture and factory farming practices"; universal health care emphasizing prevention, nutrition, and wellness; and the conversion of agriculture, transportation, and utilities to responsible, renewable, and conservation-based energy practices.

Further Reading
Organic Consumers Association. www.organicconsumers.org.

See Also: Consumers

Taylor Reid

Organic Farming Research Foundation

The Organic Farming Research Foundation (OFRF) was founded by organic producers in 1990 as a spin-off from California Certified Organic Farmers, "to foster the improvement and widespread adoption of organic farming practices." The organization has pursued this mission by conducting a competitive grants program for organic agricultural research, disseminating research results, providing education and advocacy in the public

policy arena, producing research and analysis products, and cultivating collaborative networks of producers and scientists. The organization's board
always has a constitutional majority of certified organic producers.

Since the first grants were made in 1992, OFRF has awarded more than
$2 million for over 240 projects. The goal of OFRF's grant program is to
generate practical, science-based knowledge to support and improve modern organic farming systems. OFRF-funded projects emphasize collaboration between farmers and researchers, so studies conducted on-farm or in
certified organic settings are important. In addition, the outreach and dissemination of project results are key. In 2005, OFRF evaluated the impact
of its grant-making program on farmers, research institutions, and the base
of knowledge about organic production systems. The results are compiled
in the online publication "Investing in Organic Knowledge."

Every OFRF-funded project is required to have an outreach and education component that disseminates the results to the grower and research
communities. This often includes field days, farm tours, grower conference
presentations and publication in grower newsletters. The results of funded
projects are published in the free OFRF newsletter, the *Information Bulletin*.

Since 1995, OFRF has maintained an active public policy program, to
educate the public and policymakers about organic farming issues and to
increase public institutional support for organic farming research and education funding. Its main focus has been the establishment and growth of a
USDA competitive grants program for organic research and education. The
program's objectives also include encouraging land grant universities'
organic research and extension programs developed with the participation
of organic farmers; cultivating state and federal policies that help to ensure
the economic viability of organic family farmers; and supporting organic
farmers' rights to grow and sell their products without the threat of pesticide and genetically modified organism (GMO) contamination.

OFRF encourages organic farmers to participate in the policy process
by subscribing to its Organic Farmers Action Network (OFAN). OFAN
subscribers receive free policy updates and tools for communicating with
representatives in Congress.

OFRF original research products about organic farming in the U.S.
include a series of *National Organic Farmers' Surveys* collecting information
about organic farmers' research and information needs, their experiences in
the organic marketplace, effects of GMOs on organic production and markets,
organic farmer demographics, and much more. There is also an ongoing
inventory of programs related to organic farming at the land grant universities. The online report "State of the States: Organic Farming Systems
Research at Land Grant Institutions" is a valuable resource for understanding where organic farming research takes place.

In 1995 and 1996, OFRF conducted a study to identify and catalogue
federally supported agricultural research pertaining specifically to the
understanding and improvement of organic farming. This led to the 1997
publication of *Searching for the 'O-Word': Analyzing the USDA Current*

Research Information System for Pertinence to Organic Farming, OFRF's seminal work identifying the need for federal support for organic programs.

NETWORKING

OFRF is coordinating efforts to develop a national research agenda and a farmer-scientist network for pursuit of multidisciplinary research and extension on working organic farms. The Scientific Congress on Organic Agricultural Research (SCOAR) has conducted a series of national and regional meetings with farmers and scientists to discuss and design a plan for basic, applied, and developmental organic research.

OFRF also takes part in other networking in support of organic and sustainable farming. Examples include the Sustainable Agriculture Working Group and the Organic Agriculture Consortium.

Further Reading
Organic Farming Research Foundation. www.ofrf.org.

See Also: Research

Mark Lipson

Organic Foods, Benefits of

Organic foods come from organic farms, and those farms use methods and materials designed not only to grow healthy crops and livestock, but also to care for the air, water, and soil on the farm and beyond. To best understand the benefits of organic foods and food production, it is important to understand what sets organic farms apart.

Organic farming—and therefore organic food—helps limit exposure to toxic chemicals from synthetic pesticides and fertilizers that can end up in the ground, air, water, and food supply, and are associated with health consequences, from asthma to cancer.

In the United States alone, more than one billion pounds of pesticides are released into the environment each year. As a result, consumers are exposed to them daily in the food they eat, the water they drink, the air and dust they breathe, as well as on surfaces inside their homes, at work, and in public places. Extensive pesticide residue testing by the U.S. Department of Agriculture has found that nonorganic fruits and vegetables are three to more than four times more likely on average to contain residues than organic produce, eight to eleven times more likely to contain multiple pesticide residues, and contain residues at levels three to ten times higher than corresponding residues in organic samples.

Organic agriculture minimizes children's exposure to toxic and persistent pesticides in the soil in which they play, the air they breathe, the water they drink, and the foods they eat. Research at the University of Washington showed that children eating primarily organic diets had significantly lower levels of organophosphorus (OP) pesticide metabolite concentrations than did children eating conventional diets. In fact, concentrations of dimethyl metabolites, one OP metabolite group, were approximately six times higher for the children eating conventional diets. Other studies indicate that chronic low-level exposure to OP pesticide may affect neurological functioning, neurodevelopment, and growth in children. Choosing organic products, therefore is one way to reduce dietary exposure to pesticides, and can bring exposure for children down to a range of negligible risk.

Growing crops in healthy soils results in food products that offer healthy nutrients. There is increasing evidence that organically grown fruits, vegetables, and grains offer more of some nutrients, including vitamin C, iron, magnesium, and phosphorous, and less exposure to nitrate and pesticide residues than their counterparts that are grown using synthetic pesticides and fertilizers. For example, research results on select crops indicate that organic oranges may have more vitamin C, organic apples may have more fiber, and organic pears, peaches, and oranges may have more antioxidants.

In addition, there have been some promising results concerning the nutritional qualities of organic meat and milk. A European research team led by Swiss scientist Lukas Rist has found that mothers consuming mostly organic milk and meat products have about 50 percent higher levels of rumenic acid, a conjugated linoleic acid, in their breast milk.

A three-year study in the United Kingdom, sponsored by the Organic Milk Suppliers' Co-operative, found organic milk contained 68 percent more omega-3 fatty acids, on average, than conventional milk. Findings from the study, conducted independently by the Universities of Liverpool and Glasgow during 2002 and 2005 centering on a cross-section of UK farms over a twelve-month production cycle, have been published in the *Journal of Dairy Science*.

Findings released in 2003 from studies at the Danish Institute of Agricultural Sciences suggest that organic milk is higher in conjugated linoleic acids (CLA) than conventional milk. The studies also showed that feeding clover to dairy cows led to increased levels of CLA in the milk. Previous research at the University of Wisconsin–Madison found that dairy cattle that graze produce higher amounts of CLA in their milk than those eating grain, hay, and silage. In animal studies, CLA has been linked to preventing cancer in rats and atherosclerosis in rabbits.

At the 2005 international congress Organic Farming, Food Quality, and Human Health, Professor Carlo Leifert of Newcastle University reported findings that organically produced food had higher level of specific antioxidants and lower mycotoxin levels than conventional samples, and that grass-based organic cattle diets reduce the risk of *E. coli* contamination whereas grain-based conventional diets increase the risk.

Because organic agriculture respects the balance of microorganisms in the soil, organic producers use composted manure and other natural materials, as well as crop rotation, to help improve soil fertility, rather than synthetic fertilizers that can result in an overabundance of nitrogen and phosphorous in the ground. As a result, organic practices protect groundwater supplies and avoid runoff of chemicals that can cause "dead zones" in larger bodies of water. Excess nitrogen in drinking water has been linked to certain forms of cancer.

Toxic chemicals are contaminating groundwater on every inhabited continent, endangering the world's valuable supplies of freshwater. Several water utilities in Germany now pay farmers to switch to organic operations because moving farmers to organic practices costs less than removing farm chemicals from water supplies.

Organic farm practices recognize and respect the powerful nature of antibiotics. As a result, organic practices in the United States and for products sold in the United States prohibit the use of hormones, antibiotics, or other animal drugs in animal feed for the purpose of stimulating the growth or production of livestock. If an antibiotic is necessary to restore an animal to health, that animal cannot be used for organic production or be sold, labeled, or represented as organic. Thus, organic practices avoid the abuse of antibiotics that could have profound consequences for treatment of diseases in humans.

Nearly 40 percent of the world's agricultural land is seriously degraded, undermining both present and future production capacity, according to scientists at the International Food Policy Research Institute (IFPRI). Land degradation can have significant on- and off-site effects on income and environmental quality, and can take a number of forms, including soil nutrient depletion, agrichemical pollution, and soil erosion. With organic farming's focus on building healthy soil, choosing organic foods can help safeguard food-production resources for the future. Organic farming practices have also been shown to sequester carbon in the soil and may therefore serve an important role in mitigating climate change.

Because U.S. national organic standards and industry practices do not allow the use of genetic engineering in the production and processing of organic products, organic agriculture gives consumers who wish to avoid genetically modified foods a choice in the marketplace. Organic foods give consumers who wish to avoid products produced from cloned animals a choice since cloning is not allowed in organic livestock production. In addition, use of irradiation and sewage sludge as fertilizer are prohibited in organic production.

Choosing organic products offers many people peace of mind that they are making choices that support their health and the health of the planet. Organic products are traceable back to the farm where they were grown. In the United States, and in many other countries, organic production and labeling is regulated by law. Organic products sold in the United States can be traced back to the farms where they were produced, and those farms must meet or exceed the national organic standards. All of these features of organic production give shoppers confidence about organic foods.

Further Reading

Benbrook, Charles. "Minimizing Pesticide Dietary Exposure Through the Consumption of Organic Food." *Organic Center State of Science Review*, May 2004. www.organic-center.org.

Lu, Chensheng, Kathryn Toepel, Rene Irish, Richard A. Fenske, Dana B. Barr, and Roberto Bravo. "Organic Diets Significantly Lower Children's Dietary Exposure to Organophosphorus Pesticides." *Environmental Health Perspectives* 114(2) February 2006: http://dx.doi.org (doi: 10,1289/ehp.8418).

Organic Trade Association. "Benefits of Organic." www.ota.com/organic.html.

Quality Low Input Food. "QLIF Congress 2005." www.qlif.org/qlifnews/april05/con2.html#Anchor-Produc-58519.

Rist, Lukas, and A. Mueller, C. Barthel, B. Snijders, M. Jansen, A. P. Simoes-Wust, M. Huber, I. Kummeling, U. von Mandach, H. Steinhart, and C. Thijs. "Influence of Organic Diet on the Amount of Conjugated Linoleic Acids in Breast Milk of Lactating Women in the Netherlands." *British Journal of Nutrition* 97(4) 2007: 735–43.

Sampat, Payal. "Deep Trouble: The Hidden Threat of Groundwater Pollution." *Worldwatch Paper* 154, December 2000.

Scherr, Sara J., and Satya Yadav. "Land Degradation in the Developing World: Issues and Policy Options for 2020." In *The Unfinished Agenda: Perspectives on Overcoming Hunger, Poverty and Environmental Degradation*, International Food Policy Research Institute, 2001.

Weyer, Peter. "Nitrate in Drinking Water." "Agricultural Safety and Health Network." www.agsafetyandhealthnet.org/Nitrate.PDF.

Wiebe, Keith. "Resources, Technology, and Public and Private Choices." In *Who Will Be Fed in the 21st Century? Challenges for Science and Policy*, International Food Policy Research Institute, 2001.

See Also: Health Concerns; Diet, Children's; Antioxidants

Holly Givens

Organic Labels

The purpose of the U.S. Department of Agriculture (USDA) organic label is to provide consumers with reliable and consistent information about how products are grown and processed. The organic label may only be used on products that have been produced and processed in accordance with the standards of the National Organic Program (NOP). In general, for food to be labeled organic, it must be grown and processed without the use of most synthetic chemicals, irradiation, genetically-modified seeds, and sewage sludge, and be produced in systems that rely on crop rotation to break pest cycles and replenish soil fertility. Livestock must be fed organic feed, and the use of synthetic hormones and routine administration of antibiotics is prohibited.

The NOP provides three types of organic labels to help consumers differentiate between products that are fully organic and those that contain some organic ingredients. The labeling options include: "100% organic," "organic," and "made with organic ingredients." In all cases, the name and contact information for the certifying agent responsible for determining that the product qualifies for the organic label must be included on the package. Records must be kept by farmers, processors, distributors, retailers, and certifying agents to ensure traceability and verification of label claims.

100% ORGANIC

The "100% organic" label applies to single ingredient products, such as raw fruits and vegetables, that have been produced under an organic system. The label also applies to multi-ingredient products in which each individual ingredient, as well as the food additives and other processing aids excluding water and salt, have been produced in accordance with the NOP. For such products, the "100% organic" label may be placed on any part of the product package. Further, the processor may choose to include the USDA seal.

ORGANIC

The term "organic" applies to products that contain between 95 and 100 percent organic ingredients. The percentage of organic ingredients in a multi-ingredient product is determined by weight, excluding water and salt. In such products, the organic ingredients must be produced and processed according to the NOP. Up to 5 percent of nonorganic ingredients in such products must be either nonagricultural products such as flavoring, coloring, and food additives or agricultural products that are not commercially available in an organic form. Even the nonorganic ingredients, however, must be produced without use of genetic modification, irradiation, and sewage sludge. The organic label may be placed in any location on the product. The USDA seal is allowed in any location on the package. Because products in this category can include nonorganic ingredients, the ingredient panel must indicate those ingredients that are organically produced to provide full disclosure to consumers.

MADE WITH ORGANIC INGREDIENTS

The "made with organic ingredients" label is for multi-ingredient products in which 70 to 95 percent of the ingredients are organic. The percentage of organic ingredients in the product is determined by weight, excluding water and salt. The organic ingredients in such products must be produced and processed in accordance with the NOP. As with the organic label, all nonorganic ingredients must be produced without use of genetic modification, irradiation, and sewage sludge. However, the nonorganic ingredients may be produced using synthetic substances (insecticides, herbicides, and fungicides) and other practices that are excluded from organic production. Products with 70 to 95 percent organic ingredients may list up to three organically produced individual organic ingredients or whole food groups on the front product labeling. Examples of food groups include fruits, vegetables, grains, dairy products, beans, poultry, meats, oils, nuts, seeds, herbs, beans, fish, spices, and sweeteners. For example, a label may state "made with organic bananas, strawberries and mango" or "made with organic fruit." If the label states that it is made with an organic food group, all ingredients in that group must be organic. Because products in this category contain less organic content, use of the USDA organic seal is prohibited. Additionally, organic ingredients must be individually identified in the ingredient panel.

LESS THAN 70 PERCENT ORGANIC INGREDIENTS

Products with less than 70 percent organic ingredients may not be advertised as organic. The use of the USDA seal or the logo of any certifying agent is strictly prohibited. If a product does contain some organic ingredients, the producer may identify them as such on the ingredient list. To avoid misrepresentation, no other use of the word "organic" is allowed on such products. For such multi-ingredient products, the items listed as organic must still be produced and handled in full compliance with the NOP. The nonorganic ingredients in such products are not held to any additional standards; they may be produced using genetic engineering, sewage sludge, irradiation, synthetic substances, and any other production or processing practices that are not otherwise in violation of the law.

In the case of products with any of the aforementioned organic labels, an item cannot be indicated as organic if both the organic and nonorganic forms of the ingredient are used in the product. Additionally, a product cannot claim to use the organic form of an ingredient "when available" if the nonorganic form is used in the product.

Further Reading

Agricultural Marketing Service. "National Organic Program." U.S. Department of Agriculture. www.ams.usda.gov/nop/.

See Also: Certification, Organic; USDA National Organic Program; USDA National Organic Standards Board; APPENDIX 2: Organic Foods Production Act of 1990

Kathleen Merrigan

Organic Trade Association

The Organic Trade Association (OTA) has approximately 1,700 members and represents the organic business community in North America. Since its establishment in 1985 (originally called the Organic Foods Production Association of North America), the OTA has shaped both the regulatory and market environment for organic farming and products in North America and around the world.

With professional staff in both the United States and Canada, OTA represents businesses across the organic supply chain, including farmers, processors, distributors, and retailers. OTA's members cover all types of organic products, including foods, beverages, textiles, personal care products, pet foods, and more. OTA's mission is to promote and protect the growth of organic trade to benefit the environment, farmers, the public, and the economy. OTA envisions organic products becoming a significant part of everyday life, enhancing people's lives and the environment.

OTA advocates for and protects organic standards so consumers can have confidence in certified organic production. With input from its membership, OTA develops organic standards for emerging product areas. OTA monitors the work of government agencies, takes positions on legislation that affects organic agriculture and products, and represents the industry to regulators, elected officials, and international bodies. OTA is an advocate for organic production in the U.S. Farm Bill, and is a leader in developing Canada's new mandatory national regulations, to be implemented in June 2009.

OTA founded the All Things Organic Conference and Trade Show in 2001. All Things Organic is the largest business-to-business trade show and conference in North America, focusing exclusively on organic products and organic trade issues. OTA also connects buyers and sellers of organic products and services, from farm to retail, through its fully searchable membership directory, the Organic Pages Online. OTA also publishes the Organic Export Directory Online in six languages for people around the world interested in purchasing U.S. organic products.

OTA works on many fronts to support the transition to organic farming, processing, and handling. OTA's HowToGoOrganic.com Web site is a clearinghouse of resources for farmers and businesses interested in becoming organic or creating new organic businesses.

OTA is a primary source for fact-based information about organic products and processes throughout North America. OTA shares the benefits of organic with the public and helps expand markets for organic products. It does this through press releases and events; a media newsletter, "What's News in Organic"; and a consumer Web site, the O'Mama Report. OTA also directly promotes organic products for retail via its cooperative marketing programs, including Go Organic! For Earth Day and work with *Taste for Life* magazine. OTA also promotes September as Organic Harvest Month.

Further Reading
Organic Trade Association. www.ota.com.
All Things Organic Conference and Trade Show. www.organicexpo.com.

Holly Givens

Organic Valley

Cooperative Regions of Organic Producers Pools (CROPP) Cooperative is a unique cooperative that is owned by organic farmers. Their well-known brand of organic dairy products is called Organic Valley. Starting in Wisconsin in 1988 as a regional food initiative, CROPP Cooperative has evolved into a national cooperative with regional pools of organic farmers/owners. The original cooperative was started with the concept of producing organic produce in the hilly part of Wisconsin near the Mississippi River and delivering that food to the region defined by Minnesota's Twin Cities and the Chicago area. The founders of the cooperative saw organic foods as an opportunity to redesign the broken conventional model of farm sustainability to ensure a viable living for family-sized farms. In the first year, the cooperative diversified when seven dairy farmers began to develop standards for organic milk production and CROPP Cooperative began producing organic cheese to supply a natural food distributor.

A critical foundation for CROPP Cooperative is to partner with other stakeholders when possible for operation functions such as milk hauling or manufacturing and to keep the cooperative focus on serving organic farmers and the organic marketplace. Economic sustainability of organic farming and stable prices are foundation principles that are used in all pricing decisions of the cooperative. Diversity of people is another key component of the success of the early years that produced today's dynamic team of employees and farmers serving a common vision. The founding mission

Environmentally friendly headquarters of Organic Valley Cooperative in La Farge, Wisconsin. (AP Photo/ Tara Walters)

statement and goals combined with the definition of organic have been guiding lights throughout CROPP Cooperative's development. For the founders of the cooperative, the term "organic" has always had a broad meaning that goes beyond farm practices to how to do business with the many stakeholders, how to use sustainability practices, how to treat employees, and how to think about what organic business means.

After several years, it was decided that it was necessary to be in the branded business to better ensure reaching the core goal of economic sustainability for farmers. The Organic Valley brand name was coined by an employee while bouncing along the hills of Wisconsin in CROPP Cooperative's first milk truck. The tagline of "Family of Farms" and the signature red barn were added to Organic Valley to become the foundation of all brand use.

Fifteen years later when it was finally legal to sell organic meats, the cooperative developed the Organic Prairie brand name, which also uses the "Family of Farms" tagline. CROPP Cooperative is intent that these brands represent the highest integrity and quality possible in an organic brand. Today, Organic Valley is the one of the top brands in Natural Food outlets and Organic Prairie is a well respected organic meat brand.

From humble beginnings, CROPP Cooperative has succeeded beyond the hopes of the original founders to become a cooperative with over 1,300 members in thirty-two states producing organic milk, produce, eggs, pork, beef, turkeys, broilers, juice, and soy. CROPP Cooperative is a leader in the organic movement, member farmers are pioneers in innovative organic farming practices, and employees have cocreated a productive, flexible work environment.

The cooperative has a profit goal of around 2.25 percent, of which the last 1 percent goes to employees, farmers, and the community. Employees and farmers each receive 45% of this profit share, and the remaining 10% goes to community giving. With projected sales for 2009 at over $600 million the current challenge is how to build an employee and farmer culture that can maintain the cooperative long-term viability and dedication to the mission. Such questions as "how to be big and act small" or "how to be a national business that acts regionally" or "how to inspire employees to embrace the mission" or "how to be big and maintain farmer trust" or "how to be successful big-name brands and maintain the real integrity that built the brands" are just a few of the challenges that are being discussed by both farmers and employees.

CROPP Cooperative set out to start a business with a foundation of economic and environmental sustainability. The founders did not care how unique their vision was or how many said it could not be done. Because they had clear values and mission, they were willing to fail rather than compromise those core values. They were unique in that although they were heavily mission-oriented, they realized that to succeed they could not rely on a sense of "rightness" but had to succeed first as a business so that they could fulfill their mission. As a farmer-owned business, educating farmers about the marketplace realities has been critical so they understand the evolving production standards that the cooperative has adapted and better understand the costs and forces in the marketplace that affect potential farmer pay price.

CROPP Cooperative is a unique cooperative that represents the largest block of organic farmers in the world. It is a "social experiment disguised as a business."

See Also: Livestock Production, Organic; Milk, Organic

George Siemon

P

Pest Control, Biological

Organic farming relies on natural methods of pest control, rather than the use of synthetic chemicals. The search for predators, parasitoids, and pathogens to control pest arthropods (insects) or weeds is a technique known as classical biological control. There have been numerous successes starting dramatically with the control of the invaded cottony-cushion scale and the Klamath weed where suppression has remained at or above 95 percent for over a hundred years, and without requiring any other control measures. In the numerous cases in which the success has been less dramatic, it has been necessary to integrate other control measures. Here considerable scientific knowledge is required, which all too often is unavailable. In this case, integrated pest management is another recourse. When agricultural scientists working out of colleges and universities are directly and continuously involved, there is a greater likelihood of success.

The addition of new biotic mortality factors to the pest's ecosystem is carefully scrutinized by regulatory agencies, which strive to eliminate the establishment of potentially harmful organisms. Researchers continuously seek more effective guidelines for judging a natural enemy's capabilities before importation in order to accelerate control success rates and to reduce project costs. The manner by which control is achieved varies considerably among projects and the various countries that use the technique; there is a continuing debate on proper procedures for selection of natural enemies and regulation of their importation.

The primary goals of federal, state, or university importation programs are the same: the collection, safe transport, and quarantine processing, leading ultimately to field colonization of candidate biological agents. However, there are differences in the methods that are used by each entity. The U. S. Department of Agriculture Animal and Plant Health Inspection Service (APHIS) either on its own initiative or in concurrence with established dicta (as from the Environmental Protection Agency) issues regulations regarding the importation and quarantine handling of biological agents that the USDA Agricultural Research Service (ARS), individual states, and universities are required to follow.

Of mutual concern to the explorer and government regulatory agencies and quarantine personnel are the identification of target species and their hosts, acquisition of permits to import the material collected, packaging and labeling, method of shipment, clearance at the port of entry by customs and agricultural inspectors, and the quarantine facility itself.

Most of the technical and biological considerations relative to acquiring and shipping biological agents remain much the same as those described long ago for entomophagous arthropods or weed feeders by Bartlett and van den Bosch (1964), and more recently by Boldt and Drea (1980), Coulson and Soper (1988), Klingman and Coulson (1982); and for phytophagous (weed feeders) organisms by Schroeder and Goeden (1986).

State departments of agriculture or universities usually send out members of their staff as explorer/collectors, who do not always have access to laboratory facilities while in the field. Therefore, shipments sent to their quarantine laboratories may contain more than one targeted pest species and more than one natural enemy of each of these. They must then be segregated in quarantine and studied through one generation (for newly introduced species) before they can be released. Unsolicited extraneous material accidentally included may require further study in quarantine. If so, specific arrangements must be made with APHIS Plant Protection and Quarantine (PPQ) regarding the handling of such material. USDA collectors abroad can use all available U.S. governmental facilities (embassies, agricultural attachés, commissary, vehicles, communication facilities, etc.) to expedite their missions. Thus far, U.S. state and university collectors abroad have only infrequently been able to avail themselves of similar federal cooperation even though their missions were financed by public funds and their efforts could accrue to the benefit of agricultural crop production on a regional if not national scale in the United States.

International geopolitical and socioeconomic unrest may greatly affect the success or failure of foreign exploration missions. Colleagues in such areas, or intermediary organizations that charge a fee for service, such as the Commonwealth Institute for Biological Control, Silwood Park, United Kingdom, may be able to supply the desired beneficial organisms. Nevertheless, experience has shown that when explorers who are directly responsible for a project's success or failure physically participate in the collecting process, there is a greater chance of making successful introductions.

The purpose for exploration is to search for, import, and colonize natural enemies of pests from areas where the pest is indigenous, or at least present in low numbers because its natural enemies keep it under control. The need for exploration is to protect the environment from needless or questionable use of chemical pesticides, especially those with long half-lives or broad-spectrum toxicity that can adversely affect nontarget species and beneficial organisms and ultimately the food chain within a wide range of biologically diverse species.

The basic goal is to import species or strains that are preadapted to areas targeted for colonization. Large founder numbers are desirable to keep the gene pool as large as possible. Although traditionally used for homopterous pests of perennial crops, classical biological control has been increasingly considered for nonhomopterous and annual pests in agricultural, urban, and glasshouse environments. Extra agricultural uses in medical, forest, and household entomology are expanding.

Environmental concerns and laws, public opinion, and resistance of arthropod and weed pests to chemical pesticides are increasingly forcing a consideration and implementation of nonchemical solutions for pest problems. Classical biological control is a powerful and proven tool. However, federally mandated regulations may greatly slow the process of importation and colonization of new natural enemies far beyond sound biological protocols that have served this discipline and society for over a hundred years.

Natural enemies for use in biological control may be categorized into separate risk groups. Parasitic and predaceous arthropods fit into the lowest risk category, but are the most difficult to study and to assess for potential success. The policy of certain countries to require intensive studies on native organisms before allowing them to be exported is especially devastating to the deployment of biological control. Yet progress is being made with increased attention to basic ecological and behavioral research. The rate of biological control successes may drop initially as the style of "educated empiricism" becomes more widely adopted, as has apparently already begun. Success rates could be expected to increase as the database enlarges and intercommunication possibilities expand. Certainly the trend will ever more propel the activity of exotic natural enemy importation into a solid scientific base. However, there are special difficulties of establishing integrated control in crops for which eye appeal has excessively been used as a measure of quality.

Further Reading

Bartlett, B. R., and R. van den Bosch. "Foreign Exploration for Beneficial Organisms." In *Biological Control of Insect Pests and Weeds*, P. DeBach and E. I. Schlinger, eds., 489–511. London: Chapman & Hall, 1964.

Boldt, P. E., and J. J. Drea. "Packaging and Shipping Beneficial Insects for Biological Control." *FAO Plant Protection Bulletin* 28(2) 1980: 64–71.

Coppel, H. C., and J. W. Mertins. *Biological Insect Pest Suppression.* New York: Springer-Verlag, 1977.

Coulson, J. R., ed. "Use of Beneficial Organisms in the Control of Crop Pests." Entomological Society of America Publication Proceedings of the Joint American-Soviet Conference, Washington, D.C., Aug 13–14. 1979: 62.

Coulson, J. R., and R. S. Soper. "Protocols for the Introduction of Biological Control Agents in the United States." In *Plant Quarantine*, R. Kahn, ed., 1–35. Boca Raton, FL: CRC Press, 1988.

DeBach, P. "Successes, Trends, and Future Possibilities." In *Biological Control of Insect Pests and Weeds*, P. DeBach, ed., 673–713. New York: Reinhold, 1964.

Hall, R. W., and L. E. Ehler. "Rate of Establishment of Natural Enemies in Classical Biological Control." *Bulletin of the Entomological Society of America* 25 (1979): 280–82.

Hall, R. W., L. E. Ehler, and B. Bisabri-Ershadi. "Rate of Success in Classical Biological Control of Arthropods." *Bulletin of the Entomological Society of America* 26 (1980): 111–14.

Klingman, D. L., and J. R. Coulson. "Guidelines for Introducing Foreign Organisms into the United States for Biological Control of Weeds." *Weed Science* 30 (1982): 661–67.

Legner, E. F., and T. S. Bellows Jr. "Exploration for Natural Enemies." In *Handbook of Biological Control: Principles and Applications*, T. W. Fisher and T. S. Bellows, eds., 87–101. San Diego: Academic Press, 1999.

Pimentel, D., L. McLaughlin, A. Zepp, B. Lakitan, T. Kraus, P. Kleinman, F. Vancini, W. J. Roach, E. Graap, W. S. Keeton, and G. Selig. "Environmental and Economic Impacts of Reducing U.S. Agricultural Pesticide Use." In *Handbook of Pest Management in Agriculture*. Vol I. 2nd ed., D. Pimentel, ed., 679–718. Boca Raton, FL: CRC Press, 1991.

Schroeder, D., and R. D. Goeden. "The Search for Arthropod Natural Enemies of Introduced Weeds for Biological Control—In Theory and Practice." *Biocontrol News and Information* 7(3) 1986: 147–55.

See Also: Agroecology; Insects, Beneficial; Integrated Pest Management; Pesticides

E. F. Legner

Pesticides

Many substances that people use or encounter every day, including cleansers, polishes, solvents, and fuels, are toxic. Pesticides, however, whether natural or synthetic, are among the relatively few chemicals that are used specifically because they are poisons and are intended to kill something. Pests—broadly defined when one considers pesticides—include insects, weeds, fungi, bacteria, rodents, and other organisms that compete with humans for resources (e.g., food, fiber, shelter) or transmit pathogens that cause disease. Specific categories of pesticides that target these different kinds of pests include insecticides, herbicides, fungicides, bactericides, and rodenticides. In the United States, all products, natural or synthetic, that are sold for the purpose of killing a pest, even the bacteria and fungi in swimming pools, must be approved (registered) by the U.S. Environmental Protection Agency (EPA).

Within each pesticide category—insecticide, herbicide, fungicide, and so on—individual chemicals vary in general chemical structure, mode of action, derivation, environmental fate, and toxicity. Insecticides serve as an example. The major chemical classes of insecticides, based on the elements and molecules used to make them, include organochlorines, organophosphates, carbamates, pyrethroids, and neonicotinoids. In addition to these five major classes, several more chemical groups of insecticides currently are represented by only a few chemicals each.

A pesticide's "mode of action" is the way in which it poisons or kills pests (and sometimes other organisms). The organophosphate, carbamate, and neonicotinoid insecticides disrupt the function of the neurotransmitter acetylcholine, either directly or by blocking the enzyme that breaks it down. Acetylcholine functions at gaps called synapses between nerve axons and muscle cells that receive nerve impulses, so these three classes of insecticides are called synaptic poisons. Organochlorine and pyrethroid insecticides, along with plant-derived pyrethrins, interfere with nerve impulse transmission along a nerve axon and are called axonic poisons. Although many synthetic insecticides poison the nervous system of insects and other animals, insecticides derived from the bacterium *Bacillus thuringiensis* cause breakdown of the insect gut, the botanical insecticide rotenone interferes with cellular respiration in a wide range of animals, and several synthetic "growth regulator" insecticides interfere with molting in all arthropods.

A pesticide's chemical structure and mode of action govern its acute toxicity to target and nontarget organisms, including humans, and its breakdown and deactivation. These traits define the relative safety or toxicity of pesticides to humans and determine the likelihood of rapid development of pesticide resistance in target pests. An insecticide that targets a relatively specific insect hormone that regulates molting is, in general, less likely to be acutely toxic to humans than an insecticide that interferes with neurotransmission in insects and humans alike (as do the organophosphates, carbamates, neonicotinoids, and nicotine). Pest populations that have developed resistance to a certain pesticide are more likely to also be resistant to a pesticide with a similar chemical structure or a similar mode of action (even if one of the pesticides is naturally derived and the other is synthetic).

Attention to the derivation of a pesticide usually focuses on whether it is manufactured synthetically or derived from a naturally occurring substance or organism with minimal industrial processing. This distinction may be somewhat arbitrary, but it is a key factor for the approval of pesticides for use in certified organic production systems. Very few synthetic chemicals are on the National Organic Program's List of Approved Substances, and these are only allowed for extreme situations, with special consideration. Insecticides that occur naturally or are derived from natural products with minimal processing include most microbial insecticides (*Bacillus thuringiensis*, Entrust, Mycotrol, and preparations of other viral, bacterial, fungal, and protozoan pathogens), plant oils, insecticidal soaps, pyrethrum and its component pyrethrins, neem, and sulfur. Nicotine and rotenone also are plant-derived insecticides, but they are no longer approved in certified organic production, and two more plant-based insecticides, ryania and sabadilla, are no longer registered for use in the United States. Almost all of the hundreds of other insecticides sold in the United States are synthetic chemicals. They are not inherently more or less toxic than naturally occurring compounds, but they are produced only by human-controlled chemical reactions.

Concern about the environmental fate of pesticides usually focuses on two characteristics—persistence over time and water solubility. A pesticide residue is the amount that remains in the environment (including plant parts that will be harvested for food uses) after application. Residue levels diminish over time as a result of reactions catalyzed by light, water, and enzymes produced by living organisms. In general, users of pesticides want residues to last long enough to kill target pests but not persist so long as to pose undue risks to nontarget organisms, such as humans, pets, birds, fish, or even beneficial insects, fungi, and more. Persistence of pesticides is expressed in terms of their "half-life" under specific conditions—the time it takes for a given amount of the chemical to decrease by one-half as a result of chemical breakdown. Several organochlorine insecticides of the 1950s and 1960s had half-lives of a few years in soil. In contrast, the half-life of pyrethrins on foliage in direct sunlight may be only a few minutes. Depending on specific uses, pesticides with half-lives of two to three days to a few weeks provide effective control and pose minimal risks in terms of environmental contamination. In general, most naturally derived pesticides degrade more rapidly than most synthetic pesticides. There are exceptions; plant-derived rotenone persists for a few days on plant foliage, about the same length of time as synthetic carbaryl (Sevin) and longer than Malathion (an organophosphate). Most pesticides are not highly soluble in water, but those that are should not be used in locations where they might leach into groundwater, as the likelihood of contamination is great in such circumstances. Compounds that persist for a long time and are highly soluble in water present the greatest risk for environmental contamination. Regulatory guidelines prevent such compounds from being approved for use in vulnerable settings.

The toxicity of pesticides is their defining characteristic. Insecticides are used to kill insects, and insects are animals with many physiological similarities to humans. It should not be surprising that many insecticides can injure humans as well. Measures of toxicity indicate the likelihood that a chemical will cause injury or death if encountered in a specific way—by skin contact (dermal), ingestion (oral), or inhalation. Toxic effects may result from acute poisoning (single or short-term exposures) or chronic, long-term poisoning. Acute toxicity of pesticides is usually expressed in terms of an LD_{50}—a dose that is lethal to 50 percent of the animals in a test population (generally laboratory rats or mice). That dose is expressed as milligrams of toxicant per kilogram of body weight of the test animals (mg/kg). A compound with a lower LD_{50} is more toxic than one with a higher LD_{50}. Although dermal, oral, and inhalation LD_{50}s are useful indicators of acute risks of poisoning for handlers of pesticides, they reflect only the lethality of the compound; they do not reflect other risks of exposure—eye injury, throat and lung irritation, skin burns, neurological injury, and other risks. In addition, LD_{50}s describe the lethality of only the active ingredient; that compound is diluted with formulation ingredients to produce a usable product, and addition of those formulation ingredients may decrease or increase the lethality of the final product. Finally, long-term, low-dose exposure to certain chemicals is known to cause chronic effects—cigarette smoke, benzene, and asbestos are all known to be carcinogenic. Other compounds are known to cause birth defects or neurological damage. Laboratory tests to determine whether or not individual pesticides, natural or synthetic, cause cancer or other chronic disease as a result of low-dose, long-term exposures are required for all pesticides as part of the U.S. EPA registration process. Designing tests to detect long-term effects that may occur in a small but meaningful portion of the human population (say even 1 in 10,000, or 30,000 Americans) and avoid high probabilities of producing false-positives or false-negatives has proven to be a monumental task. Consequently, precise answers about the long-term risks of intermittent low-dose exposures to pesticides, especially as residues on food in the type of exposure that many Americans experience, remain elusive and a matter of disagreement among pesticide proponents and opponents. Minimizing human exposure to pesticides remains the best way to minimize risk.

Further Reading

Bloomquist, Jeffery R. "Insecticides: Chemistries and Characteristics." http://ipm world.umn.edu/chapters/bloomq.htm.

Silva-Aguayo, G., "Botanical Insecticides." http://ipmworld.umn.edu/chapters/Silvia Aguayo.htm.

Ware, George G. and David M. Whitacre, "Introduction to Insecticides, 4th ed." from *The Pesticide Book*, 6th ed. 2004. http://ipmworld.umn.edu/chapters/ware.htm.

See Also: Agrichemicals; Agroecology; *Bacillus thuringiensis* (Bt); Fungicides; Gulf Hypoxia; Herbicides; Insects, Beneficial; Integrated Pest Management; Neem; Pest Control, Biological; Soaps, Insecticidal; Water Quality

Richard Weinzierl

Policy, Agricultural

Agricultural policy includes the full range of programs funded through the United States Department of Agriculture (USDA) to enhance national food security. Agricultural programs may be viewed by taxpayers, farmers, or food stamp recipients as farm subsidies or welfare payments; however, the political justification for government involvement in agriculture is invariably couched in terms of national food security. Government programs administered by the USDA include those designed to stabilize agricultural production and farm income, conserve soil and natural resources, provide food for low-income consumers, ensure public food safety, provide nutrition education, fund agricultural research and education, support rural community development, and promote international trade. Government payments to producers of agricultural commodities and food stamps for low-income consumers are the most widely recognized and most controversial of agricultural policies, and they typically claim a large portion of the total USDA budget. Agricultural programs have become increasingly controversial in recent years as the connection between farm programs and national food security has become more ambiguous and tenuous.

The initial involvement of government in U.S. agriculture began with publicly funded agricultural education and research in the late 1800s. The Moral Act of 1862 made federal land-grants available to the states for the purpose of establishing state colleges to provide higher education in the agricultural and mechanical arts; President Abraham Lincoln established the USDA in the same year. The Hatch Act of 1887 provided federal funds for agricultural experiment stations in each state. Two years later, the USDA was elevated to cabinet status. The Smith-Lever Act of 1914 established the cooperative extension service, extending the benefits of U.S. agricultural research and educational programs to farm families and rural residents. The political leaders understood that the security and economic progress of the nation at that time depended on the resilience and productivity of the nation's farms.

Agricultural productivity increased and agriculture prospered during the early 1900s and through World War I. However, while the rest of the economy "roared" through the 1920s, the agricultural economy languished, primarily because of production stimulated by wartime demand and later declines in demand for U.S. exports to Europe. Many farmers accumulated large debts, and President Herbert Hoover established the Farm Board in 1929 in an attempt to slow farm bankruptcies during the banking and stock market crises. The New Deal policies of President Franklin D. Roosevelt, who was elected in 1933, greatly expanded the role of government in agriculture during the Great Depression. About one-fifth of all Americans earned their living from farming at the time, and most rural communities relied on agriculture for their economic well-being. Consequently, policies to bolster farm income provided a badly needed stimulus to the whole U.S. economy, in addition to ensuring the continued production of food and fiber needed to feed growing numbers of urban Americans. The foundation for today's USDA supply, price, and income stabilization programs, commonly referred to as "commodity programs," was established by the Agricultural Adjustment Act of 1933.

Agricultural prices are inherently unstable because of unpredictable weather conditions and the necessity for farmers to base their production decisions on prices expected months or years in the future. High prices tend to create expectations of profitable prices in the future, which tend to bring about surplus production and unprofitable prices, rather than the profitable prices expected. Low prices then trigger cuts in production, sometimes after lengthy time lags, which result in scarcity and more profitable prices than expected, thus starting a new cycle of surplus and scarcity. Unusually good or bad growing conditions also trigger new cycles of production and price instability. A wide variety of agricultural programs have been devised to help stabilize agricultural production.

First, "parity prices" were established for a number of agricultural commodities to reflect the prices necessary for commodities to maintain their purchasing power. For example, wheat prices maintained at parity levels would always allow a bushel of wheat to contribute the same amount to an average standard of living, the parity price of wheat being adjusted to reflect changes in prices of all other goods and services.

To keep prices from dropping significantly below parity, the government would purchase and store surplus commodities and would then sell commodities from government storage to keep prices from rising significantly above parity. During World War II, government stocks were depleted and U.S. farmers had profitable markets for all they could produce. The government resorted to food rationing, making government stabilization programs both unnecessary and impractical. Government programs focused on maximizing food production to enhance food security.

After World War II, new mechanical and chemical technologies, developed or refined through public agricultural research, made it both possible and profitable for farmers to farm more land and produce more wheat, corn, cotton, rice, peanuts, tobacco, and virtually everything else. Parity prices then became highly profitable prices and the government could no longer afford to buy and store enough commodities to keep commodity prices at parity. In the 1954 Farm Bill, the government resorted to giving away surplus commodities as food aid to other countries (Food for Peace) and paying farmers to divert land from production to conservation uses (the Soil Bank). During the 1960s, Congress expanded food assistance programs by distributing surplus commodities to low-income Americans, the forerunner of Food Stamps and other special food assistance programs. The national school lunch program, initiated by President Harry Truman in 1946, was expanded to include breakfasts by President Lyndon Johnson in 1966. These food distribution programs provided the obvious public benefit of promoting healthier and more productive citizens, but they also helped stabilize agricultural prices and production.

During the 1970s, growing export markets for agricultural commodities resulted in dramatic increases in prices of agricultural commodities, which reduced the costs of commodity programs but increased the costs of food assistance. Farmers borrowed heavily during the 1970s to buy farmland, machinery, and buildings to expand production. A significant part of this expansion was financed by government subsidized loan programs, initiated during the 1930s, to ease the financial burden of the Great Depression. The initial purpose of Farm Credit Service was to stabilize production; access to subsidized loans facilitated overproduction instead. In the early 1980s, a global recession brought a sharp drop in export demand at a time when farmers were financially committed to production increases. Consequently, commodity prices plummeted and once again farmers were caught with large debts and unprofitable prices.

In 1973, the government began providing direct "deficiency" payments to farmers to make up for differences between market prices and support prices. Rather than buying commodities to support prices, the USDA began to offer nonrecourse loans to farmers, allowing farmers to store commodities to benefit from any later price increases. To ease the government burden, farmers were required to remove land from production to qualify for commodity payments.

The 1985 Farm Bill was called the Food Security Act of 1985. It included the Conservation Reserve Program (CRP), which paid farmers to idle environmentally sensitive farmland for ten to fifteen years, and the "sodbuster" and "swamp-buster" programs to protect highly erodible farmland and wetlands. The Soil Conservation Service had been established in 1935, during the Dust Bowl years, to reimburse farmers for conservation practices. For the first time, however, the 1985 Farm Bill *required* farmers to implement soil and water conservation plans to qualify for commodity payments. The primary purpose of these conservation provisions was to reduce agricultural surpluses; however, the 1985 Farm Bill empowered an emerging environmental movement in agriculture. The cost of government commodity programs increased dramatically during the 1980s. However, increases in market prices during the late 1980s and early 1990s reduced the government's costs of commodity payments, making it easier to fund conservation and natural resource programs.

Government commodity payments were "decoupled" from current production during the 1990s, representing a dramatic change in farm programs. Export markets were again providing U.S. farmers with profitable market prices and the 1990s appeared to be a good time to begin phasing out commodity-based payments to farmers. The Freedom to Farm Act of 1996 was designed to systematically reduce commodity payments over a five year period by basing initial commodity payments on historic production levels for specific farms and reducing payments each of the following four years. However, payments for the initial year were inflated dramatically to gain political support for the phase-out process. Furthermore, when export markets faltered and agriculture commodity prices fell, Congress responded with generous annual "emergency farm payments," ballooning government farm program costs to record levels. "Freedom to farm" was dubbed "freedom to fail." The 2002 Farm Bill essentially institutionalized the earlier "emergency payments" into "counter-cyclical" payments that were designed to increase as prices fall and decrease as prices rise.

Over the decades, government involvement in U.S. agriculture has become firmly entrenched politically, although the taxpaying public is becoming increasingly skeptical regarding whether farm programs are benefitting the general public or large landowners and agribusiness. During the 1990s, the emphasis of government programs in coping with periodic overproduction shifted from managing market supplies to stimulating increases in farm export demand. A variety of export promotion and enhancement programs were initiated to help make U.S. farmers more competitive in an increasing global marketplace for food. The previous decoupling of government payments from current production facilitated the integration of agricultural commodities into international trade negotiations under the new World Trade Organization (WTO); fixed payments would not distort trade patterns. Growing conflicts in global trade negotiations caused the United States to turn to regional trade agreements, such as the North American Free Trade Agreement (NAFTA) and the Central American Free Trade Agreement (CAFTA), but international trade continues to be the government's means of choice for coping with the instability in agricultural production. Increasing exports during times of domestic surplus moderate price declines, and decreasing exports during times of domestic scarcity moderate price increases. However, relying on global markets to stabilize U.S. agricultural production and prices also means relying on the global markets for U.S. food security. Thus, it has become increasingly difficult to justify continuing large, commodity-based payments to U.S. farmers under the guise of national food security.

POLICIES FOR AGRICULTURAL SUSTAINABILITY

Concerns for agricultural sustainability emerged from the 1985 Farm Bill in a provision to enhance "agricultural productivity." This provision was supported by a coalition made up of conventional farmers who were concerned about the farm financial crisis and the economic viability of farming. They were joined by organic farming advocates, who were concerned about the impacts of conventional agriculture on soil productivity and environmental quality, and rural advocacy organizations, who were concerned about economic and environmental impacts of agriculture on rural communities. The program emerging from this provision was called Low Input Sustainable Agriculture (LISA). The program was retained and expanded in the 1990 Farm Bill. Its name was changed to Sustainable Agriculture Research and Education (SARE), and it has continued to receive funding in each farm bill since, including the Food, Conservation, and Energy Bill of 2008.

Sustainable agriculture was defined by the LISA and SARE programs as integrated systems of crop and livestock production that conserve natural resources and protect the environment, maintain the viability of the agricultural economy, and preserve a desirable quality of life for farmers, rural residents, and society as a whole. Thus, sustainable agricultural policies seek to expand and integrate the government's previous conservation and natural resources programs with programs that stabilize and enhance farm income and programs that maintain the traditional culture of family farms and rural communities. The National Organic Program (NOP) expanded this agenda in 2002 by providing USDA standards for certification of organic foods, accompanied by modest funding to support research on organic farming methods. Various other government programs have provided indirect support for sustainable agriculture, including a small program to promote farmers markets and other local foods programs. However, USDA funding for sustainable and organic agriculture remains less than 2 percent of the USDA agricultural research and education budget.

Ever since their inception, agricultural policies have been promoted to the public as being necessary to save the family farm. In the 1930s, saving the family farm was equivalent to saving U.S. agriculture and was clearly critical to agricultural sustainability and national food security. Between the 1930s and the early 2000s, however, the number of farms in the United States dropped from more than six million to around two million and the percentage of Americans earning their livelihood from agriculture dropped from 20 percent to less than 1 percent. More than 95 percent of U.S. farms are still owned and operated by families rather than stock corporations. However, today's large mechanized, chemical-intensive crop farms and large concentrated animal feeding operations (CAFOs) hold little resemblance to the family farms of the 1930s or even of the 1950s.

The chemical and biological wastes from today's large farming operations have depleted the natural fertility of soils and turned agriculture into the number one source of nonpoint-source (diffuse) water pollution in the nation. The demise of traditional family farms has decimated many rural communities. As farm families have left rural communities, fewer people are left to support local businesses, to attend local schools and churches, and to serve in various positions of local civic and political responsibility. In addition, larger, specialized farming operations often bypass local communities in marketing and purchasing of farm inputs. In their quest for economic survival, many operators of larger farms have felt compelled to sign comprehensive contractual agreements with multinational agribusiness corporations to gain access to the capital and technologies needed to compete in global markets. The first obligation of stock corporations is to their stockholders, not to any particular nation. Although past agricultural policies have successfully promoted economic efficiency, they have done little to ensure agricultural sustainability or long-run national food security.

The 2008 Farm Bill has added yet another dimension to farm policy, with greatly expanded government subsidies for biofuels, particularly ethanol. However, diversion of cropland from food crops to fuel crops has been at least in part responsible for rapidly rising food prices and increasing food *insecurity* both in the United States and globally. The initial public reaction has been one of even greater skepticism regarding the legitimacy of current agricultural policies in light of government's inherent responsibility for national food security.

Further Reading
Cochrane, Willard. *The Curse of American Agricultural Abundance: A Sustainable Solution.* Lincoln: University of Nebraska Press, 2003.
Cochrane, Willard, and Ford Runge. *Reforming Farm Policy: Toward a National Agenda.* Ames: Iowa State University Press, 1992.
Faeth, Paul. *Paying the Farm Bill: Agricultural Policy and the Transition to Sustainable Agriculture.* Washington, D.C.: World Resources Institute, 1991.
Knutson, Ron, J. B. Penn, and Barry Flinchbaugh. *Agricultural and Food Policy.* 6th ed. Upper Saddle River, NJ: Prentice Hall, 2006.
Pollan, Michael. *The Omnivore's Dilemma: A Natural History of Four Meals.* New York: Penguin, 2006.
Spitze Robert, Daryll Ray, and Milton Hallberg, eds. *Food, Agriculture, and Rural Policy into the Twenty-First Century: Issues and Trade-Offs.* New York: Westview Press, 1994.

See Also: 1990 Farm Bill, Title XXI; 2008 Farm Bill; Agricultural Subsidies; Country-of-Origin Labeling; Environmental Protection Agency; Policy, Food; APPENDIX 2: Organic Foods Production Act of 1990

John Ikerd

Policy, Food

"Food policy" is an abstract concept that specifically encompasses the policy, legal, and regulatory arena that governs or influences the food system. The modern food system, the policy foundation of which was largely established in the Great Depression era, is a global system that includes the production, processing, delivering, marketing, consumption, and disposal of agricultural products. Food policy is influenced by a complex web of international, federal, state, and local laws and trends that have evolved over several decades and have helped give rise to the legal disciplines of agricultural law and food law.

The farm bill is a major component of food policy in the United States. The term "farm bill" is the generic name for the federal omnibus legislation, typically enacted approximately every five years since the 1930s. Each farm bill usually includes provisions covering multiple areas, including commodity programs, trade, conservation, credit, agricultural research, food stamps, nutrition, forestry, and marketing.

The economic conditions of the Great Depression era and President Franklin D. Roosevelt's New Deal legislation combined to create the farm commodity programs in their current form. These programs are designed to support prices and incomes predominantly through the use of supply controls. This type of program was maintained up through enactment of the Federal Agriculture Improvement and Reform Act—commonly referred to as the 1996 Farm Bill—when commodity programs transitioned toward a more market-oriented economic system. Declining prices and decreased exports in years following enactment of the 1996 Farm Bill resulted in significant emergency support legislation for producers. The 2002 Farm Bill, formally known as the Farm Security and Rural Investment Act of 2002, was enacted in some measure to help avoid those expenses. The 2002 Farm Bill maintained features of the commodity programs contained in the 1996 Farm Bill, but added additional price support measures as well. The most recently enacted farm bill is the Food Conservation and Security Act, generally referred to as the 2008 Farm Bill.

In addition, food policy is shaped by a multitude of other federal laws, such as the Federal Insecticide, Fungicide, and Rodenticide Act (FIFRA); the Federal, Food, Drug, and Cosmetic Act (FFDCA); and the Clean Water Act (CWA). FIFRA governs the registration of pesticide products to ensure that certain health, safety, and environmental criteria are satisfied. FIFRA is administered by the Environmental Protection Agency (FDA), a reflection of the complex web of federal regulations that impact the nation's food system. FIFRA was originally enacted in 1947 and has been amended on multiple occasions, including amendments by the Food Quality Protection Act (FQPA), which established, among other changes, new standards for chemical residues in foods, updated the scientific basis for rule-making, and reformed pesticide registration. The FQPA also modified EPA regulatory practices for pesticides by including a new safety standard applicable to all pesticides used on foods.

The FFDCA was originally enacted in 1938. According to the *Glossary of Agricultural Production, Programs, and Policy*, which is published at the National Agricultural Law Center at the University of Arkansas, FFDCA provides "the basic authority that expanded the Food and Drug Administration's consumer protection capabilities that are intended to ensure that foods are pure and wholesome, safe to eat, and produced under sanitary conditions; that drugs and devices are safe and effective for their intended uses; that cosmetics are safe and made from appropriate ingredients; and that all labeling and packaging is truthful, informative, and not deceptive."

The Clean Water Act (CWA) is the name commonly used to describe the 1972 legislative amendments to the Federal Water Pollution Control Act of 1948. The CWA was enacted "to restore and maintain the chemical, physical, and biological integrity of the Nation's waters." The CWA is a relatively complex statute that addresses water quality standards for surface waters and requires that a National Pollution Discharge Elimination System (NPDES) permit be obtained in order to discharge point-source pollutants into navigable waters. The CWA also authorizes state-directed planning for control of nonpoint-source pollution into navigable waters. Point-source water pollution is typically from industry or urban water treatment plants, which dump water from a point (or single source) into a waterway, whereas non–point source water pollution is water that flows off fields, lawns, golf courses, and other diffuse areas.

Federal marketing orders also comprise an important component of food policy. USDA is authorized by the Agricultural Marketing Agreements Act of 1937 to issue marketing orders and agreements for a variety of agricultural commodities and their products, which include marketing orders for milk, fruits, and vegetables. According to the National Agricultural Law Center at the University of Arkansas, marketing orders "promote orderly marketing and help to collectively influence the supply, demand, price, or quality of particular commodities. A marketing order is requested by a group of producers, and must be approved by the USDA and through a referendum by a required number of the commodity's producers (usually two-thirds) in specified areas. Conformance with the order's provisions is mandatory for all producers and handlers covered by the order. It may limit total marketing or shipping, prorate the movement of a commodity to market, or impose size and grade standards."

The National Organic Program (NOP) is the program administered by the Agricultural Marketing Service, an agency within the USDA that establishes strict standards for organic certification and labeling. The NOP was statutorily authorized by the Organic Foods Production Act of 1990. Under the program, any farm, wild crop harvesting, or handling operation that wants to represent an agricultural product as organically produced must comply with the extensive program standards, which include operating under an organic system plan approved by a USDA-accredited certifying agent and using materials specifically permitted by the National List of Allowed Synthetic and Prohibited Non-Synthetic Substances. The regulations that implement the NOP were originally published in December 2000 and became effective in October 2002.

Further Reading

Culver, Chuck. "Glossary of Agricultural Production, Programs, and Policy." National Agricultural Law Center. www.nationalaglawcenter.org/glossary/index.phtml.

Knutson, Ronald D., J. B. Penn, and Barry L. Flinchbaugh. *Agricultural and Food Policy*, 6th ed. Upper Saddle River, NJ: Prentice Hall, 2006.

National Agricultural Law Center. www.nationalaglawcenter.org.

Pederson, Donald B., and Keith G. Meyer. *Agricultural Law In a Nutshell*. St. Paul, MN: West Publishing, 1995.

United States Department of Agriculture. www.usda.gov.

See Also: 1990 Farm Bill, Title XXI; 2008 Farm Bill; Agricultural Subsidies; Country-of-Origin Labeling; Environmental Protection Agency; Food Policy Council; Policy, Agricultural; USDA National Organic Program; USDA National Organic Standards Board; U.S. Food and Drug Administration; APPENDIX 2: Organic Foods Production Act of 1990

Harrison Mauzy Pittman

Political Ecology of Food

Political ecology openly addresses how power relationships of the political economy affect environmental decision-making. Political ecologists recognize that formal governments and informal governance influence how the environment is perceived and thus what approach is taken to deal with the natural environment. The term "political ecology" was coined by Eric Wolf in 1972, but political ecology studies first flourished in the 1980s.

Political ecology developed largely from Marxist traditions concerned with material equity. Changes in the physical environment affect varied stakeholders in a myriad of ways. For some the change brings benefits, others experience worsening quality of life. The unequal distribution of benefits and harm relates to politics because the distribution of natural resources is a reflection of power between stakeholders and can change the power relationships between stakeholders. Political ecology research addresses a multitude of subject matter because the approach is applicable to any human-environment topic.

The first step of political ecology research is to identify root causes of inequality expressed through relationships with the physical environment. Political ecologists will work to identify how people use natural resources, how people conceptualize their relationship with the natural environment, and then how they negotiate and at times struggle over the distribution of resources. The struggle over the distribution of natural resources can be expressed through formal government action or coordinated social movements. In other cases the struggle is more subtly experienced through everyday interactions with neighbors or even within families. Once the root causes of inequality are better understood, the second step of political ecology research is to explore how more equitable relationships can be established.

Sometimes the political ecology implications of food systems are clearly evident, such as when collectivized farms were privatized after the collapse of the Soviet Union. Changes in inputs, outputs, farming techniques, and links to markets were directly influenced by changes in the formal political system. In other cases the link between political power and food production and distribution are less clear. Informal ad hoc governance is more difficult to identify and may change rapidly without notice. For example, if men in largely agricultural areas began to migrate to cities to find higher-paying jobs, the power dynamics in rural communities would change. Women left behind in the communities would take on new roles in the community and in households and would likely still maintain at least part of the agricultural operation. The social and political infrastructure that was in place when men played a larger role in agriculture may not be well suited to the different roles and needs of women in the community.

Political ecology addresses food first, by offering a critique of how sustainable, organic, and local food can be and has been posited as apolitical. Once the initial analysis demonstrates how power relationships between stakeholders has led to less than ideal practice, suggestions and dialogue to bring about more equitable and sustainable practice can be developed. Concepts from political ecology should be used to understand decision-making at every point along the food commodity chain, such as determining who decides how food is grown, where food is distributed, which food can be purchased, and regulations for production. It is fruitful to explore how a political ecology approach can inform understanding of processes and provide guidance on more equitable distribution of benefits and costs of promoting organic, local, and sustainable foods.

The process of formalizing organic standards through government regulation would clearly benefit from a political ecology analysis. If one's goal is to expand the production of organics, a complicated political process may deter investment in organic practice. But, rigorous inspections, transparency in all stages along the food commodity chain, and clear homogenous labeling may increase consumer confidence in the quality of the organic product. Political ecologists can help to identify how governments can work to maximize farmers' access to the tools needed to convert to organic production and benefit from such production.

If one moves to a wider international or global scale, obstacles to conversion to organic agriculture often change and intensify. In the latter half of the twentieth century, Green Revolution programs promoted the use of pesticides. Now some programs promote conversion back to organic agriculture. Monetary support for both movements has largely come from the developed world. One criticism of the Green Revolution is that some programs did not adequately recognize differential power relationships in areas that received technological inputs. An issue is whether a similar pattern of benefits and costs could take place with a conversion to organic production.

A political ecology perspective also offers a critical standpoint on local foods. The process of defining what is or is not local can be disputed, and those disputes may be resolved based on differential power dynamics between stakeholders. On an individual basis, one may be able to informally and easily identify what comprises local. Once an informal system of promoting local foods becomes a more formal, government-supported system, local territories become more difficult to delineate. The formal government definition of "local" will likely rely on extant political boundaries like municipal, county, state, or international boundaries. People that live near the edge of these political boundaries may not have greater access to locally produced food under government-supported local food initiatives. Likewise, the local food areas will also influence which markets are open to producers and distributors.

Political ecologists would also identify a larger ideological problem related to local foods: the local trap. Localization of food networks will not necessarily lead to empowerment for all and ecologically sound practice. The success of localized food production and distribution will vary greatly according to the local area. Those that wish to promote local food consumption quickly recognize that the physical geography of an area will limit the diversity of food that could be grown. Localization of food pro-duction could potentially increase inequality based on the agricultural rich-ness of the area. Beyond the physical environment, one should also consider how local power relationships will be affected in such a system because inequality and exclusion can occur at any geographic scale.

Arguments that organic or local food systems are universally and essen-tially good can be dangerous. Political ecologists recognize that the concept of nature is a construct based on human ideologies that vary over time and place. Even movements that have been largely conceptualized as apolitical, such as promotion of agriculture intended to reproduce natural ecological systems, must be critiqued. Political ecology asks how conceptualizations of nature are socially constructed because understandings of natural systems are not objective. Knowledge of the environment is informed by experi-ences, goals, and power dynamics. Identification of exactly which elements of the natural system will be reproduced and which will not is essential. Political ecologists point out that the best practice in managing agricultural systems will vary not only according to conditions in the physical environ-ment, but also according to stakeholder perceptions. Therefore, a singular best practice in agriculture does not exist, even within a small local area.

Sustainable food is an example of a concept that is often thought of as universally good. After all, everyone needs food to survive. However, the meaning of "sustainability" varies greatly and sustainability, organic, and local food movements are not apolitical. There are many different ways that sustainable food can be promoted and each approach will have different repercussions on those involved in production, distribution, and consump-tion of food. Political ecologists would question what is and is not sustained by current food production systems. Efforts could be made to sustain envi-ronmental quality in food production and distribution. What sort of envi-ronment should be maintained—one that maximizes average yields of food crops, one that maximizes biodiversity, or one that is resilient to exogenous change—would still need to be determined.

Respect for cultural systems associated with food could also fall under the umbrella of sustainable food. Indeed, arguments in support of organic and local food production often link such a system to the maintenance of an area's cultural heritage and the strengthening of a community's social ties. Political ecologists point out that encouraging organic production or sales to local markets will affect power relations between actors. A call to revitalize relationships between the consumer and producer does not hold the same romanticism for all people, even in the United States. For exam-ple, for some ethnic and minority groups, farming and close relationships with the land evoke images of slavery, forced removal from ancestral areas, and low-wage migrant labor, not a bucolic place of peace where hard work is rewarded.

Political ecologists would not argue that efforts for environmental or cultural sustainability are necessarily bad, or that such efforts are necessar-ily good. Rather, political ecology demands that the differential effects of such agricultural systems on people with varying degrees of power need to be understood. If one is able to identify stakeholders that will experience significant harm or will see few benefits, perhaps changes could be made for greater inclusivity.

Further Reading

Eaton, Emily. "From Feeding the Locals to Selling the Locale: Adapting Local Sustainable Food Projects in Niagara to Neocommunitarianism and Neoliberalism." *Geoforum* 39 (2008): 994–1006.

Peet, Richard, and Michael Watts. *Liberation Ecologies: Environment, Development, Social Movements.* New York: Routledge, 1996.

Robbins, Paul. *Political Ecology.* Malden, MA: Blackwell Publishing, 2004.

See Also: Food Security; Food Sovereignty; Development, Sustainable; Food Deserts; Free Trade; Globalization

Julie Weinert

Precautionary Principle

The precautionary principle is a system designed to deal with environmental uncertainty. The precautionary principle is designed to prevent stagnation due to uncertainty by creating a buffer that can be used to mitigate against unforeseen negative consequences of an action. The goal of the precautionary principle is to create a system that allows action while hedging that action to mediate any unexpected outcomes. Several forms of precautionary principle exist, with varying degrees of action allowed when faced with uncertain outcomes.

European Union (EU) laws accept the precautionary principle as a basic tenet. The EU acknowledges that there is uncertainty in understanding some of the issues involved in human health and the environment. EU laws do not wait until harm is proven. Rather, the burden of proof is up front: prove that that new production/action/food is safe before it is allowed. The United States does not follow the precautionary principle. Thus, new products/actions/foods may come on the market and remain there until they are proven unsafe (which can take years or decades, as the data must be gathered and analyzed for a lengthy time). This is a fundamental difference between how the Europeans and Americans view public health and environmental issues. Most notable is the difference in how genetically modified organisms (GMOs) have been regulated between the two continents. Except in very carefully regulated situations, the use of GMOs have not been permitted in Europe because there are no long-term scientific studies to verify their safety. In the United States, however, GMOs have been on the market since the corporations introduced them in the 1990s. Their long-term effects remain unknown.

Another variation, the "participant pays" precautionary principle deals with this issue of uncertainty by levying a tax on the action. For example, there is some degree of uncertainty over the impacts of new GMOs. According to the participant pays precautionary principle, those wishing to introduce these new species into the wild should have to secure a bond that could be used to offset negative impacts if they develop.

This precautionary principle's strength lies in its transparency. Those seeking to engage in an action understand the costs, usually some kind of bond. Although the tax may seem to raise transaction costs, it may actually lower these costs by eliminating the need for expensive models that precisely determine actual costs.

The drawbacks to the precautionary principle system revolve around the size of the tax. If the tax is too high, then transaction costs will be raised and some beneficial transfers will be unnecessarily prevented, resulting in continued market stagnation.

If the precautionary principle tax is too low, then downstream parties may not be adequately protected because there is an inadequate bond available to offset the unforeseen negative impacts of an action. The precautionary principle relies on accurate estimates of potential harm resulting from the proposed action. Rarely is there experience to draw from in determining a proper tax rate, so adoption of this system would require experts to make a "best estimate" of potential harm. From this baseline, using adaptive management techniques, states could modify the tax over time if empirical evidence suggests a different tax level would be more appropriate. It may be wise to start with a tax rate that errs on the side of caution— that is, a tax that is too high, to ensure that downstream parties are not injured, and that could be incrementally lowered over time if necessary.

Further Reading
Costanza, Robert. "The 4P Approach to Dealing with Environmental Uncertainty." *Environment* 34 (1992): 9.

See Also: Genetically Engineered Crops; Roundup Ready Soybeans; Water Quality

Michael Pease

Prices

A price premium measures the differential between the price of a certified organic product and the price of the conventional version of the same product. The premiums paid by consumers are a measure of the value that consumers assign to the organic certification. Producers, manufacturers, and retailers are interested in price premiums because they face higher costs and greater financial risks in supplying organic products. Statistics that show increasing sales with positive price premiums encourage supplier confidence that investments in organic production will pay off.

A price premium is the percent difference in organic and conventional prices, calculated as:

$$price\ premium = \frac{(price\ of\ organic - price\ of\ conventional)}{price\ of\ conventional} \times 100$$

That is, a price premium is the ratio of the difference in organic and conventional prices to the conventional price for the same item, multiplied by 100 to obtain a percentage. The price of an organic item is usually higher than the price of a similar conventional item, so the price premium is usually a positive number. However, the calculation allows for the price premium to be zero, meaning there is no difference between an organic and conventional version of the item, or negative, meaning the organic version is less expensive than the conventional version.

For the calculation of a price premium to be meaningful, the organic and conventional version of the item must be as close to identical as possible so that the comparison is truly about the organic certification and not other characteristics. Most agricultural products are described by federal grading standards (quality characteristics) as well as units (for example, 10-pound sacks, cartons of 12 dozen) that make standard comparisons possible. Processed food and fiber products are compared on the basis of package sizes and product descriptions (for example, 1-pound loaf of 100% whole wheat bread, 24-ounce jar of spaghetti sauce). Unit conversions made to enable price comparisons to be made between organic and conventional products should be done cautiously, because prices do not usually scale up linearly according to units. For example, a flat of 12 pints of strawberries would be unlikely to have the same price per pint as a single pint of strawberries sold alone. Nor would a 12-ounce jar of spaghetti sauce be half as expensive as a 24-ounce jar.

Price premiums may be calculated for all levels of the marketing channel, which stretches from the farm to the consumer. The points in the marketing channel at which price premiums are calculated are farmgate, wholesale (which includes brokers, handlers, and certain other intermediaries), and retail. The price premium is based on the amount paid by the next entity downstream in the marketing chain. Farmgate price premiums are determined by what wholesalers pay, wholesale price premiums depend on retailers, and retail price premiums are set by consumers' willingness to pay extra for organic.

Using average U.S. prices for 2005, the box shows price premiums calculated at three levels of the marketing channel for fresh carrots. Most wholesalers and retailers handle both organic and conventional products, so the calculation of price premiums is straightforward, and a single data base may be used. Most farmers have either an organic or a conventional system, so there is no single data set that tracks both price series together.

Farmgate Price Premium (US$ per 48-pound sack)

	Organic	Conventional	Price Premium
Mean	$17.72	$10.22	73.3%
Standard Deviation	$0.83	$0.38	

Sources: For organic prices, *Organic Food Business News* (discontinued in 2007). For conventional prices, National Agricultural Statistics Service, U.S. Department of Agriculture, as reported in *ERS Vegetables and Melons Yearbook.*

Wholesales Price Premium (US$ per 24-count box of 2-pound film bags)

	Organic	Conventional	Price Premium
Mean	$32.66	$12.97	151.8%
Standard Deviation	$0.82	$0.33	

Source: U.S. Department of Agriculture, Agricultural Marketing Service. Using data from Boston Wholesale Market.

Retail Price Premium (US$ per pound, packaged)

	Organic	Conventional	Price Premium
Mean	$1.26	$1.10	14.5%
Standard Deviation	$0.65	$0.51	

Source: Biing-Hwan, L., T. A. Smith, and C. L. Huang. "Organic Premiums of U.S. Fresh Produce," *Renewable Agriculture and Food Systems* 23 (2008): 208–16.

For each level in the marketing channel, price premiums were calculated from the prices received at that level. At each level, the units were the same for the organic and conventional carrots. While the prices themselves are not directly comparable across levels due to differing units, the price premiums are comparable because percentages are unitless numbers. Farmers pack carrots at harvest in 48-pound sacks. Wholesalers repack the carrots into 2-pound film bags and sell these in boxes of 24 bags each. Retailers offer the carrots in 1-pound (or larger) bags for purchase by consumers.

The values in the example are consistent with the relative premiums usually obtained at the different levels. For 2005, the largest price premium was obtained at the wholesale level, at which organic price was on average 150 percent (two and a half times) the conventional price. Farmers received an average price premium of 73 percent and retailers about 14 percent. Wholesalers absorb the most risk in the organic marketing channel because the channels are not well developed and supply is not always sufficient to meet demand faced by the wholesalers. Price premiums for wholesalers are often significantly larger than for other levels. Retailers face resistance to excessive price premiums from consumers, who usually state that on average they are willing to pay 10 to 35 percent more for organic products than conventional products, so premiums are usually lowest at this level.

Price premiums may go up or down over time as organic prices or conventional prices or both change. What happens on the farm affects price premiums at all levels. Organic commodities grow best in the same regions as conventional commodities. For example, both conventional and organic grains are grown mainly in the Midwest and Great Plains. This means that weather-related problems affect both organic and conventional crops to about the same degree. Price increases due to supply disruptions of this type would theoretically keep price premiums about the same. However, because organic crops are often planted on fewer acres to begin with, the supply effect may be magnified for organic crops and prices may increase more than for conventional crops. This would cause the price premium to increase.

For perishable items, prices are higher at the beginning and end of the season and lowest during the middle of the season. Again, this affects both conventional and organic perishable products, but organics tend to be more affected due to limited supply. If supply is greater than demand in a given period, price premiums may be negative because organic price may be lower than conventional price. This happens for fresh fruits and vegetables, especially in smaller geographic markets where sales are direct to consumers and demand is not high enough to absorb all the organic produce offered. As more farmland is certified organic, the amplitude of this seasonality effect on organic prices will be dampened and farmgate price premiums should become more stable.

Beyond the farmgate, price premiums are affected by inconsistency in supply chains, contract obligations, and sourcing needs. Where very limited production capacity currently exists, price premiums are highest. Demand for organic animal products is one of the fastest-growing segments of the market. For fresh brown eggs, wholesale price premiums ranged from 100 to 400 percent from 2004 to 2006, and price premiums for whole broilers in the same time frame were 170 to 260 percent. Ingredients for which availability is limited but is critical to the final product, such as organic vanilla for baked goods, may command price premiums of several hundred percent.

Wholesalers must accumulate enough product from farmers to fulfill their contracts with manufacturers and retailers. If the contracted supply is not enough, wholesalers must bid for product against other buyers (other wholesalers, manufacturer agents, and retailers who buy direct from farmers). With more limited supply, greater costs of transportation and handling may be incurred for wholesalers to accumulate the volume needed. Setting prices to cover these costs causes greater divergence from conventional prices, and results in price premiums at the wholesale level. A better developed supply chain may alleviate some of this pressure on organic wholesale prices.

Regardless of how much organic farmland is certified or how well developed the supply chain becomes, organic production and wholesaling will probably continue to cost more than conventional. Farms and firms that produce, process, or handle organic foods must be certified as in compliance with federal regulations. The costs of certification (recordkeeping, inspections, etc.) and the costs of compliance are accounted for in pricing organic products. Cost-based pricing almost ensures that organic price premiums at these levels will be positive, if smaller, in the long run, despite occasional instances where excess supply drives price down in the short run.

At the retail level, consumer demand and retailer response drive organic price premiums. Consumer surveys routinely show that relatively higher price for organic is the primary deterrent to purchase. For conventional supermarkets and discount retailers, such as Walmart, price and product selection are the strategic advantages, and price premiums have to be smaller for organic products to sell. Organic products that are sold in natural products supermarkets tend to have the highest price premiums, compared with conventional supermarkets and discount mass marketers. Consumers who shop at natural products supermarkets tend to be willing to pay a higher price for the perceived advantages of organic foods.

Price premiums may be negative or zero (organic price lower than or equal to conventional price) at the retail level for several reasons. A retailer may stock the organic version of a product offered by manufacturers that already supply conventional lines. If there are several such manufacturers, this may result in several organic brands being offered. The competition is heightened when stores offer their own private label organic products. If there are many brands of an organic product to choose from, particularly if there is a store brand, the competition may drive down the price premium. This has been observed for coffee and baby food, for example.

Organic foods are subject to the same types of retail strategies that are applied to conventional foods. A retailer may deliberately offer an organic item at a lower price as a "loss leader"—a product on which a revenue loss is absorbed in order to increase consumer traffic in the store. This strategy was used by some western U.S. retailers with organic milk around 1998, when organics were first being introduced to conventional supermarkets. In response to the economic downturn in 2008, natural products stores such as Whole Foods began weekly discounting on a rotating list of products to generate interest and convince consumers that they could still afford to eat organic foods.

Regardless of the size of the price premium, it is still the price received that determines how rapidly the organic supply chain will expand. Even a large price premium is no guarantee that a farmer, wholesaler, or retailer can cover the costs of organic production, processing, and handling. However, price premiums do provide a simple statistic to measure the relative strength of the organic market.

Further Reading

Biing-Hwan, L., T. A. Smith, and C. L. Huang. "Organic Premiums of U.S. Fresh Produce." *Renewable Agriculture and Food Systems* 23(2008): 208–16.

Dimitri, C., and K. M. Venezia. *Retail and Consumer Aspects of the Organic Milk Market.* LDP-M-155-01. Washington, D.C.: U.S. Department of Agriculture, Economic Research Service, 2007.

Oberholtzer, L., C. Dimitri, C. Greene. *Price Premiums Hold on as U.S. Organic Produce Market Expands.* VGS-308-01. Washington, D.C.: U.S. Department of Agriculture, Economic Research Service, 2005.

Oberholtzer, L., C. Greene, E. Lopez. *Organic Poultry and Eggs Capture High Price Premiums and Growing Share of Specialty Market.* LDP-M-150-01. Washington, D.C.: U.S. Department of Agriculture, Economic Research Service, 2006.

See Also: Agriculture, Organic; Marketing

Luanne Lohr

Processors, Local Independent

Processing is the method of adding value to a product and preparing it for retail consumption. Local independent processing facilities went through a period of decline as a result of the structural changes that have occurred in conventional agriculture over the past sixty years. As production started to grow into industrial scales, new technology allowed processing facilities to scale up to match the industrial levels of production. Few small-scale or independent processing facilities have remained in business, and most can no longer afford to do their own processing. Many farmers entered into contracts with large processors (agribusiness) who required certain volumes and consistency of product. Processing has become a vertically integrated, highly consolidated sector of the food system. The farmers who do small-scale farming, alternative agriculture, or direct sales have few options for sending their products to be processed into a value-added product ready for retail sale.

The lack of local independent processing facilities is recognized as a major obstacle to creating a local or regional food supply chain in many places in North America. Many producers face barriers in locating adequate local processing facilities. These barriers often are related to the lack of nearby processing facilities, particularly abattoirs, fruit/vegetable processing facilities, and grain mills. The distance to a facility and the facilities' volume requirements may be prohibitive to smaller-scale producers. Even where local facilities exist, many are not set up to return the product to the farmer. This means the farmer is not able to retail directly to the consumer and therefore loses the premium from a value-added product. There are several strategies that farmers are using to overcome these barriers. Some are creating cooperatives and building or purchasing local processing facilities. Other producers are building small-scale processing facilities on farms and buying more raw products from local producers to keep the facilities running full-time.

Producers who have attempted and succeeded in building their own processing facilities have faced a different set of challenges. Regulations for processing are geared toward large commercial operations. Currently, there are no scale specific regulations that make on-farm processing an easy option. Many producers have had to operate their facilities while meeting federal standards meant for large-scale facilities. They have struggled through lack of information and have found little assistance in resolving the issues. Starting a processing facility requires a ready product, a large capital investment, and a market demand.

Further Reading

Atkins, Peter, and Ian Bowler. *Food in Society: Economy, Culture, Geography.* London: Hodder Arnold, 2001.

Billie Best, Board of Directors, Regional Farm and Food Project, Open Letter. April 2007. www.farmandfood.org/newscommentary/articles_past/MeatProcessing Policy-Apr27-2007.pdf.

Ollinger, Michael, Sang V. Nguyen, Donald Blayney, Bill Chambers, and Ken Nelson. *Structural Changes in the Meat, Poultry, Dairy and Grain Processing Industries.* Washington, D.C.: United States Department of Agriculture, Economic Research Report, 2005. www.ers.usda.gov/publications/err3/err3.pdf.

See Also: Cooperatives, Grocery; Local Food; Marketing, Direct

Shauna M. Bloom

Production, Treadmill of

The treadmill of production is a theoretical framework originally developed in the late 1970s by Allan Schnaiberg, who sought to explain why social and ecological phenomena were misaligned, leading to social and environmental degradation, particularly in the postwar period. The only viable solution, according to the treadmill approach, was a wholesale reorganization of the very structures of industrial society by actively taking power away from the elite class, which places economic growth via the expansion of production above the needs of the environment.

In the postwar period, both communist and capitalist firms (supported by governmental policies, university research, and union interests) intensified the rate and scale of industrialization to replace manual labor with more economically efficient technologies. The mechanization of the production process increasingly forced firms to augment the scale of production to increase profits. Thus, industrial production generated an ever-accelerating treadmill, requiring ever more energy and raw materials to increase profits. More recently, Schnaiberg and collaborators have emphasized the transnational character of the treadmill, where the industrial treadmills of the North increasingly rely on the natural resources, labor pools, economies, and waste depositories of the global South.

For agriculture, the Green Revolution exemplifies the treadmill process, whereby higher yielding hybrid crops promoted by the West and exported to the global South require ever-increasing chemical inputs and mechanization to sustain (and increase) agricultural production. Such production changes are seen as fundamental causes of environmental degradation by treadmill theorists. Since Rachel Carson's 1962 book *Silent Spring*, a solution to this aspect of the treadmill has been for national, and now transnational, social movements to raise concern about the social and environmental destruction caused by conventional agriculture.

In terms of organic agriculture, treadmill theorists would be skeptical of the potential of organic movements operating at the local scale to create systemically sustainable agriculture. Although social movements have been successful in obtaining new potentially sustainable organic production practices, organic growers often must expand production beyond the local scale to keep up with competing firms. Organic growers often find themselves needing to compromise their values, shifting their production goals of sustainable reproduction to ever-increasing profit, thus replicating the system they sought to challenge. Small-scale organic growers are often bought out by larger firms. Agri-industrial firms often succeed in taking over niche markets previously controlled by growers concerned primarily with ecological, not economic, sustainability.

There are specific cases that demonstrate how organic farming can remain loyal to its original mission of locally produced food, stressing the ecological sustainability of reproduction, and maintaining a separation from the conventional agricultural complex. Treadmill theorists, however, would argue that focusing on specific cases of success veils the larger trend taking place, that is, the continuation of an unsustainable agricultural treadmill in need of fundamental social-environmental restructuration.

Further Reading

Gould, Kenneth A., David N. Pellow, and Allan Schnaiberg. *The Treadmill of Production*. Boulder: Paradigm Press, 2008.

Pellow, David, Allan Schnaiberg, and Adam Weinberg. "Putting the Ecological Modernization Thesis to the Test: The Promises and Performances of Urban Recycling." In *Ecological Modernization Around the World: Perspectives and Critical Debates*, A. P. J. Mol and D. A. Sonnenfeld, eds., 109–37. London: Frank Cass, 2000.

See Also: Agriculture, Conventional; Factory Farming; Food Dollars

Brian J. Gareau

R

rBGH

Recombinant bovine growth hormone (rBGH), also known as recombinant bovine somatotropin (rBST), is a genetically engineered version of a cow's naturally occurring growth hormone. It is injected into the animal to increase milk production by 10 to 15 percent. Monsanto Corporation holds sole patent rights to the drug, which is marketed under the brand name Prosilac. Since its commercial introduction in February 1994, rBGH has been engulfed in controversy and widespread opposition from consumers, animal rights advocates, and farmers. Canada, the European Union, Australia, New Zealand, and Japan currently restrict the drug's use out of concerns for its effects on the environment, the dairy economy, and human health. Nevertheless, rBGH is widely used in the United States.

Opposition to rBGH centers broadly on three concerns. First, opponents question the need to increase milk production in the United States when the country consistently maintains a surplus. Contrary to initial claims of the biotechnology industry and the U.S. Department of Agriculture, there is little or no income difference between farmers that use rBGH and those that do not. Nor has the price of milk products fluctuated with the drug's introduction.

Second, animal rights advocates and environmentalists argue that rBGH adversely affect cows' health. Treated animals exhibit 25 percent more mastitis (udder infections), increased rates of reproductive problems (including infertility, ovarian cysts, still births, and birth defects), and a 50 percent increase in the risk of lameness (hoof or leg problems). The rapid expansion of milk production can also result in painful udder growth and negatively impact a cow's lifespan. Moreover with increased infection rates, farmers tend to increase antibiotic use, the residues of which are excreted in milk and fecal matter, potentially increasing antibiotic resistance in pathogens.

Third, despite Monsanto's insistence that the product is safe for human and animal consumption, multiple scientific studies link rBGH to increased levels of Insulin-like-Growth Factor One (IGF-1) in milk. IGF-1 is a naturally occurring human hormone that plays an important role in childhood development and cell growth. Elevated levels of IGF-1 are known to promote cancer cells, particularly in breast, colon, and prostate tissue.

In response to growing opposition, major food manufacturers and retailers such as Kraft, Walmart, Whole Foods, Tillamook Country Creamery, Ben and Jerry's, Dean Foods, Starbucks, and Kroger Grocery banned or limited the use of milk from cows treated with rBGH. In 2006, California Dairies Incorporated (CDI), the United State's second largest dairy cooperative imposed a surcharge on rBGH milk, arguing that the premium is necessary to cover the added cost of selling the product. A year later, the CDI banned rBGH milk completely. With a shrinking market, fewer farmers are choosing to inject cattle, and more are transitioning to organic methods which offer a better financial return.

Further Reading

DuPuis, Melanie. *Nature's Perfect Food: How Milk Became America's Drink*. New York: New York University Press, 2002.

Tokar, Brian, ed. *Redesigning Life: The Worldwide Challenge to Genetic Engineering*. London: Zed Books, 2001.

See Also: Factory Farming; Genetically Engineered Crops; Milk, Organic; U.S. Food and Drug Administration

Robin Jane Roff

Research

The key issue related to organic and sustainable agriculture research concerns how much is being spent and on what (and by whom). The first part of the issue is more difficult to determine than might be expected because research support includes not only money, but also allocation of facilities and personnel. Research into organic and sustainable agriculture in the

Researchers educate farmers during on-farm agricultural field days. Courtesy of Leslie Duram.

United States started with on-farm experimentation of alternative systems in the late 1970s and early 1980s as energy prices rose and availability of fossil fuels became less assured. Farmers were looking for ways to both save money and reduce fossil fuel energy consumption. The emphasis of this research was on input self-sufficiency and substitution away from fossil-fuel-based inputs, such as pesticides and fertilizers, toward tillage, variety selection, nitrogen-fixing plants, crop rotation, and compost for plant protection and nutrition. Organic farming associations were formed in part to share information about and moral support for alternative cropping and animal systems.

In 1985, when federal funding for organic and sustainable agriculture research was first authorized, the types of projects expanded and university-trained researchers began to apply scientifically rigorous methods to production problems. From applied questions about what works on the farm, research has expanded to address economic and sociological questions as well as to provide answers to specific pest management, nutrition, and systems problems. The majority of this research since 1985 has been funded by the U.S. government.

Federal funding for agricultural research is particularly important because it signals to university and private sector researchers the key areas that the agricultural sector, and by extension the U.S. Department of Agriculture, consider priorities. Agricultural research is an investment in improving current systems and developing new systems to address emerging challenges in food and fiber production. Agricultural research is conducted by federal agencies, by agricultural-degree granting institutions, and by nonprofit organizations.

The majority share of federal research financing has shifted from formula funding, which allocates money to state-specific research projects conducted by agricultural universities, to competitive grants funding, which requires that project proposals submitted by government agencies, universities, and nonprofits be judged against each other for the same pool of money. The amount of money allocated for agricultural research is set in five-year cycles within the comprehensive federal agriculture and nutrition legislation, known commonly as the Farm Bill. The final amount authorized for federal agricultural research is subject to the appropriations process and executive approval, as are all laws that involve spending federal funds.

Various organizations with the U.S. Department of Agriculture dispense research funds or participate in research collaborations, but the majority of competitive grants are administered by the Cooperative State Research, Education, and Extension Service (CSREES) based on a priority setting process that takes place within the USDA. A number of programs not specifically targeted at organic agriculture support organic research, including the Small Business Innovative Research Program, which is aimed at private sector market research and commercialization, and the Higher Education Program, which has supported development of university-level degrees and certificate programs in organic agriculture.

The first federal funding made available for organic and sustainable competitive grants was through the Low Input Sustainable Agriculture (LISA) program established by the 1985 Farm Bill with almost $4 million in research funding distributed through four regional Sustainable Agriculture Research and Education (SARE) offices. Subsequent support in the 1990 Farm Bill expanded research funding and established programs for extension service training and producer initiated grants in sustainable agriculture. The extension service is an outreach organization run through the land grant universities and funded primarily by state governments that provides research-based technical support.

After the Organic Farming Research Foundation's (OFRF's) watershed report disclosing the low level of federal funding for organic research, the first organic-targeted research funding was authorized in the 1998 Farm Bill. The 2008 Farm Bill provides $78 million in mandatory funding to be allocated to the USDA's Organic Agriculture Research and Extension Initiative (OREI) over four years. This represents a fivefold increase over the 2002 farm bill. There may also be additional appropriations of up to

$25 million per year for additional research into organic methods of production. Under this bill, the organic program is specifically directed to investigate seed breeding for organic systems and to study the conservation outcomes of organic practices. These are new directions for research that could greatly help farmers, consumers, and the environment.

Nonprofit organizations typically use donor money or endowments to fund projects that match the organization goals and mission. The Organic Farming Research Foundation is in Santa Cruz, California, and has funded farm-centered research projects since 1992, with about $1.3 million provided over that period. Donor-driven research funding may flow through nonprofits, such as the $450,000 provided to OFRF by Stretch Island Fruit Company in 2008 to support organic fruit research. Priorities for larger self-sustaining philanthropics such as the W. K. Kellogg Foundation and the Charles Stewart Mott Foundation have shifted away from sustainable agriculture and toward community empowerment and food security, making federal funding even more critical.

Nonprofits themselves compete with universities and government agencies for research funding, often leading to the development of new coalitions (government-university, university-nonprofit) with broader agendas and wider support. Funding may bypass university and government researchers altogether as nonprofits team with for-profit research firms or develop their own teams to conduct mission-specific research. An example is the $750,000 awarded in 2008 by the USDA Foreign Agricultural Service to the Organic Trade Association and to Sustainable Strategies, a for-profit food and agriculture consulting firm.

Federal agencies, private nonprofit organizations, and universities support organic and sustainable agriculture research through allocation of personnel and facilities. Prior to 1990, organic and sustainable agriculture were not considered viable production systems by many researchers. As of 1997, OFRF could identify only thirty-four projects in U.S. universities that were "strongly organic." By 2003, scientists at land grant universities in forty-four states were conducting organic research. By 2008, more than two hundred projects conducted by researchers in land grant universities used "organic agriculture" in the descriptions of their federally funded projects. Much of this shift was due to the availability of federal funding and state agricultural industry demands, which drive university research priorities. Within federal agencies, organic research was initially limited by a small constituency to lobby for funding, disbelief among administrators and researchers that organic agriculture offered enough benefits to justify the cost of research, and hostility toward organic agriculture by powerful conventional farm organizations. As organic production systems became more widespread and consumer demand increased, the organic agribusiness sector became more organized and large-scale conventional agribusinesses (food manufacturers, retailers, and exporters) became involved in distribution. This burgeoning market growth and the passage of federal organic regulations under the Organic Foods Production Act in 2000 provided the impetus for a change of policy within the USDA that broadened the involvement of several agencies.

The need for data to conduct domestic market analysis was recognized in 1998 by the Economic Research Service (ERS), but collection costs were limiting. Most of the research on organic production and markets prior to 2002 was conducted on data provided by private organizations, such as the OFRF National Organic Farmers Survey, the Organic Business News price reports published by Hotline Printing and Publishing, and the Natural Marketing Institute, or from nonsystematic surveys of small samples of producers, manufacturers, and consumers. As of 2007, the Agricultural Marketing Service was reporting wholesale fruit and vegetable prices in Boston and San Francisco as well as initiating organic poultry, egg, and grain price reports. Organic producers were included for the first time in the 2002 national Census of Agriculture conducted by the National Agricultural Statistics Service. The first organic commodities in the Agricultural Resource Management Survey of production costs were dairy, in 2005, and soybeans, in 2006. Organic handlers in the supply chain were first sampled by the ERS in 2004. The 2008 Farm Bill mandates $1.25 million per year and authorizes up to $5 million per year for price data collection and analysis. The ERS will likely expand its series of market analysis publications to include more organic reports as these data become available to researchers.

The bulk of scientific research conducted by the federal government is through the Agricultural Research Service, which operates research stations throughout the United States that routinely collaborate with land grant university scientists on projects in nutrition and food safety, plant and animal production and protection, and natural resources and sustainable systems. In 2005, approximately 8 percent of 2,340 ARS scientists indicated an interest in working on organic agriculture systems, and about 4 percent were actively engaged in such research.

The private sector is highly competitive in organic research in two areas—long-term production systems and market research. Most of the cropping trials that have survived over this time are projects that began under the auspices of individual visionaries who believed in organic and sustainable systems early on and had the money to support decades of research. The most famous is the Farming Systems Trials research at the Rodale Institute (Kutztown, Pennsylvania), started in 1981 and still running. Sustainable agriculture centers in Kansas (The Land Institute), Oklahoma (the Kerr Center for Sustainable Agriculture), and Wisconsin (Michael Fields Agriculture Institute) conduct research on a range of plant and animal production systems that emphasize ecological principles examined with sound scientific methods. Nutrition and food quality have recently taken center stage in the private research sector with the founding of the Organic Center in Enterprise, Oregon. With the most significant researchers in these fields on its Board of Directors and Scientific Committee, the Organic Center has produced an influential body of literature on organic food quality, pesticides and health, and nutritional benefits of organic food since it was started in 2002.

University researchers typically receive funding to conduct a survey one time, and can at best develop a snapshot of a market at a particular point in time. Most companies (food manufacturers and retailers) need information that is updated. The for-profit market research sector makes repeated surveys of the same individuals (called panels) or conducts national surveys of the same type of questions repeated every one to three years. The exceedingly high cost of such surveys can be recouped with sales of reports that cost from several hundred to several thousand dollars. Market research on consumers, suppliers, and the marketing chain is provided by companies such as Mintel International Group, the Organic Monitor, Natural Marketing Institute, HealthFocus, SPINS/Nutrition Business Journal, and the Hartman Group. Organic retail price data are collected weekly from supermarket scanners by SPINS/A. C. Nielsen for all major marketing areas and sold by individual SKU (stock keeping unit code—a unique identifier for each product) or accumulated into reports or charts. These types of data products give businesses the most current information on which to make decisions.

The perceived disadvantage of university research—slowness due to adherence to strict standards of conduct and peer-reviewed publication—is also its greatest advantage. The research plan for competitively funded research must be vetted at several levels within the institution as well as by a review team at the funding agency. The research must be rigorously conducted, from data collection to analysis, and the results subjected to review at presentations and in written form. University researchers have time to be certain their results are statistically correct, and generally cannot obtain more funding for a project unless previous work has yielded significant results. The outputs of university research are often immediately useful to organic and sustainable farmers and agribusinesses, particularly if the research team includes extension specialists. More importantly, university research contributes to the body of public knowledge about agricultural and economic systems, and often is a training tool for the next generation of agricultural researchers who may be in the public or the private sector.

The universities that have generated the most research on organic and sustainable agriculture research are University of California, Davis; University of California, Santa Cruz; Iowa State University; University of Minnesota; University of Wisconsin; Michigan State University; Washington State University; Ohio State University; West Virginia University; Rutgers University; and North Carolina State University. University scientists study everything from inputs selection and management to maximize organic output to basic research on soil chemical and biological processes to optimal marketing chains. Much of the research on organic systems has positive implications for conventional agricultural systems, which is beginning to be recognized and supported by nonorganic farmers. Since land grant universities receive the majority of their funding from state governments, research programs tend to be responsive to the needs of the state and the region. In states such as California, where the organic farming and agribusiness sector is ingrained and extensive, university research has a longer history of working to solve problems faced by organic farmers.

Training students in organic agriculture is the means to ensuring that research on organic and sustainable systems will continue. The heart of this type of education is multidisciplinary training. The Center for Agroecology and Sustainable Food Systems at the University of California, Santa Cruz, is one of the oldest undergraduate training programs in organic farming in the United States. Students study agronomic principles, economics, and ecology, and conduct on-farm internships before conducting research. The Agriculture, Food, and Environment Program at Tufts University in Boston provides graduate training in a combination of agricultural science, policy, and nutrition courses. The only current degree program in Organic Agriculture Systems in the United States began in 2007 at Washington State University in Pullman. Cross-disciplinary instruction, practical experience, and systems design and analysis training figure prominently in all these programs.

Further Reading

Bull, C. T. "US Federal Organic Research Activity Is Expanding." *Crop Management.* 2006. www.cropmanagement.org.

Dimitri, C., and L. Oberholtzer. *Market-Led Growth vs. Government-Facilitated Growth: Development of the U.S. and EU Organic Agricultural Sectors.* Washington, D.C.: United States Department of Agriculture, Economic Research Report, 2005. www.ers.usda.gov/Publications/wrs0505/.

The Organic Center. State of Science. 2009. www.organiccenter.org.

Organic Farming Research Foundation. Organic Provisions in the 2008 Farm Bill *May 20, 2008.* http://ofrf.org/policy/federal_legislation/farm_bill/080520_update.pdf.

Scientific Congress on Organic Agricultural Research (SCOAR). Organic Farming Research Foundation. 2008. http://ofrf.org/networks/scoar.html.

See Also: Education and Information, Organic; Organic Farming Research Foundation; Rodale Institute; USDA Agricultural Research Service; USDA Economic Research Service; Appendix 4: Recent Growth Patterns in the U.S. Organic Foods Market, 2002; Appendix 3: U.S. Organic Farming Emerges in the 1990s: Adoption of Certified Systems, 2001, 2001.

Luanne Lohr

Restaurants

The restaurant industry employs over 13 million people and is expected to increase in size to 15 million by 2018. This makes the restaurant industry one of the largest employers in the United States. The overall economic impact of the restaurant industry exceeded $1.5 trillion in 2008, including sales in related industries such as agriculture, transportation, and manufacturing.

Manufacturing accounts for about 80 percent of industrial energy use. Within this group, the food industry is the fifth largest consumer of energy. As a whole, the restaurant industry consumes almost one and a half times more energy per square foot than other commercial businesses. The average restaurant generates 100,000 pounds of garbage and consumes excess water each year. Over the last few years, there has been growing interest among foodservice operations to develop environmentally sustainable operations. Using locally grown meat and produce, running energy-efficient equipment, monitoring water usage, and implementing a waste management program can help a restaurant save thousands of dollars while greatly reducing its impact on the environment.

Whatever format the foodservice operation takes—whether fast-food, quick-service, casual dining, or upscale—environmental sustainability is becoming an important issue. As customers are becoming more aware of "green practices," they are demanding that the businesses they support be accountable for their impact on the environment.

QUICK-SERVICE RESTAURANTS

Quick-service restaurants make up the largest segment of the industry. Customers patronizing quick-service restaurants expect low-priced, adequate-quality food delivered relatively quickly. Competition is fierce in this sector, because there is usually another quick-service restaurant nearby to meet consumers' needs. Most customers expect fast, friendly service, but their service expectations don't generally go beyond the basics—low price, adequate-quality food, fast service. Going beyond the customer's basic expectations could provide a point of differentiation and provide an opportunity to exceed customer's expectations. One fast-food restaurant in Portland, Oregon, does just that. Burgerville is famous to the locals for serving traditional fast food in which the ingredients come from local or regional sources. Burgerville is committed to fresh, local, sustainable business. The fast-food chain restaurant serves the typical menu of hamburgers, French fries, milkshakes, and onion rings. Burgerville's hamburger meat is ground from pasture-fed, antibiotic-free beef. The fries come from locally produced potatoes or sweet potatoes. The milkshakes are made with local berries, and the onions are a regional variety known as Wala Wala, sourced from neighboring Washington state. Burgerville not only purchases local ingredients to improve environmental sustainability, it also uses local businesses and is committed to renewable energy, composting, and recycling.

UPSCALE DINING RESTAURANTS

Upscale dining restaurants provide a venue for high customer expectations. Patrons expect excellent food and service in an unrushed environment. Highly trained, professional, attentive wait staff anticipate customer needs.

Since 1971, Chez Panisse, an upscale, full-service restaurant in Berkeley, California, has served the best local and organic cuisine in season, produced in an ecologically sound manner. Chez Panisse and their suppliers are careful not to harvest more than is necessary in order to sustain the natural resources for future generations. This embodies the idea of sustainability, cultivated at a time before it was popular. The menu at Chez Panisse is based on fresh, seasonal, local foods and is *prix fixe*—it includes a single multicourse meal for a fixed price. Founder Alice Waters's philosophy, in essence, began the "slow food movement" and was the beginning of what is now known as "California cuisine."

ORGANIC RESTAURANTS

Organics has finally reached the mainstream consumer and is making inroads in the restaurant business as well, although organic restaurants are still few at this time. One leader in Washington, D.C., is blazing the trail. Restaurant Nora became the first USDA certified organic restaurant in 1999. Restaurant Nora ensures that 95 percent of the ingredients it uses are from organic distributors. In a high-volume restaurant, this can be a difficult task. Besides the need for a consistent customer base, there is a need for reliable and consistent organic ingredients. Although the organic market has reached the mainstream, with a 20 percent market growth for home purchases, organic producers and suppliers are fewer than conventional suppliers, and their product yields are smaller. Despite this fact, organic restaurants such as Restaurant Nora are popping up throughout the country, supported by an educated and affluent clientele desiring high-quality, sustainable food from their dining experience.

Further Reading

Burgerville. www.burgerville.com.
Local Harvest. "Organic Restaurants." www.localharvest.org/restaurants/.
Restaurant Nora. www.noras.com.
Tzeng, G. H., M. H. Teng, J. J. Chen, and S. Opricovic. "Multicriteria Selection for a Restaurant Location in Taipei," *International Journal of Hospitality Management* 21(2) 2002: 171–87.

See Also: Fast Food

Sylvia Smith

Rodale Institute

The Rodale Institute, a brainchild of Jerome Irving Rodale, was first inaugurated in 1947. Rodale, a highly successful publishing entrepreneur, was the foremost North American pioneer of organic agriculture. In the 1930s, Rodale was given less than two years to live by his family physician. After transforming his diet, however, he lived an additional forty-plus years. During this time he became a leading spokesman and postwar proponent of organic food and agriculture.

The mission of the Rodale Institute is the promotion and development of organic foods and agriculture through the use of scientific research, education, and information outreach. Its motto, "Healthy Soil = Healthy Food = Healthy People," stresses the connection of health outcomes to health of natural resources used in agriculture and food production.

The Rodale Institute is a nonprofit and nongovernmental organization organized under the 501(c)(3) section of the Internal Revenue Service Code.

Nestled in the rolling hillsides of the Lehigh Valley, Pennsylvania, the Institute comprises 331 acres certified organic production and a research center reflecting the rich Mennonite farm heritage of the area. Unlike the corn belt "English farming," the Mennonites stood steadfastly by traditional mixed cropping. This contrasts with the monotony of the seemingly endless fields of corn that are typical of much of North America's postwar corn belt.

In 1971, the Rodale family purchased the Rodale Institute from the Siegfried family. Archaeological evidence indicates continuous support of Native Americans for over 9,000 years on the site. In the last Siegfried generation, the family farm had been converted into one huge corn field with massive gully erosion on the hillsides. Ardath Rodale brought the farm, which became a living laboratory to help complete Robert Rodale's dream that through organic agriculture it would be possible to resurrect land under the worst deteriorated conditions.

It is not accidental that in 1981 the Rodale Institute research scientists chose to compare conventional, chemically based corn and soybean farming with a mixed cropping system based on small grains, cover crops, and corn and soybean without chemical production methods. This landmark study in collaboration with USDA Agricultural Research Service (ARS) scientists is called the Farming Systems Trial.

By 1984, the corn yield in organic mixed cropping systems had equaled that of conventional corn and soybean monocultures. Since 1985, grain yields were found similar in both conventional and organic corn and soybean overall. However, under the low input organic mixed agriculture approach, corn and soybean yields were 30 to 70 percent higher in drought years than conventional production. In addition, the typical organic yields on poor soil have now exceeded those of conventional yields for corn and soybeans in Berks County, Pennsylvania. These results prove the wisdom of the regeneration strategy that Robert Rodale championed.

Robert Rodale coined "regeneration" to define continuous land improvement using natural biological processes. There was no need to rest land. However stripped the land, it renews itself through the natural biological processes similar to building natural resources by biological succession.

After twenty-one years of the 1981 Farming Systems Trial, the longest side-by-side comparison of organic and conventional farming in North America, mixed organic systems had increased soil carbon or organic matter by about 1 percent per year, reaching the current situation of about a 30 percent increase. This significant level of soil improvement was not found in the conventional corn and soybean system. Moreover, it has now been documented that the added carbon in soils results in 25 to 50 percent more water penetrating and percolating the soil. This transformation explains the superior performance of organic crops in drought years.

The importance of the Rodale findings is enormous in their potential impact. Because organic agriculture stores carbon effectively, it is a powerful tool to fight global climate change from elevated greenhouse gas content. Because it stimulates percolation of water that feeds underground and stream water systems, it is crucial to recycling and purifying source water for human use. Because it uses no chemical poisons, it improves water quality. Because it improves soil fertility and biology, it improves the diets of our domesticated animals.

The Rodale Farming Systems Trial proves beyond any scientific doubt that organic agriculture is a powerful tool in reclaiming the environment. Currently, about 75 percent of all private land is under agriculture. By shifting to regenerative organic agriculture, a better future can be assured, supporting farmers making transitions from a chemical dependent conventional farming treadmill to natural farming systems. Organic agriculture can not only prevent unnecessary environmental damage but can also promote the growth of natural resources, especially soil, water, and air. The Rodale Institute has been a leader in proving that organic farming has firm economic, environmental, and energy justification.

Present priorities of the Institute include combating global warming and changing agriculture and food policies to promote agriculture and supply healthy and abundant food, especially to those most in need.

See Also: Agriculture, Organic; Research; Appendix 1: Report and Recommendations on Organic Farming, 1980

Paul Reed Hepperly

Roundup Ready Soybeans

Roundup Ready (RR) soybeans are one of a family of genetically engineered crops designed to resist specific herbicide types, in this case Monsanto Corporation's broad-spectrum glyphosate, Roundup. Similar technologies include Calgene's (now Monsanto) BXN cotton and Bayer Crop Science's Liberty Link (LL) crops, most notably their LL Rice varieties, which sparked international controversy in 2006 when unapproved lines were discovered in the global supply of long-grain rice. To date the biotechnology industry has introduced herbicide tolerant varieties of canola, sunflower, cotton, corn, rice, and soybeans. Monsanto is currently developing Roundup Ready wheat.

Herbicide tolerance is marketed primarily as a means of simplifying weed management. Whereas in conventional practice farmers must turn to hand or machine tilling to remove weeds after crops germinate, with herbicide tolerant varieties farmers can spray fields as needed throughout the growing season. Farmers in the United States and Canada have taken great advantage of this easier production process and rapidly adopted Roundup Ready varieties and other similar crops since their introduction in 1996. As of 2008, 63 percent of U.S. corn acres and 92 percent of soybean acres were planted to herbicide tolerant varieties.

Like all genetically engineered (GE) crops, Roundup Ready soybeans are not without problems. RR soybeans are one of the most widely used GE products and are the focus of much critique. Opponents of agricultural biotechnology challenge herbicide tolerance, arguing that by removing biological constraints on pesticide application the technology leads to dramatic increases in chemical use. Out-breeding between GE crops and wild relatives has also created new strains of herbicide tolerant weeds (so-called superweeds) that require highly toxic pesticides to control or additional manual tilling. In canola, a plant that is far more susceptible to out-crossing than soybeans, "stacked" superweeds—varieties with tolerances to more than one herbicide—have emerged on the Canadian prairies. Controlling herbicide tolerant varieties in crop rotations is similarly difficult. "Volunteers" from the previous season must be removed manually or with older non-glyphosate-based pesticides that pose a greater risk to ecological health. If not detected early, these volunteers may contaminate the current crops. Whether tolerant to one or multiple chemicals, tolerant plants pose a significant ecological and economic problem and reverse much of the initial labor-saving benefits of these types of GE crops.

Further Reading

Duke, Stephen, ed. *Herbicide-Resistant Crops: Agricultural, Environmental, Economic, Regulatory, and Technical Aspects.* Boca Raton: Lewis Publishers, 1996.

National Agricultural Statistics Service. *2008 Acreage Report.* U.S. Department of Agriculture. 2008. http://usda.mannlib.cornell.edu/usda/current/Acre/Acre-06-30-2008.pdf.

See Also: Bt Corn; *Diamond v. Chakrabarty*; Genetically Engineered Crops; Precautionary Principle; Substantial Equivalence; Terminator Gene

Robin Jane Roff

S

Seafood, Sustainable

Sustainable seafood generally refers to fish that have been caught or farmed in an environmentally responsible manner. Terms like "eco-friendly," "environmentally preferable," or "ocean-friendly" seafood are also commonly used. With more than 75 percent of the world's fisheries either fully fished or overexploited, and a global aquaculture industry that now accounts for almost half of all seafood, there is a growing interest in learning more about the origin and production methods of the seafood we eat.

HISTORY

Dolphin-Safe Tuna

Perhaps the first example of a sustainable seafood campaign, the dolphin-safe label was created in response to public concern over the practice of harassing or killing dolphins in the process of catching tuna. This led to a major shift in how these fisheries operated.

Single-Species Campaigns

Another early approach to sustainable seafood awareness was employed by environmental nonprofit groups like Natural Resources Defense Council,

National Environmental Trust, and SeaWeb. They developed campaigns with catch slogans around the plight of a given overfished species, like North Atlantic swordfish ("Give Swordfish a Break"), Patagonian toothfish ("Take a Pass on Chilean Seabass"), or Caspian Sea sturgeon ("Caviar Emptor"). Chefs and retailers were encouraged to stop offering these seafoods, and consumers were urged not to buy them. These initiatives were successful, but had limited reach.

Business Partnerships

More recently, environmental organizations have partnered with large retailers and foodservice companies to implement more sustainable purchasing policies for seafood. With the help of the World Wildlife Fund and others, Walmart announced that it would carry only wild fish certified to the Marine Stewardship Council standard. Compass Group—the largest foodservice company in North America—no longer sells seafood from the Monterey Bay Aquarium's "red list." In partnership with the Environmental Defense Fund, Wegmans supermarkets developed purchasing policies for more sustainable farmed shrimp and salmon. However, a recent Greenpeace report found that although many businesses were making progress in the area of seafood sustainability, there was still much work to be done.

CONSUMER GUIDES

Perhaps the most well-known sustainable seafood tools today are pocket guides or wallet cards, distributed by a number of conservation organizations. These were first developed approximately ten years ago by groups like the Monterey Bay Aquarium, Audubon Society, and Environmental Defense Fund, and encourage individual consumers to make choices that can shift the seafood marketplace.

These guides are extraordinarily useful in that they are updated frequently, offer a large list of choices, and are available in print, online, or on mobile devices. They have even been expanded to cover sushi. The downside of these guides is that they only work when consumers refer to them when buying or ordering fish.

These stoplight-type guides usually offer seafood choices in three categories: "green" (fish that come from abundant, well-managed fisheries or eco-friendly farms); "yellow" (fish that are farmed or fished using methods that cause some environmental concerns); and "red" (fish that come from depleted populations or are fished or farmed in a destructive manner).

In determining the sustainability of farmed fish, several factors are considered. First is the feed: some farmed fish are fed large quantities of wild-caught fish, leading to a net loss of fish protein from the ocean. For example, three to six pounds of wild fish are needed to produce one pound of farmed salmon. More "vegetarian" species, such as tilapia and catfish, require very little wild fish in their feed, and mollusks (e.g., clams, mussels, oysters, scallops) require none.

CRITERIA USED TO DETERMINE ENVIRONMENTAL RATINGS FOR
WILD FISH

1. Basic biology. Small schooling fish like sardines and herring grow fast and repro-
    duce quickly, making them more resilient to heavy fishing pressure. Slow-growing
    species like sharks, rockfish, and orange roughy mature late in life and are there-
    fore much more susceptible to overfishing.
2. Status of the population. How abundant a given species is at any one time.
3. Bycatch. Some fishing methods are less selective than others. For example, bottom
    trawls, dredges, and longlines have high levels of incidental catch. This bycatch
    often includes threatened species like sea turtles, marine mammals, sharks, and
    seabirds. Traps and pole-and-line fishing have substantially lowered bycatch.
4. Habitats and ecosystem effects. Fishing gear itself, especially those types used for
    bottom-dwelling species, can damage sensitive ocean habitats. This is common for
    trawled species like cod, flounder, monkfish, and rockfish. More sustainable meth-
    ods include traps and pole-and-line gear.
5. Effectiveness of management. Eco-friendly fisheries usually have scientifically
    determined catch limits, and a management system that responds appropriately
    to natural and human-made fluctuations in the fishery. Good management also
    requires that bycatch and habitat damage be minimized.

Second is risk of escape: farmed fish are often grown where they are not
native, or are genetically distinct to their wild counterparts. If they escape,
they can harm wild populations through interbreeding or competition for
resources. This is common with salmon netpens, where thousands of fish
can escape at a time. Closed systems, like ponds and tanks, reduce the risk
of escape.

Third is risk of disease and parasites: crowded conditions make stressed
fish more susceptible to diseases and parasites, which often require chemical treatments. This is common in shrimp and salmon farming, where excess
chemicals—or the diseases and parasites themselves—can spread to the
adjacent ecosystem.

Fourth is pollution and ecosystem impacts: some types of fish farms,
especially those for shrimp and salmon, allow fish waste and uneaten feed
to pass untreated to surrounding waters. Operations that capture and treat
waste, such as recirculating tank systems, are more sustainable options.

ECO-CERTIFICATION

Wild Fish

The Marine Stewardship Council is the most widely accepted independent eco-certification for wild-caught fish. Their blue-and-white check
logo can be found on seafood products around the world and is gaining
popularity in the United States with large retailers like Walmart and Whole
Foods. Under this program, fisheries are assessed against three main principles: sustainable fish stocks, minimizing environmental impact, and effective management. Thirty-five fisheries worldwide have already been
certified, and another seventy-nine are undergoing assessment.

Friend of the Sea is another, smaller, eco-certification program for wild fish, but their environmental assessments are far less detailed than those of the MSC.

Farmed Fish

There is no definitive eco-label for farmed fish. Friend of the Sea has begun to certify fish farms on a small scale, and the Global Aquaculture Alliance, an industry trade association, offers self-certification for its members. The World Wildlife Fund, an environmental nongovernmental organization, is holding a series of multistakeholder dialogues around the world to develop credible environmental standards for fish farming. Although this process is encouraging, it will be several years before standards are developed for the most popularly farmed species.

The closest thing currently to a true farmed fish eco-label may be private organic certification in Europe, most notably by Naturland (Germany) and the Soil Association (United Kingdom). However, standards differ between labels, and some are more stringent than others. Currently there are no USDA organic standards for farmed fish.

HOW AND WHERE TO BUY

Sustainable seafood is available at restaurants, retailers, and fish markets across the country. Consumers can check with nonprofit groups such as the Seafood Choices Alliance, Chefs Collaborative, or the Green Restaurant Association to support local eco-friendly seafood businesses. They can also order online from a company like Ecofish, which sells 100 percent environmentally responsible seafood.

The consumer guides can help shoppers identify sustainable seafood options when buying fish in a store or restaurant. However, sustainability information is rarely posted on commercially available seafood products. Two notable exceptions are FishWise, a program based in California that labels fish at point-of-sale with sustainability and health information. Washington State's Department of Health has also begun a similar pilot program, identifying fish that are good or bad for the environment, and high or low in mercury and omega-3 fatty acids.

A new law passed in 2004, called Country of Origin Labeling (COOL), requires supermarkets to label unprocessed seafood (fresh or frozen, not canned, smoked, or otherwise "altered") with its country of origin and whether it is farm-raised or wild-caught.

Monterey Bay Aquarium provides a scientifically based assessment of the sustainability of various fish species for consumers at their Web site (www.montereybayaquarium.org/cr/seafoodwatch.aspx). Shoppers can download information or even sign up to have the seafood watch sent to their cell phones. The Monterey Bay Seafood Watch Guides are regionally specific, with eight unique lists. Shoppers in the Northeast, for example, have a guide that is distinct from the list for shoppers in the West or Southeast. Up-to-the-minute information on the sustainability level of various fish species can help concerned shoppers make the best choices.

QUESTIONS TO HELP DETERMINE FISH AND
SEAFOOD SUSTAINABILITY

1. Is this fish farmed or wild?
2. What country is it from?
3. How was it caught or farmed?
4. Does the restaurant or market offer any eco-friendly fish and seafood options?

Further Reading

Costello, Christopher, Steven D. Gaines, and John Lynham. "Can Catch Shares Prevent Fisheries' Collapse?" *Science* 321(5896) 2008: 1678–81.

Ecofish. www.ecofish.com.

Environmental Defense Fund. "Seafood Selector." www.edf.org/seafood.

Greenpeace. "Carting Away the Oceans." www.greenpeace.org/usa/press-center/reports4/carting-away-the-oceans.

Marine Conservation Biology Institute. "Shifting Gears: Addressing the Collateral Impacts of Fishing Methods in U.S. Waters." www.mcbi.org/publications/pub_pdfs/ShiftingGears.pdf.

Marine Stewardship Council. www.msc.org.

Monterey Bay Aquarium. "Seafood Watch." www.seafoodwatch.org.

National Oceanic and Atmospheric Administration. "FishWatch: U.S. Seafood Facts." www.nmfs.noaa.gov/fishwatch/.

Seafood Choices Alliance. www.seafoodchoices.org.

Sustainable Fishery Advocates. "Fish Wise." www.sustainablefishery.org.

United States Department of Agriculture. "Country of Origin Labeling." www.ams.usda.gov/cool.

Washington State Department of Health. "Healthy Fish Eating Guide." www.doh.wa.gov/ehp/oehas/fish/fishchart.htm.

See Also: Fish, Farm-Raised; Fish, Line-Caught; Tuna, Dolphin-Safe

Tim Fitzgerald

Seeds, Extinction of

Agriculture output has benefitted from human-splicing of crop genes. Taking several varieties of a crop and introducing them together to create a crop that yields specific characteristics has occurred for centuries. For example, modern corn varieties have more ears of corn per plant, larger ears of corn, with larger kernels than natural varieties. The bioengineered process has recently increased in frequency with the introduction of genetically modified crops. Many of these crops are now considered "monocrops" as the number of varieties of a given food have been reduced.

Native species are increasingly threatened for several reasons. They are increasingly being removed or reduced in abundance to make room for modified crops with their higher outputs. Native crops are also being modified in their natural environment by cross-pollination with modified crops, producing yet another new crop variety. Last, seeds are not being collected and saved as was customary practice in previous generations.

A downside to the hybridization of plants is a loss of the original genetic materials and seeds. Natural crop varieties developed over millennia via the process of natural selection. During this evolutionary process, surviving plants developed immunities and resistances to various environmental threats such as climatic extremes, pests, and invasive species. Because many environmental threats are localized and require specific defenses against threats such as fungi or pests, many hybrid species do not have these resistances. Genetically modified species can be treated with pesticides and herbicides to withstand many of these threats, but in many cases the application of these substances is usually via a wide broadcast technique that can have negative impacts on the local ecosystem.

Native seeds are also viewed as important for regeneration of crops in the event of a large-scale extirpation. Collection of native seeds is an ongoing process undertaken globally by local agricultural cooperatives, individual farmers, and universities. Many collected seeds in developing nations are not stored properly (in a climate-controlled facility with low humidity) or are at risk of disruption during times of political strife. A seed storage facility known as the Svalbard International Seed Vault in Norway, which opened in 2008, aims to become the world's largest collection of native seeds in a climate-controlled environment well below freezing. The goal of the program is to protect as much seed diversity as possible and to be able to replant native species if necessary after a cataclysmic event.

See Also: Biodiversity; Seeds, Heirloom Varieties of

Michael Pease

Seeds, Heirloom Varieties of

Heirlooms are varieties of vegetables, fruits, and grains that were naturally emergent, historically cultivated, and developed in specific regions. Heirloom varieties disappeared rapidly in agricultural production with the advent of industrialization and hybridization of varietals. With the advent of the organic and sustainable agricultural movement and, equally important, the organic consumer movement in the United States, the identification and preservation of heirloom varieties has once again become widespread and systematic. Heirloom seeds are also growing in retail presence, and most often are cultivated organically through open pollination methods for use in small-scale agriculture and home gardening.

Heirloom seed revival in the United States has been primarily guided by two organizations. Through the 1970s, retention of heirloom varieties, predominantly of vegetables, had generally been the work of local gardeners.

However, the Seed Savers Exchange (SSE) first emerged in 1975 in Decorah, Iowa, headquartered at Heritage Farm. The farm not only serves as the primary seed bank of over twenty thousand heirloom varieties, but also maintains heritage livestock from the region. SSE is membership-based, whereby members receive newsletters and information on planting and cultivating techniques, and more recently, have broad access to materials in the farm's extensive seed bank. SSE has also expanded internationally, with the largest memberships found in Europe, Australia, and New Zealand.

In 1989 another leader in this area began operation in the southwest United States—Seeds of Change. Like SSE, Seeds of Change began with some of the varieties grown and tended by home gardeners, but went further in enlisting gardeners in future seed production. Kenny Ausubel, Alan Kapular, and like-minded others in the Santa Fe, New Mexico area, established a cooperative network with gardeners, consumers, and chefs. This network served to cultivate and preserve many heirloom varieties and became an early source point for heirloom seed retail. The company, its founders, and most of its customers are specifically "value-driven"—they are propelled by strong belief in the importance of retaining biodiversity and the need for a return to organic production systems.

Heirloom seeds are generally produced through open pollination and organic methods. Open pollination includes pollination by wind, insects, or other natural mechanisms. The end result of open pollination is biological diversity, as the distribution of pollen is unpredictable. In contrast, much of modern hybrid seed production is based on closed pollination, or the controlled manipulation of pollination to preserve and limit specific traits. Large-scale agricultural production depends on the controlled development of hybrid varieties to select for transportability and weather resilience.

Heirloom varieties are noted for their variety of appearance and flavor, which is typical of open pollination processes. The most commonly known forms of heirloom for new consumers are tomato and apple varieties. However, heirloom seeds are increasing in both production and popularity, and many more are emerging with the growth of farmers markets, community-supported agriculture, and other forms of small-scale production.

Further Reading

McDonald, Miller B., and Lawrence Copeland. *Seed Production: Principles and Practices*. New York: Chapman and Hall, 1997.

Stickland, Sue, foreword by Kent Whealy. *Heirloom Vegetables: A Home Gardener's Guide to Finding and Growing Vegetables from the Past*. New York: Fireside Books, 1998.

Weaver, William Woys, foreword by Peter Hatch. *Heirloom Vegetable Gardening: A Master Gardener's Guide to Planting, Growing, Seed Saving, and Cultural History*. New York: Henry Holt, 1997.

See Also: Seeds, Extinction of

Meredith Redlin

Shiva, Vandana (1952–)

Vandana Shiva is a leader in promoting environmental and human rights. She was born in 1952 in Dehradun in the northern Indian state of Uttarakhand. Shiva earned her PhD in physics from the University of Western Ontario in 1978 but shortly thereafter chose to pursue a life dedicated to environmental and social activism. In 1982 Shiva founded the Research Foundation on Science, Technology, and Ecology, located today in New Delhi, to support research on ecology, sustainability, and social issues. Navdanya ("nine seeds") was established by Shiva in 1991 to support seed biodiversity and local food sovereignty.

Shiva's first introduction to environmental activism was through the Chipko Movement in northern India in the 1970s. Poor rural women formed the backbone of the movement because environmental degradation caused by commercial forestry negatively affected women in the area to a greater degree than it affected men. The Chipko women were the original "tree huggers." Shiva eventually became an early proponent of ecofeminism and co-wrote the influential book *Ecofeminism* published in 1993.

For Shiva, human rights and environmental rights are inseparable. Shiva has documented links between farmer suicides in India and the introduction of agricultural modernization of the Green Revolution. Dependency on multinational corporations for expensive seeds and synthetic fertilizers and pesticides increased some farmers' debt and led to desperation. So-called modernized agricultural technology robbed Indian farmers of their dignity and degraded the environment at the same time.

Shiva opposes neoliberal globalization because of its negative impacts on food security and sovereignty for developing countries. Shiva is a vehement critic of the World Trade Organization's Agreement on Trade-Related Aspects of Intellectual Property Rights (TRIPS) because multinational corporations have used TRIPS as a tool to appropriate and profit from millennia of indigenous knowledge of seeds. When multinationals patent crop DNA, the indigenous people that had safeguarded and bred seed varieties for generations find that they must pay to use some of those same seed varieties. Shiva refers to this practice as biopiracy. Biopiracy not only leads to dependency on corporations for seeds and inputs, but can also reduce biodiversity. If farmers are no longer allowed to save and exchange seeds without the threat of being sued, fewer varieties of crops will be grown.

The institutions that Shiva has founded aim to honor human dignity by promoting a safe and healthy living and working environment. Shiva does not recommend a romanticized return to premodern agriculture as the solution to problems wrought by the globalized multinationals-dominated agriculture of today. Rather she argues that present-day solutions must be beneficial to all, especially the poorest of the poor. To achieve these goals Shiva believes that indigenous knowledge must be respected and advanced. One method employed is the establishment of community seed banks to promote conservation of seed diversity and sharing of knowledge in local communities. Shiva is also supportive of using local contextualized knowledge of organic farming to promote health for the natural environment, farmers, and consumers.

Further Reading
Meis, Maria, and Vandana Shiva. *Ecofeminism*. London: Zed Books, 1993.

See Also: Development, Sustainable; Green Revolution

Julie Weinert

Slow Food

Slow Food is both an organized global movement and an overall cultural response to a fast food mentality and an increasingly unstable and unsustainable global market economy.

The Slow Food movement put down its organizational roots in 1986 when Italian Carlo Petrini led a protest to a McDonald's fast food restaurant opening near the Spanish Steps in Rome, Italy. Petrini has since become a figurehead for the international Slow Food movement, which was officially constituted with a founding manifesto signed by fifteen countries in Paris in 1989. Since then, it has grown to more than 83,000 members in 122 countries. In the United States, there are nearly 120 Slow Food chapters in forty-five states.

The Slow Food movement encourages practices that focus on a variety of means to preserve the integrity of culturally and biologically diverse local food systems, such as forming heirloom seed banks, promoting local and traditional cuisine, and improving local infrastructures in the areas of processing, storage, and distribution. The movement also educates consumers about the health risks of fast food, the environmental and biological limitations of vast industrialized monocultures, factory farms, and the dangers posed by an over-reliance on too few plant varieties. Slow Food chapters also work to influence sustainable food policies in agriculture and lobby for support of organic and ecological farming practices.

Apart from specific organizational goals, the Slow Food movement has come to represent a more general desire for change in the current global food system on social, environmental, financial, and even aesthetic fronts. Socially, the Slow Food movement recognizes the value of food as a way to build stronger bonds with families, friends, and communities. It stresses the importance of knowing where one's food comes from and who grew it.

Environmentally, there is much to change by redirecting industrial agriculture's over-reliance on chemical fertilizers and cheap fossil fuels. Pesticides and herbicides also have well-documented negative environmental impacts from toxic ground water to hypoxic zones.

Financially, proponents of the Slow Food movement believe that a focus on local food production and infrastructure will help to revitalize rural economies by providing a living wage to farmers and farm workers. Currently, subsidies for commodity crops in Europe and the United States provide artificial price supports for large-scale growers, which most small to mid-sized growers cannot compete with realistically.

Nutritionally, locally produced fruits and vegetables are fresher and more healthy than their far-flung or heavily processed counterparts. Local food also supports a greater diversity in plant varieties because the current industrial food system values fruits and vegetables that emphasize durability over freshness and taste.

Aesthetically, the Slow Food movement wants to see more people savor the flavor of fresh, locally produced food and enjoy the time it takes to share and prepare meals with families, friends, and communities.

Further Reading
Petrini, Carlo. *Slow Food Nation: Why Our Food Should Be Good, Clean, and Fair.* New York: Rizzoli Ex Libris, 2007.
Slow Food International.www.slowfood.com.
Slow Food USA. www.slowfoodusa.org.

See Also: Fast Food; Local Food

Jerry Bradley

Soaps, Insecticidal

Insecticidal soaps, and all soaps in general, are made from the salts of fatty acids. Fatty acids are the principal components of the fats and oils in plants and animals. Oleic acid, present at high levels in olive oil and in lesser amounts in other vegetable oils, is more active as an insecticide than most other fatty acids, so many commercial insecticidal soaps contain potassium oleate (the potassium salt of oleic acid).

Although soaps have been mixed with water and used as sprays for insect control for many years, their mode of action is not completely understood. At least a portion of their toxicity to insects results from their dissolving or disrupting the waxes in insects' cuticles (body coverings), resulting in dehydration. In addition, soap solutions enter insects' spiracles (air tube openings) and damage cell membranes in the tracheae and smaller tubes that allow oxygen and carbon dioxide exchange within the insect body.

Many insecticidal soaps are approved for use in organic pest control because they are naturally derived and lethal only to the insects that they contact before sprays dry on plants or other surfaces. Soap sprays and their residues are not significantly toxic by contact to humans or other mammals, though like all soaps, they may cause drying or irritation of skin, and they are irritating to the eyes and other mucous membranes (nose, mouth, and genital openings).

In general, soaps are most effective against soft-bodied, immobile insects and related arthropods that do not fly away ahead of applications. Targets include aphids, mites, and immature whiteflies, leafhoppers, thrips, and scales (crawlers only). Soaps are less toxic to and less effective against more mobile insects with thicker, heavier cuticles, such as adult grasshoppers, beetles, wasps, bees, and flies.

Because soaps may dissolve or disrupt the waxy cuticles of plants, phytotoxicity (plant injury) may result in some species. Where this occurs, browning or curling of leaf edges is a common symptom, and leaf drop may follow. This is one reason to choose EPA-registered insecticidal soaps instead of attempting to prepare an insecticide from household products; the soaps that are formulated and registered specifically for their insecticidal value generally cause minimal injury to common plants to which they might be applied. Plants with very hairy leaves may be more susceptible to injury from soap sprays because the hairs help to hold spray droplets on leaves for a longer time. In general, the risk of plant injury from soap sprays can be reduced by rinsing any remaining soap droplets from plants 20 to 30 minutes after application, when most insect control benefits have been achieved but before severe plant injury occurs.

Do-it-yourself references may suggest making one's own soap sprays or other concoctions from common household cleansers or chemicals. Although common sense suggests that a 1 percent Ivory soap solution applied to houseplants should present no meaningful hazard, other detergents, cleansers, and solvents that may kill insects also may be quite toxic to humans, especially if applied to food crops or left in places where children or pets might contact or ingest them. Making one's own insecticides to use as "soap" sprays—even if the idea seems rather simple—usually produces substances equally toxic as or more toxic than many synthetic commercial pesticides.

See Also: Agriculture, Organic; Pesticides

Richard Weinzierl

Soil Health

In 1938 the USDA monograph of *Soil and Men* sponsored by Henry Wallace, President of Pioneer Hi-bred and then Secretary of Agriculture, developed a compendium highlighting the roles and importance of soil for present and future generations. This consciousness was born out of the 1930s Dust Bowl era and the world economic depression. Nevertheless, however remarkable the soil consciousness of the 1930s was, it was soon replaced by amnesia in

"The nation that destroys its soil destroys itself."
—Franklin D. Roosevelt

regard to the crucial importance of soil to food and agriculture.

Even as soil erosion in the United States continues at an unsustainable rate of over 4 tons per acre, top soil continues to be the most abundant United States export. World soils, besides being plagued by erosion of enormous acreage, are lost to salinity, low fertility, acidity, lack of top soil, and low organic matter. Production technologies have stressed fertilizers, pesticides, and genetic advancement of seed varieties, but little work has focused on the core issues of soil quality and health.

In organic agriculture, soil health is considered a prerequisite for plant and animal health. This paradigm is trumpeted by the Rodale Institute motto that "Healthy Soil = Healthy Food = Healthy People." Perhaps the scientist who most focused the connection of soil health to plant and animal health was William Albrecht, past head of the Soil Science Department at the University of Missouri.

He stated that "creation starts with a hand of dust." Albrecht used extensive U.S. Navy Dental records to show soil fertility was related to dental caries. Under areas of lesser soil mineral reserves, such as the southeastern United States, dental cavities were significantly increased, by 40 percent over the heartland.

Unlike many plant breeders who suggested only improved genetics can improve our food system, Albrecht was adamantly opposed to genetics as a substitute for soil fertility. As he quipped, "Do not ask that seed pedigree can substitute for good nutrition." He understood and explained that good nutrition was a reflection of quality of diet and that "good horses require good soil."

On the other side of balanced fertility proponents such as Rodale or Albrecht was Justus Liebig, who saw fertilizer salts as the foundation of modern high-yield agriculture. The lowest stave in a barrel was depicted as the chemical limiting input for plant productivity to optimize crop production. After World War II, there was an explosion of pesticide and fertilizer use in the development of the so-called Green Revolution to increase crop production and avoid food shortages. Although the short-term effects of fertilizers and pesticides were spectacular for their effects on production, their impact on the environment, health, and long-term productivity were neither desirable nor sustainable.

In the chemical age, soil was no longer considered the bastion of agricultural productivity. It became a victim of the Better Living Through Chemistry movement. By the 1960s Rachel Carson's book *Silent Spring* ushered in a new consciousness of the detrimental effects of chemically based agriculture. Insect pests were killed with the new chemistry, but so were the birds and the bees and even the humble earthworms that till the soil. It became abundantly clear that things had to evolve from chemical dependency.

During the 1970s the Oil Crisis portended that energy resources were limited. Moreover, chemical agriculture depended heavily on these, and their expense was skyrocketing. At this same time, monies devoted for medical treatments of chronic degenerative disease skyrocketed and the cost for industrial food sources declined sharply. Major diseases such as obesity, hypertension, diabetes, and osteoporosis have clear links to nutrition, and improved nutrition coming from the soil is increasingly seen as one alternative to quick fixes from synthetic chemical fertilizers, pesticides, drugs, and surgery.

This situation been referred to as the end of the Oil Age of Peak Production with the beginning of the Soil Age, in which renewable, sustainable, biological-based options will predominate. As an example of this drastic change, in 1960 20 percent of the Gross National Product was devoted to food production. By 2008 this same contribution was estimated at just 6 percent. Meanwhile the medical proportion of the GNP went from 6 percent in 1960 to over 20 percent currently. This switch indicates the future and current well-being related to food and agriculture choices.

In response, organic agriculture and food has become the fastest growing sector of the world food system. Over the last thirty years, organic agriculture has been growing at double-digit annual rates.

Although Justus Liebig and his followers fought the early battles in the war for the food system, by the turn of the nineteenth century alternative voices of change grew. F. H. King, in his book *Farmers of Forty Centuries*, marveled at the ability of soils from the Orient to maintain their health and productivity from management of organic amendments only, without use or dependency on fertilizers.

By the 1930s, Sir Albert Howard became the foremost proponent of the Law of Return, citing recycling as the most effective and natural mechanism for promoting soil productivity. Howard is credited with the Indore method of composting based on layered mixing of high carbon and high nitrogen plant and animal waste products. Jerome Irving Rodale trumpeted the value of compost in *Pay Dirt;* by the 1990s compost was a large, mainstream business.

In 1981, the long-term Rodale Farming Systems Trial showed that organic agriculture was not only competitive with conventional, chemical-based corn and soybean production systems, but could reduce cost, reduce environment degradation, and improve soil organic matter. In the 1990s Robert Rodale challenged the academic community to develop tests and assessments of soil quality or health. John Doran from USDA Research Service in Nebraska was among the pioneers to heed the call for a better understanding and measurement of soil quality and health.

Currently soil health is considered as multi-phasic measurement

"The land ethic simply enlarges the boundaries of the community to include soils, waters, plants, and animals, or collectively: the land.

A land ethic, then, reflects the existence of an ecological conscience, and this in turn reflects a conviction of individual responsibility for the health of land."

—Aldo Leopold, renowned environmental scientist. From *A Sand County Almanac*. New York: Oxford University Press, 1949.

of the physical structure, chemical balance, and biological vigor and integrity of the soil system. A 1999 study showed that organic farms were 22 percent higher in soil organic matter and 20 percent higher in soil nitrogen. Eighty percent of organic farms were closer to neutral acidity pH and had higher water holding capacity, higher weight of soil microbes, and higher respiration (i.e., were more alive).

Criteria that were identified in the study's soil health or quality index included soil respiration, water infiltration into soil, bulk density of soil, soil solution electro-conductivity, soil pH, soil nitrate, soil aggregate stability, earthworm activity, and soil water quality. The importance of biological indicators of soil health is gaining in its recognition and measurement.

In the late 1990s Sarah Wright from the USDA identified that mycorrhizal fungi were crucial to the development of crumblike structure through the action of beneficial glycoproteins that sheath these beneficial organism. In addition, the resistance of these compounds is being implicated in carbon accumulation in soils.

Soil carbon accumulation is a good indicator of the development of increased biological activity, higher water holding capacity, and better physical structure of the soil. In this way it is very important in understanding soil health, quality, and its measurement.

Whether from mycorrhizae or other sources, carbon in the soil has its root in the photosynthesis in plants and its repatriation into the soil. Carbon dynamics are so important because an increase in soil carbon resources will mean a decrease in greenhouse gas carbon dioxide. Although many scientists believe nitrogen to be the most planetary limiting factor for food production, water and energy sources are very important competitors for this role and position. Whereas nitrogen fertilizer holds no water and takes large amounts of energy to make, organic amendment improves the water percolation, retention, and crop use, and biological nitrogen fixation can supply all the nitrogen needs at the same proportion of the energy used in chemical nitrogen fixation by the Bosch Haber process. Unlike ammoniated nitrogen, organic forms are less prone to acidify the soil requiring lime applications to remedy.

The use of chemical fertilizers, which were believed to maintain critical soil organic matter through stimulated crop growth, has recently been studied. Nevertheless, about one hundred years of data show a continuing decline of soil organic matter reserves. The use of so-called modern agriculture techniques are known from several long term studies to reduce soil organic stocks by about 1 percent for every year they are practiced for up to fifty to seventy-five years before a new low soil organic matter is achieved.

The Rodale Institute has shown that organic agriculture can not only stave off the decline of soil organic matter but can promote its accumulation. Soil organic matter is not only a key component of soil health and crop productivity but a key to reversing global greenhouse gases that threaten the climate.

Further Reading

Albrecht, William. *Soil Fertility and Animal Health*. Austin, TX: Acres, 1975.

Hepperly, P. R., D. Desmond, C. Cook, and John Haberern. "Organic Farming Sequesters Atmospheric Carbon and Nutrients in Soils." The Rodale Institute White Paper 22 Year Farming System Trial Data. Kutztown, PA, 2003. www.newfarm.org/depts/NFfield_trials/1003/carbonwhitepaper.shtml.

Howard, Sir Albert. *An Agricultural Testament*. Rodale Press, Emmaus, 1943.

King, F. H. *Farmers of Forty Centuries*. Rodale Press, Emmaus, 1911.

Liebig, M., and J. Doran. "Importance of Organic Production Practices on Soil Quality Indicators." *Journal of Environmental Quality* 28 (1999): 1601–9.

Pimentel, D., P. Hepperly, J. Hanson, D. Douds, and R. Seidel. "Environment, Energy, and Economic Comparisons of Organic and Conventional Farming Systems." *Bioscience* 55(7) 2005: 573–82.

Rodale, J. I. *Pay Dirt*. Rodale Press, Emmaus, 1959.

U.S. Department of Agriculture. *Soils and Men: Yearbook of Agriculture 1938*, Washington D.C.: USDA, 1938.

See Also: Agriculture, Organic; Agriculture, Sustainable; Agroecology; Compost; Crop Rotation; Vermiculture

Paul Reed Hepperly

Substantial Equivalence

Substantial equivalence is a food safety concept that was first described in a document published by the Organization for Economic Cooperation and Development in 1993. It stated that

> For foods and food components from organisms developed by the application of modern biotechnology, the most practical approach to the determination is to consider whether they are substantially equivalent to analogous food product(s) if such exist.

> The concept of substantial equivalence embodies the idea that existing organisms used as foods, or as a source of food, can be used as the basis for comparison when assessing the safety of human consumption of a food or food component that has been modified or is new.

Once the concept of substantial equivalence was established, whenever official approval for the introduction of genetically modified (GM) foods has been given in Europe or the United States, regulatory committees have invoked it to justify and support their actions. This means that if a novel food, in particular one that is genetically modified, can be characterized as substantially equivalent to its "natural" antecedent or more traditional counterpart, for example, as safe as its traditional counterpart, it can be assumed to pose no new health risks and hence to be acceptable for commercial use. That is, it is generally regarded as safe or GRAS.

Such a regulatory approach is considered permissive toward technology commercialization and is consistent with U.S. regulatory policy that does not emphasize the process by which a technology is developed but rather emphasizes its final characteristics. In addition, U.S. regulatory science is oriented toward controlling the error of concluding falsely that a new technology will likely result in negative health or environmental outcomes (also known as Type 1 error). Therefore, a genetically engineered soybean is GRAS if its chemical composition, as measured by variables such as the amounts of protein, carbohydrate, vitamins and minerals, amino acids, fatty acids, fiber, ash, isoflavones, and lecithins, is substantially equivalent to a soybean produced through classical breeding techniques. The concept of substantial equivalence has been criticized as overly vague to the benefit of biotechnology firms. Also, it has been argued that scientists cannot predict or measure biochemical or toxicological effects of novel foods, including GM foods, from knowing their chemical compositions.

Further Reading

Jasanoff, S. *Designs on Nature: Science and Democracy in Europe and the United States.* Princeton, NJ: Princeton University Press, 2008.

Kessler, D. A., M. R. Taylor, J. H. Maryanski, E. L. Flamm, and L. S. Kahl. "The Safety of Foods Developed by Biotechnology." *Science* 256 (1992): 1747–49.

Millstone, E., E. Brunner, and S. Mayer. "Beyond 'Substantial Equivalence.'" *Nature* 401 (1999): 525–26.

Organization for Economic Cooperation and Development. *Safety Evaluation of Foods Derived by Modern Biotechnology.* Paris: OECD, 1993.

Welsh, R., and D. Ervin. "Precautionary Technology Development: The Case of Transgenic Crops." *Science, Technology and Human Values* 31 (2006): 153–72.

See Also: Bt Corn; *Diamond v. Chakrabarty*; Genetically Engineered Crops; Roundup Ready Soybeans; Terminator Gene

Rick Welsh

Suburban Sprawl

"Suburban sprawl" has more than one definition. As it relates to farmland use, it is generally thought of as low-density residential and commercial development characterized by large lot sizes (over one acre) and rural land conversion. As large-lot housing development has grown steadily over the last couple of decades, resulting in increasing farmland loss, the public's interest in issues surrounding sprawl has also increased.

Developers often prefer agricultural lands because they are flat and contain well-drained soils. The USDA's Economic Research Service estimates that between 1997 and 2001, on average, 2.1 million acres of land were converted to urban uses annually, a number that has increased over time. In addition, research has shown that those lands with the best soils for agricultural use are the ones being developed faster. The loss of farmland is also geographically skewed. Although the total amount of cropland has remained relatively stable over the last fifty years, regions such as the Northeast and Southeast have had larger percentage losses of cropland than many other regions, most likely due to urban use.

Conversion of farmland can have many effects on a farming community and the viability of the agricultural sector. One of the most important issues is the increase in land values as demand for the land grows. Although increasing land values may benefit farmers wanting to sell land for retirement or to raise capital for the farming enterprise, the cost to farmers looking to expand their operations or new farmers trying to enter the sector can become prohibitive. In addition, urbanization can result in fragmented landscapes in which the remaining farmland is noncontiguous. In these cases, the increasing proximity between suburban residents and farms can raise the threat of neighbor complaints or lawsuits about the consequences of normal farm operations (e.g., farm odors from livestock, noise of moving machinery, dust). It can also make once normal operations—such as moving equipment on public roads—difficult for farmers. Related to this is the loss or merger of agricultural input businesses (e.g., equipment dealers, bank lenders, chemical dealers) in communities as farm numbers dwindle; these changes can make it more difficult for the remaining farms to access needed inputs. Yet another impact is what some researchers call the impermanence syndrome—farmers in areas under development, perceiving the impermanence in agriculture in the community, cease to maintain soil fertility, drainage tiles, fences, livestock herds, and other essential elements of viable farm businesses.

Some researchers argue that conversion of prime farmland results in less productive farming because crop production is forced onto lands of lower quality. However, others argue that the remaining operations—or those farms on the urban edge—are not necessarily less productive. In fact, the increasing proximity to urban and suburban populations provides unique opportunities for some of these farmers. Research has shown that farms near metro areas are generally smaller, more diverse, and more focused on high-value production; these farms make up a third of the number of all U.S. farms and a third of the value of U.S. farm output. These farmers generally shift to more capital intensive commodities, and add businesses to their farm that take advantage of the proximity to urban populations. For example, the access to urban populations provides farmers with direct markets, such as farmers markets or farm stands, which can often result in a higher value for their agricultural products.

The percent of farmland being converted to other uses is small compared to the overall number of farmland acres in the country, leading some researchers to assert that there is little threat to the agricultural sector and its production capacity. However, although the overall food security of the nation is unlikely to be impacted by future farmland loss, specific sectors of agriculture can be greatly affected. For example, research by the American Farmland Trust found that in 1997 those counties subject to the most urban influence accounted for 63 percent of all of the nation's dairy products (by value) and 86 percent of all of its fruits and vegetables.

Public concern for the issue of farmland loss comes from a number of directions. For some, access to open and scenic landscapes is an important component of farmland protection. Related to this is the retention of the farming culture of the community. Some are also concerned about the costs of development to communities, including often the increased costs of public services (e.g., roads, water) that cannot be fully covered by the residential use taxes. Finally, the environmental effects of development compared to farmland use are of concern to some.

Federal, state, and local laws have been enacted in response to the growing public interest in the loss of farmland. All fifty states have at least one farmland protection program. Some of the key federal, state, and local agricultural protection programs include agricultural protection zoning (APZ), agricultural use-value tax assessments, and the purchase or transfer of development rights (PDR or TDR programs).

In general, APZ programs are county and municipal ordinances that designate areas where farming is the desired land use, discouraging other uses. Density of development is set, for example, one house per twenty or forty acres. In addition to limits on residential development, some APZ ordinances also contain limits on subdivision, site design criteria, and right-to-farm provisions. Although APZ has the potential to protect large tracts of farmland from residential and commercial development, many researchers and advocates are wary of it as the sole tool for farmland protection because it is easily modified by changing majorities in local government legislative bodies. Farmers can also be opposed to these programs if it affects their ability to sell their land at higher prices.

Agricultural use-value tax assessment programs (or differential assessment laws) direct local governments to assess agricultural land at its value for agriculture, instead of its full fair market value, which is generally higher in urban-influenced counties. Owners of farmland require fewer local public services than residential landowners, but they often pay a disproportionately high share of local property taxes. Differential assessment programs are meant to help bring farmers' property taxes in line with what it actually costs local governments to provide services to the land. Most states have these programs.

As part of state and local PDR or TDR programs, easements (usually permanent) are sold and attached to the property's deed and stipulate one or more restrictions on the use of the land, such as limiting the number or size of houses built on the property. Although the use of the land is legally restricted to agricultural use (e.g., production of crops, livestock, or livestock products), the purchaser of development rights normally cannot compel land to be actively farmed or ranched. The buyers are usually governmental entities or private land trusts in PDR programs, whereas the cost of development rights in TDR programs is usually borne by a developer who seeks to transfer the development rights from one parcel of land to another—developers "buy" development rights from agricultural land owners to be used for growth in designated development zones or land selected for its capacities (sewer, water, roads) to sustain denser growth. At the federal level, the Farmland Protection Program was enacted in 1996 to provide matching funds to state, local, and tribal governments to purchase conservation easements. Although changes have been made to the program through the years, funding for what is now called the Farmland Protection Program (in the 2008 Farm Bill) is set to increase from $48 million in 2007 to $200 million by 2012.

Although PDR and TDR programs are increasingly popular with the public, they have mixed reviews in their effectiveness. One of the goals of the programs often is to keep farmland affordable for both existing and new farmers; however, there is little evidence that these programs result in lower land values. These programs also tend to be expensive, and some have questioned whether enforcement of what is supposed to be a permanent restriction on development will continue over the long term. Finally, another concern is whether these farms are being preserved as "working farms" or farm businesses, rather than open space with minimal farm income being generated.

Little is still known about the relationships between farm viability in areas that are experiencing high farmland conversion, and the complex environment in which these farms operate, including the supply of land, labor, and agricultural inputs, and the marketing opportunities for farmers on the urban edge. On top of this, policies working to address the loss of farmland are being developed and implemented at local, state, and the federal levels, creating an assortment of possible models and responses to the problem. A number of case studies in counties across the United States have recently become available as researchers look at the specifics of these issues.

Further Reading

American Farmland Trust. Farmland Information Center. www.farmland.org/resources/fic/default.asp.

Bryan, Christopher Rex, and Thomas Johnston. *Agriculture in the City's Countryside.* Toronto: University of Toronto Press, 1992.

Daniels, Tom. *When City and Country Collide: Managing Growth in The Metropolitan Fringe.* Washington, D.C.: Island Press, 1999.

Heimlich, Ralph, and William Anderson. *Development at the Urban Fringe and Beyond: Impacts on Agricultural and Rural Land.* Washington, D.C.: U.S. Department of Agriculture, Economic Research Service, No. AER803, June 2001.

Nickerson, Cynthia, and Charles Barnard. *AREI Chapter 5.6 Farmland Protection Programs.* Washington, D.C.: U.S. Department of Agriculture, Economic Research Service, 2006.

USDA, Economic Research Service. "Land Use, Value, and Management Briefing Room." www.ers.usda.gov/Briefing/LandUse/.

Lydia Oberholtzer

Sustainability, Rural

Rural sustainability refers to the ability of rural communities to meet the needs of those who live and work in rural areas today without compromising the opportunities of those who might choose to live and work in those areas in the future. People have material, social, and spiritual needs that must be met if they are to experience a desirable quality of life. People also benefit socially and spiritually from their direct personal relationships with other people and with nature. Furthermore, people benefit materially from the natural and human resources that support their economies. Thus, rural sustainability must be built on a foundation of ecological, social, and economic integrity if the current needs of people in rural areas are to be met without compromising opportunities for future generations of rural people.

People historically have chosen to congregate in rural areas—in clans, tribes, towns, cities, and other types of communities—to pursue specific purposes. For example, the indigenous people who occupied rural America at the time of European immigration congregated in places where they were able to hunt and gather food and materials for clothing and shelter. They lived sustainably by relying on each other for companionship and protection while harvesting only as much as could be restored by nature in the particular areas where they lived. European immigrants brought a distinctly different culture, but they settled in rural areas for the same basic purpose: to take advantage of the natural resources. They harvested wildlife, timber, and minerals and, perhaps most important, grew crops on the fertile American farmland. European settlements sprang up on the frontier wherever fur trading, logging, mining, and farming would support them. Unlike the indigenous tribes of hunters and gathers, the European settlers didn't limit their production to only what was needed to meet their own needs; they produced surpluses of fur, lumber, minerals, and food crops for both local and distant markets. This allowed the establishment of urban manufacturing centers that relied on rural areas for their raw materials, and most important, for their food.

Unlike the Native American use, the Europeans' use of rural resources was not sustainable. Eventually, the fur-bearing animals were killed off, the old growth forests were logged off, and the minerals were mined out. The most persistent and durable of the initial rural settlements proved to be the farming communities. Logging and mining towns remain in a few rural areas; however, most remaining rural communities in the United States are remnant farming communities. Today, many of these remaining farming communities are being

"I am talking about the idea that as many as possible should share in the ownership of the land and thus be bound to it by economic interest, by the investment of love and work, by family loyalty, by memory and tradition."

—Wendell Berry, renowned author, poet, and farmer, in *The Unsettling of America: Culture and Agriculture.* San Francisco, CA: Sierra Club Books, 1977.

"farmed out," like most of the fur trading, logging, and mining communities before them.

At the beginning of the twentieth century, about 40 percent of all Americans lived on farms, and well over half were residents of nonmetropolitan areas. However, the industrialization of agriculture brought dramatic changes to rural areas. The process of industrialization began with the mechanization of agriculture in the early 1900s. Horse-powered farm machinery had been around since the early 1800s, but in the early 1900s, steam power threshing machines dramatically increased the efficiency of grain harvest. Most early harvesting machines were costly and complicated, meaning most farmers had to share work with their neighbors to gain access to these new technologies, which actually strengthened the social fabric of rural communities. The United States remained largely an agrarian nation through the Great Depression of the 1930s.

U.S. agriculture changed dramatically following World War II, and the changes in agriculture spawned equally dramatic changes in rural areas. The factories that had produced jeeps and tanks during World War II began manufacturing multiple-purpose farm tractors powered by internal combustion engines. Pull-type grain combines and silage choppers soon allowed farmers to work more efficiently and more independently. Wartime munitions factories began producing chemical fertilizers rather than explosives and chemical warfare technologies were used to create and manufacture agricultural pesticides. Farmers no longer needed both crops and livestock on their farms or the diverse crop rotations they had needed to maintain soil fertility and manage pests. The new technologies freed farmers from many of the vagaries of nature, giving them greater control over their farming operations, which greatly simplified farm management. By the 1960s, individual farmers were capable of producing far more crops and livestock than had seemed possible a couple of decades before. By specializing, standardizing, and consolidating into fewer larger farming operations, the industrial model of manufacturing was used to transform U.S. agriculture.

The transformation of agriculture transformed rural America. Communities depend on people, not just production or economic activity. Larger farms meant fewer farms, as agricultural productivity had expanded far faster than increases in population or consumer demand for food. Larger farms began to bypass their local dealers in marketing their products and buying their feeds, fertilizers, and equipment; they could get a better deal elsewhere. Fewer farms also meant fewer farm families to buy cars, groceries, clothes, and haircuts in small towns. Fewer families also meant fewer people to fill the desks in rural schools, the pews in rural churches, and the waiting rooms of rural doctors, or to volunteer for fire departments or serve on county commissions. Fewer people with a purpose for being in specific rural areas meant that many rural communities were rapidly losing their purpose for being. They were not sustainable.

Some rural communities were successful in recruiting industry to replace agriculture, with the promise of a dependable local workforce willing to work hard for relatively low pay. Labor unions had made deep inroads into traditional manufacturing areas with demands for higher wages, more fringe benefits, and better working conditions. Many rural people were willing to work harder for less. Some rural areas flourished economically, at least for a time, during the continued de-unionizing and decentralizing of U.S. manufacturing. By the turn of the twenty-first century, however, globalization of the economy meant that U.S. corporations could take advantage of even cheaper labor in less-developed countries, such as Mexico, China, and India. With the continued outsourcing of U.S. manufacturing, many rural communities are again searching for a purpose for being.

A few rural areas have been able to take advantage of unique local geographic features to develop viable tourist industries linked to lakes, mountains, or scenic landscapes. Others are located sufficiently close to fast-growing metropolitan areas to become "bedroom communities" for commuters. However, many rural areas were left competing with each other for prisons, urban landfills, toxic waste incinerators, or giant concentrated animal feeding operations (CAFOs). Lacking any higher economic purpose, their large empty spaces and sparse populations have been seen as opportune places to dispose of society's "wastes." Such purposes for rural places are not sustainable.

Within a few decades following World War II, the United States was transformed into a metropolitan society. At the turn of the new century, less than 2 percent of Americans called themselves farmers and on average farm families earned around 90 percent of their incomes from nonfarm sources. Only one-fourth of Americans lived in rural or nonmetropolitan areas, and many of those commuted daily to a city to work. In the meantime, the natural resources of many rural areas were polluted and degraded by factories that abandoned rural communities for cheaper labor elsewhere. Chemical-intensive farming and large animal CAFOs also polluted local streams, groundwater, and air with their wastes and degraded the soil through over-fertilization and erosion. Furthermore, the industrialization of rural America frequently resulted in short-run economic benefits for a few local investors at the expense of the communities as a whole. This violation of the traditional rural ethic of economic equity weakened the social fabric of once close-knit rural communities, leaving them unable to function politically. Many rural communities were left without ecological, social, or economic integrity—without rural sustainability.

PURPOSE AND PRINCIPLES OF SUSTAINABLE RURAL DEVELOPMENT

Sustainable rural development must be fundamentally different from the past industrial development of rural areas, both in purpose and principles. Industrial development is driven by the purpose of productivity, which is generally interpreted as economic efficiency. The principles that guide the quest for productivity are maximization of profits and growth. Industrial development is designed to maximize individual economic well-being, under the assumption that whatever is best for individuals is also best for communities and societies. The industrial model or paradigm of development seemed logical at the time of the Industrial Revolution. The economy was dominated by small individual proprietorships, unable to extract excess profits either from their customers or their workers. Natural resources were abundant and seemed unlimited, and the world seemed a largely empty place capable of absorbing infinite quantities of wastes. However, the world today is quite different. The economy today is dominated by large multinational corporations, quite capable of exploiting both their customers and employees in their pursuit of profits and growth. The productivity of today's industrial economies depends on the continuing extraction and exploitation of natural resources that are now known to be finite and are being rapidly depleted. Furthermore, nature has a limited capacity to assimilate toxins in its soil, water, and air, and humanity is rapidly running out of safe places to dump its wastes. The twin threats of declining fossil energy supplies and global climate change are putting the sustainability of human life on earth at risk.

The purpose of sustainable development is permanence—to meet the needs of both present and future—which requires ecological, social, and economic integrity. Sustainable development must be self-renewing and regenerative rather than extractive and exploitative. Sustainable development must rely on solar energy to recycle the physical molecules and elements that make up all living beings and nonliving materials and to offset the loss of usefulness of energy whenever energy is used or reused. Only living things are capable of capturing and storing solar energy, and thus, sustainable development must be patterned on the principles of living systems.

The principles of sustainable development can be derived from the ecological, social, and economic principles of sustainable living systems. All relationships in healthy, productive natural ecosystems are holistic, diverse, and interdependent. Human relationships in healthy, productive societies are trusting, kind, and courageous. Efficient and productive economic relationships are based on value, efficiency, and sovereignty. In addition, sustainability depends on integrity, meaning completeness, strength, and soundness. The sustainability of rural areas depends on the extent to which the ecological, social, and economic principles of sustainability permeate all aspects of the community. Social integrity depends on the principles of ecological and economic integrity. Ecological integrity depends on the principles of social and economic integrity. And perhaps most important, economic integrity depends on the principles of ecological and social integrity.

STRATEGIES FOR RURAL SUSTAINABILITY

Strategies for rural sustainability are different from strategies for industrial economic development. Sustainable development must be self-renewing and regenerative and thus must mimic the development stages of life. Living organisms are conceived, born, grow, mature, reproduce, and then die. Rural development ventures are far more likely to be sustainable if they originate with local people, are nurtured by local people, and grow to maturity under the direction of local people who are committed to the future of their community.

Support for local ventures during the stages of early development should focus on empowering people to solve their own problems and to realize their unique opportunities with information and knowledge appropriate for sustainable living processes. Public policies for sustainable rural development should be fundamentally different from the industrial development policies. Access to financing, appropriate marketing infrastructure, accommodative laws, and facilitating regulations are examples of the types of public policy encouragement local entrepreneurs need to grow, develop, and become mature, productive members of their communities.

The economic developmental stages of growth and maturity require little more than encouragement from the community. Once people have achieved maturity—economically, socially, and ethically—they are then capable of making meaningful commitments to the well-being of others. Mature members of sustainable rural communities are then capable of caring for others and for stewardship of nature as a way of giving back for what they have received. They become the most productive members of the local economy who also represent the self-renewing and regenerative forces of rural sustainability.

First, however, people must have a purpose for choosing to live and work in rural areas. Obviously, employment opportunities are an important consideration in most location decisions. However, quality of life factors also affect location decisions, including open spaces, clean air and water, scenic landscapes, and a sense of belonging or neighborliness. Rural areas traditionally have been viewed as desirable places to live and raise a family if acceptable employment opportunities were available. When people are free to live wherever they choose, as when people are independently wealthy or retired, they often choose to live in rural areas. With the advent of microcomputers and the Internet, an increasing number of jobs can be done in home offices, allowing more people to choose a rural lifestyle and quality of life.

The sustainability of the local agriculture can be an important consideration for people choosing to live in a particular rural area. Organic and sustainable agriculture have the capacity of employing far more people productively than industrial agriculture. Sustainable and organic agriculture also rely more on productive soils, rather than purchased inputs, and on labor and management provided by farm families, which keeps more of the economic returns to farming in the local community. Potential agricultural employment will be even greater as communities develop local food systems, which keep more of the food marketing, processing, and distribution activities in local communities. Agriculture may become more important, but will not likely become the primary purpose for many rural communities in the future.

More important for rural sustainability, sustainable farms protect the natural environment from degradation and pollution and sustainable local food systems help build a sense of connectedness and neighborliness in local communities. Organic and sustainable farms are good places for families to live and work and are good places for families to have as neighbors. They value clean air and water and maintain diverse landscapes for personal as well as professional reasons. Many sustainable farmers market fresh, healthful food directly to local customers at farmers markets and through community-supported agriculture associations and local food retailers, which help strengthen social and economic relationships within rural communities. Healthy farms and healthy foods help build healthy communities.

Further Reading

Bernard, Ted, and Jora Young. *The Ecology of Hope: Communities Collaborate for Sustainability.* Gabriola Island, BC: New Society Publishers, 1997.

Berry, Wendell. *The Art of the Commonplace: The Agrarian Essays of Wendell Berry.* Berkeley, CA: Counterpoint, 2003.Flora, Cornelia, ed. *Interactions between Agroecosystems and Rural Communities.* Boca Rotan, FL: CRC Press, 2001.

James, Sarah, and Torbjorn Lahti. *The Natural Step for Communities: How Cities and Towns Can Change to Sustainable Practices.* Gabriola Island, BC: New Society Publishers, 2004.

Summer, Jennifer. *Sustainability and the Civil Commons: Rural Communities in the Age of Globalization.* Toronto: University of Toronto Press, 2005.

Wood, Richard. *Survival of Rural America: Small Victories and Bitter Harvests.* Lawrence: University Press of Kansas, 2008.

See Also: Family Farms; Food Dollars; Local Food; Farmers Markets; Appendix 3: U.S. Organic Farming Emerges in the 1990s: Adoption of Certified Systems, 2001

John Ikerd

T

Terminator Gene

The terminator gene, also known as terminator technology, is the most controversial genetic use restriction technology (GURT). The agricultural biotechnology industry began developing GURTs in the late 1990s to restrict the spread of genetically engineered (GE) traits to non-GE plants. GURTs reinforce patent law by disabling or limiting biological reproduction and further protecting proprietary seed lines.

Plants engineered with the terminator gene produce sterile seeds with the express purpose of preventing seed saving. Delta & Pine Land (D&PL) and the United States Department of Agriculture introduced the technology in 1998 and until 2007 held sole patent rights. In June 2007, Monsanto successfully acquired D&PL. Despite a public promise in 1999 not to commercialize the terminator gene, the company, with the backing of the USDA, reinvigorated the research program. As of 2008, however, field trials were restricted to greenhouses in the United States.

The possibility of introducing genetically engineered sterility has spurred an international controversy; widespread opposition in North America, Europe, and the developing world successfully prevented commercialization for the last decade. Owing to public outcry, the United

Nations Convention on Biological Diversity adopted a *de facto* moratorium on the technology in 2000, recommending that GURTs not be approved for field-testing until sufficient scientific evidence indicated their safety. The parties to the Convention reaffirmed this position in 2006.

Opponents to the technology argue that the terminator gene has significant consequences for farming, particularly in the developing world, where seed saving is practiced widely. The terminator gene substantially affects the cost of production by increasing the price of seeds and further indebting farmers to multinational agribusiness firms. Limiting seed saving also restricts the development of regionally and locally adapted varieties and threatens agricultural viability across the globe. Opponents also fear that the terminator trait will not be expressed in the first generation, enabling the gene to move between GE and non-GE plants. Cross-pollination between crops will endanger the seed lines of those farmers who choose not to adopt the technology. Similarly, contamination of wild plants may disrupt biological reproduction in closely related species.

In the face of international opposition, biotechnology firms have developed T-GURTS, or trait-level GURTs as alternatives to the terminator gene. T-GURTs allow farmers to manipulate access to engineered traits using agrichemical promoters, thereby allowing them to save seeds without infringing on patents. A particularly interesting trait under development would allow farmers to remove fragments of DNA introduced through genetic engineering. This "Exorcist" gene snips specific sequences between two markers and could be turned on using a promoter. Biotechnology advocates claim that the technology would allow farmers to take advantage of the benefits of GE traits but still market their products to consumers wary of GE technology.

Further Reading

Smyth, Stuart, George G. Khachatourians, and Peter W. B. Phillips. "Liabilities and Economics of Transgenic Crops." *Nature Biotechnology* 20 (2002): 537–41.

Jianru Zuo, Qi-Wen Niu, Simon Geir Maller, and Nam-Hai Chua. "Chemical-regulated, Site-specific DNA Excision in Transgenic Plants." *Nature Biotechnology* 19 (February 1, 2001): 157–61.

See Also: Bt Corn; *Diamond v. Chakrabarty*; Genetically Engineered Crops; Roundup Ready Soybeans; Substantial Equivalence

Robin Jane Roff

Trader Joe's

Trader Joe's, named after company founder Joe Coulombe, began as Pronto Market convenience stores in the Los Angeles area in 1958. The markets were renamed Trader Joe's in 1967 when Coulombe sought to distinguish the stores from competitors like 7-Eleven. Trader Joe's were decorated with a South Seas motif, including Hawaiian shirts as the uniform for employees.

The average size of a Trader Joe's is about 12,000 square feet, smaller than traditional grocery stores. Trader Joe's is a privately owned company that was bought by Aldi, a German discount grocery chain, in 1979. Over three hundred stores, located mainly in suburban strip malls, are in the United States. Trader Joe's headquarters is located in Monrovia, California.

Trader Joe's markets itself as a place to find exotic cuisine at a low price. Prices are kept low via a few different strategies. Trader Joe's is a limited-assortment store that sells about two thousand different items in each store, less than 5 percent of the number of products sold in most full grocery stores. Therefore, Trader Joe's sells only those items in high demand with a high profit margin. About 85 percent of products in the stores are private label, or store brand goods. A private label allows Trader Joe's to negotiate directly with producers rather than rely on name brand wholesalers. Limited information is made available from the company about suppliers, but many goods come from small-scale operations from around the world. Local sourcing is not a stated priority, but the company Web site indicates that local goods, in particular fresh produce, are used when possible.

Although many organic products that have been certified by an independent third party are available at Trader Joe's, organic goods and sustainable food production are not the main focus of the company. All eggs sold under the Trader Joe's label come from cage free hens. In 2001 Trader Joe's made a commitment to eliminate genetically modified food from all products. Trader Joe's direct relationship with suppliers means that this goal can be accomplished more easily than for a store that works through multiple distributors. Trader Joe's makes efforts to inform customers about the origins and health benefits of their products through distribution of pamphlets and clear product labeling. The store's niche in the grocery market is as a place where educated, but not particularly well-paid, people can go to buy a high-quality product that is not found at a traditional grocery store. Organic goods have a fairly high profile at Trader Joe's because of demand from the customer base, not because of a deep commitment to the natural environment or to the health of the local agrarian economy. The example of Trader Joe's demonstrates the power that consumers can have in promoting sustainable practice. As a result of Trader Joe's response to customer demand, the company has been lauded by environmental and animal rights groups.

See Also: Brands; Natural Food; Whole Foods

Julie Weinert

Transition to Organic

Transition to organic is the process of converting a parcel of land from conventional, chemical agriculture to certified organic farming. The transition takes three years, during which time the crops that are harvested cannot be sold as organic. In the late 1990s, there was a market for these transitional crops, but that market no longer exists. Now farmers cannot count on any premium for their transition crops, so transition crops are sold in conventional markets.

In reality, transition is more complicated than simply not applying prohibited substances for three years. In most cases those substances served a purpose, usually to control weeds, insects, and diseases, and to provide soil nutrients. Without these conventional tools, growing a productive crop can be more challenging. These issues are best addressed before transition begins. Farmers contemplating a transition to organic production need to develop a clear plan for how weeds and other pests will be controlled and how soil quality will be maintained and, it is hoped, improved over time.

Transition serves two purposes. First, it provides time for residual chemicals to be purged from the soil and farm system. Three years is not a magic number. In reality, agricultural chemicals can persist in the soil in small quantities for many years. Three years is the time period set by the National Organic Standards Board (NOSB), the group responsible for crafting the official USDA definition of "organic" and the rules organic farmers must follow to maintain certification. Three years is considered a reasonable balance between the need to remove chemicals from the system and the time a farmer is required to wait before selling a crop as organic. Transition also provides time for the farmer to "transition" his thinking. Learning to farm organically usually presents a steep learning curve for the transitioning farmer.

Farmers decide to transition to organic for many different reasons. As long as organic products command substantial premiums, economics will be a strong incentive. Many times farmers decide to eliminate chemicals for personal health or safety reasons. Others switch because they believe organic farming is better for the environment. Some farmers find the challenge of farming organically more enjoyable.

Whatever the reason, there is no one method of transition. Every farm is different, and therefore, every transition is different. The particular characteristics of the farm should be taken into consideration as the farm plan for transition and organic production is developed. Farming organically typically requires a radically different approach to solving production problems and managing agricultural systems.

In recent years there has been a flood of organic products developed for use by organic farmers. It might be tempting to approach transition as mere input-substitution—removing prohibited substances and replacing them with organic inputs approved by the Organic Materials Review Institute (OMRI) and supposedly designed to serve the

State and Private Organic Certification

In 1988 Washington became the first state to develop organic standards and implement an organic certification program. Today, thirteen states have such programs: Colorado, Idaho, Iowa, Indiana, Kentucky, Maryland, New Hampshire, New Mexico, Nevada, Oklahoma, Rhode Island, Texas, and Washington. Otherwise, organic farmers chose independent certifiers that operate nationwide. These include Quality Assurance International and the Organic Crop Improvement Association.

same functions. Though possible in theory, there are good reasons to avoid this approach. First, the per-acre cost of alternative organic inputs is usually substantially higher than conventional inputs. In addition, most organic inputs developed for pest control are simply not as effective as their conventional counterparts. Farmers attempting the input-substitution approach will find themselves, much of the time, paying more for inputs that don't work very well. There may be times when the use of manufactured organic inputs is appropriate and effective, but these tools should be viewed as a part of a multifaceted approach to organic farm management, not as silver bullets.

Farmers planning for transition might think about their farms as a biological system. The official definition of organic as determined by the NOSB denotes more than just the absence of chemicals. Words like "ecological," "biodiversity," and "biological cycles" are used to emphasize an approach to farming characterized as a living system. These concepts are at the heart of organic farming. The ecological mentality is substantially different from that which operates by industrial standards. Typically, this can mean the use of practices such as longer rotations, cover crops, grazing animals, cultivators, rotary hoes, flaming, biological control agents, and so on.

The steps to organic certification are the same for all farms. First, a suitable, USDA-accredited certifying agency must be chosen. They should have experience in the farm's enterprise type, offer all additional certifications that might be needed, be a stable business, and provide their services at a reasonable cost.

Once a suitable certifier is identified, an application is submitted to that certifier. Though all certifiers work from the same USDA organic rules, each use their own, unique process and paperwork. An application form should be requested from the certifying agency. It should then be filled out and returned during the second year of transition. Part of the application is the Farm Plan Questionnaire. This document spells out how the land and crop will be managed in compliance with NOP rules. A fee is also paid at the time of application. Fees vary according to the certifying agency and the farm being certified.

Once the application is submitted, the certifying agency conducts a completeness review. During this time the application is reviewed for completeness and to confirm the ability of the farm to be certified in accordance with National Organic Program (NOP) rules. It is not uncommon for the certifying agency to request more information from the farmer at this point in the process.

During the third growing season of transition (and every year thereafter that certification is maintained) the certifying agency sends an inspector to the farm. The inspector's job is to observe every aspect of the operation—fields, buildings, and equipment. The inspector will also look at field borders and assess the risk of contamination from neighboring fields and commingling within storage and processing facilities. While at the farm, the inspector will examine all records related to the fields and will want to create an audit trail with those records documenting organic production practices from seed to harvest and storage. When the inspection is complete, the inspector will review all noncompliance issues with the farmer. The farmer and inspector then both sign an inspection affidavit.

After the farm visit the inspector writes and submits a report to the certifying agency. With the application, farm plan, and inspector report in hand, the certifying agency conducts a final review. This is typically done by a committee of people (the inspector is not included) who make one of the following rulings: approval for organic certification, request for additional information, notification of noncompliance, or denial of certification.

In addition to increased paperwork and recordkeeping, organic certification commits the farmer to a more ecological approach to production—using farming and ranching techniques and materials that conserve and build soil resource, create minimal pollution, and encourage development of a healthy diverse agroecosystem. Certification is also a commitment to maintaining the integrity of organically labeled food, by preventing contamination of organic production with prohibited materials, preventing accidental mixing of organic and conventional products, maintaining buffers—clear separation of organic crop and nonorganic crop—and cleaning equipment if needed.

Approaches to transition vary. The best strategy will depend on the farm and the farmer. Geography can make a huge difference. A grain farmer in the Midwest will have different challenges from those encountered by a fruit tree grower in California. For farmers with large acreage or multiple fields, the question comes down to the rate of transition—all at once or field by field. Some farmers who have transitioned large acreage all at once do not recommend this strategy. In the first year of transition, weed control can present a particularly vexing problem. To many, it makes better sense to start with one field, see how it goes, and learn from the experience before adding other fields in future years. Experienced farmers will have a sense for their land, and what order fields should be transitioned.

As transition progresses, the land, as a living entity, will itself change. Soil accustomed to chemical weed control and fertility year after year must have time to adjust after those chemicals are suddenly removed. The sudden change will affect crops differently. Because of weeds and possibly low yields, first-year organic soybeans are almost not worth trying. With corn, the nitrogen demands might require many tons of compost or manure, hauled in and spread, in order to get a respectable crop.

Transition can be made easier by putting fields into a small grain—wheat or oats—then into grass/legume cover crop. The land should rest in this state for a couple years and take a hay cutting or graze if some income is needed from the field. During transition, it's important to minimize tillage. If the land is rented, landlords will not like this approach because of low or no income during these years. But neither will they have any expenses. In the third year, begin the rotation with corn, soybeans or a small grain. This proven strategy makes transition much simpler for grain farmers, and should work well for vegetable producers, as well.

Further Reading

Anderson, Dan. *Pioneering Illinois Farmer Recount Transition to Organic.* Rodale Institute. 11/16/07. www.rodaleinstitute.org/pioneering_illinois_farmer_recounts_transition_to_organic.

Fossel, Peter V., *Organic Farming: Everything You Need to Know.* Voyageur Press. Minneapolis, MN, 2007.

"How To Go Organic." 2009. www.howtogoorganic.com/index.php?page=producers.

Rodale—The Organic Transition Course. www.tritrainingcenter.org/course/.

USDA-SARE. *Managing Cover Crops Profitably,* 3rd ed. Beltsville, MD: Sustainable Agriculture Network, 2007.

USDA-SARE. "Transitioning to Organic Production." Bulletin 408. 10/2003. www.sare.org/publications/organic/index.htm.

See Also: Agriculture, Organic; Certification, Organic; Family Farms; Grassroots Organic Movement

Dan Anderson

Tuna, Dolphin-Safe

The term "dolphin-safe" appears on the labels of canned tuna found in grocery stores. This general term refers to tuna that was caught without harming dolphins. During the 1990s, the United States began to restrict imports of canned tuna caught by dolphin sets (nets that are set by using dolphins as markers for the tuna); currently, only "dolphin-safe" tuna may be imported into the United States. "Dolphin-safe" tuna is defined as tuna caught in purse seine trips (a system of nets) where there are no sets on dolphins.

In the eastern Pacific Ocean, large yellowfin tuna (*Thunnus albacares*) weigh 10 to 40 kg. These large tuna associate with several species of dolphins, including spotted dolphins (*Stenella attenuata*), spinner dolphins (*S. longirostris*), and common dolphins (*Delphinus delphis* and *D. capensis*). Indeed, only large yellowfin are found with dolphins, as smaller yellowfin cannot match the swimming speed of these marine mammals. Tuna tend to concentrate below the marine mammals, rather than with them at the surface. In the eastern Pacific Ocean, fishers can easily find a school of tuna by locating a group of dolphins.

The gear used in this fishery is called purse seines and involves encircling a school of tuna with a long net (typically 200 m deep and 1.6 kilometers long). The net is weighted at the bottom and the top is kept at the surface of the water by a series of floats. One end of the net is pulled out from the main vessel by a skiff, which encircles the school of tuna, and the bottom of the net is then closed by a purse line run through the leadline by a series of rings. The net is then hauled in, and most of the net is brought onboard, leaving a small volume of water in the net and allowing the catch to be brought onboard using a large dip net. There are several types of purse seine sets: those set on dolphins (dolphin sets); those set on floating objects (floating object sets, or FADs); and those set on a school of tuna that is not associated with either dolphins or a floating object (unassociated sets).

To decrease marine mammal bycatch, when a purse seine net is hauled in, fishers "backdown" the net by reversing the vessel once the net is closed; this causes the corks and the top of the net to sink below the surface of the water, allowing dolphins to escape. Purse seine nets are also modified with a "Medina" panel, which is located at the top of the net, and is made of smaller mesh so that dolphins do not get entangled in the gear. In addition, dolphin mortality can be reduced with the use of rescue rafts that free dolphins that may be trapped in the net.

The bycatch of dolphins in the purse seine fishery in the eastern Pacific Ocean has declined dramatically since the 1980s. This reduced mortality is due to changes in the way fishers use the gear, as well as modifications to the gear. There was, however, an increase in dolphin mortality in the mid-1980s due to the presence of international fleets, which began to replace the diminishing U.S. fleet. This increased dolphin mortality was due to a lack of crew experience and ill-equipped vessels. Trade sanctions and international cooperation have resulted in decreased dolphin mortality in purse seines worldwide. However, the number of mother dolphins killed may be an underestimate due to unobserved or unreported mortalities.

An estimated six million dolphins have been killed in the eastern Pacific Ocean yellowfin purse seine fishery since it began; dolphin mortality was at its highest from 1959 to 1971. The Marine Mammal Protection Act (MMPA) requires mandatory observer coverage of tuna fisheries, and the Inter-American Tropical Tuna Commission has worked with fishers to modify their fishing practices in such a way as to reduce dolphin mortality. The International Dolphin Conservation Program (IDCP) has also set annual mortality rates for dolphins in purse seine fisheries.

Although the use of purse seines has resulted in fewer dolphin mortalities, there have been some unintended consequences. The switch from purse seines set on dolphins to those set on floating objects has resulted in an increase in bycatch of juvenile tuna and other fishes, as well as vulnerable species such as sea turtles and sharks.

Further Reading

Environmental Defense Fund. "Seafood Selector." www.edf.org/page.cfm?tagID=16314.

Monterey Bay Aquarium. Seafood Watch. www.montereybayaquarium.org/cr/seafoodwatch.aspx.

Monterey Bay Aquarium. "Seafood Watch, Seafood Report." July 9, 2009. www.scribd.com/doc/4434501/Seafood-Watch-Yellowfin-Tuna-Report.

See Also: Fish, Farm-Raised; Fish, Line-Caught; Seafood, Sustainable

Jesse Marsh

U

Union of Concerned Scientists

The Union of Concerned Scientists (UCS) is one of the leading science-based nonprofits working for a healthier environment and a safer world. The organization, which began as a collaboration between students and faculty members at the Massachusetts Institute of Technology in 1969, is now an alliance of more than 250,000 citizens and scientists. UCS combines independent scientific research and citizen action to develop innovative, practical solutions and to secure responsible changes in government policy, corporate practices, and consumer choices.

The organization is dedicated to the transformation of the U.S. food system to a more sustainable enterprise capable of bringing healthier food to our tables and economic opportunity to rural America without compromising the environmental integrity on which our future prosperity depends. UCS works toward creating an agricultural system that is part of the solution to environmental problems like climate pollution rather than among its causes while producing high-quality, safe, and affordable food.

UCS's Food and Environment program promotes sustainable agriculture practices, including organic and pasture-based livestock systems while eliminating harmful confined animal feeding operations (CAFOs), strengthening government oversight of genetically engineered food, and minimizing antibiotic use in animal agriculture. The organization is also becoming increasingly engaged in the interface between food production and climate change.

The organization has published a number of significant reports on food and agriculture issues, including: Hogging it!: Estimates of Antimicrobial Abuse in Livestock (2001); Gone to Seed: Transgenic Contaminants in the Traditional Seed Supply (2004); Greener Pastures: How Grass-fed Beef and Milk Contribute to Healthy Eating (2006); and CAFOS Uncovered: The Untold Costs of Confined Animal Feeding Operations (2008). These and other UCS reports are available on their Web site.

To promote sustainable agriculture, UCS engages in direct lobbying on key legislation such as the farm bill and actively monitors the administration of key food and agriculture programs at the United States Department of Agriculture and the Food and Drug Administration. UCS communicates on a regular basis with over 200,000 activists to educate them about sustainable food choices, provide information on current issues, and alert them about key opportunities to become involved and make their voices heard.

UCS accomplishments include successfully shaping legislation that supports organic and sustainable agriculture, achieving a meaningful consumer label for grass-fed meat, preventing a valuable human medicine from being used routinely in animal agriculture, pressing for a ban on the outdoor production of drugs and industrial chemicals in food crops, and strengthening oversight of genetically engineered crops and cloning.

Further Reading
Union of Concerned Scientists. www.ucsusa.org.

Brise Tencer

United Natural Foods

United Natural Foods (UNFI) is the largest organic and natural foods distributor in the United States, with annual sales totaling $3.4 billion in 2008. UNFI carries more than 60,000 products, including food, nutritional supplements, and personal care items. Approximately one-third of its sales are to one customer: Whole Foods Market, the leading natural/organic food retail chain. Although 42 percent of its sales are to thousands of independently owned natural food stores, the share of sales to conventional supermarkets is increasing, reaching 23 percent in 2008. Although UNFI is most dominant in the distribution of processed organic and natural foods, its subsidiary Albert's Organics is also the largest distributor of organic produce in the country.

UNFI is the result of more than two dozen mergers and acquisitions since the 1980s. The growth in natural food sales in the 1970s created an increased need for a distribution infrastructure to supply retailers. As a result, retail food cooperatives began to establish distribution businesses with a cooperative ownership structure. By 1982 there were twenty-eight cooperative distributors operating regionally in the United States. A process of consolidation then occurred, beginning with private natural foods distributors. UNFI was formed through a merger between Mountain Peoples Warehouse and Cornucopia Natural Foods in 1996, and became publicly traded on the NASDAQ stock exchange that same year. UNFI continued to acquire private and cooperatively owned distributors as it expanded. In 2002 UNFI acquired Blooming Prairie Cooperative and merged with Northeast Cooperatives (NEC), which had already merged with another cooperative distributor, FORC. Also in 2002, one of the largest remaining cooperative distributors, North Farm, declared bankruptcy. North Farm had previously merged with Michigan Federation of Food Cooperatives and Common Health Warehouse Cooperative Association in order to achieve economies of scale and keep pace with a rapidly growing industry. By 2008, all but one of the twenty-eight cooperative distributors had either been acquired by UNFI or exited the business.

Although UNFI is primarily a distributor, it is vertically integrated with a manufacturing division and a retail division (thirteen stores located in the Eastern United States). In addition to distributing major natural and organic food brands, it has acquired or developed its own private label brands, including Woodstock Farms, Grateful Harvest, Harvest Bay, Old Wessex, Natural Sea, Rising Moon Organics, and Organic Baby.

UNFI's largest national competitor is the organic and natural foods distributor Tree of Life. It is a subsidiary of Royal Wessanen, a publicly traded company based in the Netherlands. Tree of Life has also grown through acquisitions; it purchased nine distribution companies between 1996 and 2007. UNFI also faces competition from regional distributors, such as Kehe, Nature's Best, and Distribution Plus, as well as from supermarkets that increasingly source organic and natural foods directly from manufacturers.

Further Reading
United Natural Foods, Inc. www.unfi.com.

See Also: Brands; Market Concentration

Philip H. Howard

Urban Agriculture

Urban agriculture is the growing of plants and raising of animals in and around urban or peri-urban areas. It includes related activities such as marketing and processing of the food products. Although in many places, such as the United States, urban agriculture is practiced for recreational reasons, many people around the world practice it as a way to increase food security or alleviate poverty through income generating activities. As urban populations have increased, urban agriculture activities have also expanded rapidly over the last decade or more.

The Food and Agriculture Organization (FAO) of the United Nations has defined urban agriculture as small areas within cities, such as vacant lots, community gardens, rooftops, or balconies, used for growing crops and raising small livestock for personal consumption or sale in neighborhood markets. This food production is usually small scale. Peri-urban agriculture, on the other hand, includes farms close to the urban area that are producing crops or livestock for commercial purposes. These systems generally can be characterized by closeness of markets, limited space, competition for land, and specialization of both production systems and food products. FAO estimates that worldwide, 800 million urban residents are taking part in urban agriculture, and the United Nations Development Programme estimates that 15 percent of food worldwide is grown in cities. In addition, the majority (65 percent) of urban agriculture enterprises worldwide are undertaken by women.

There are a number of impetuses for an increasing interest in urban agriculture. Food security and poverty alleviation are two major influences. Higher food prices, consumer interest in locally grown products, and poor access to healthy food in the inner cities are also important issues in the development of urban agriculture. Many countries across the world, and international organizations such as the Food and Agriculture Organization, are also increasing their support of development of urban agriculture and often link it to poverty reduction, employment generation, and food security measures. In terms of food security, urban agriculture can increase the amount of fresh food (especially fresh produce) available to food insecure households. In addition, since lower-income households spend a higher percentage of household income on food, the savings for an individual household can be substantial. In some countries, urban agriculture makes up a significant percentage of the fresh produce supply in cities. Research has also shown that in some areas, urban agriculture also contributes to improved nutrition as a result of more stable access to food.

Other interests include environmental and community aspects. Some view urban agriculture as a way to improve environmental aspects by increasing green space in cities. For instance, the increased use of city rooftops for gardens can have a positive effect on energy consumption, water pollution, and biodiversity. Much attention has also been paid to the potential reuse of urban wastes through urban agriculture. Economic development and community revitalization, such as gardening on vacant lots or unused land in cities, are also viewed as important. Others see the opportunity for small business development.

Urban agriculture can also impact the social environment. For instance, in the United States, there is an increasing interest from school programs to incorporate gardens into the curriculum and school lunch programs, linking nutrition education and food access to fresh produce. Urban gardening may also be part of a therapeutic setting, such as senior centers, prisons, and drug treatment centers.

An increasing amount of urban agriculture also takes place on the urban edge. As population growth increases the expansion of urban areas around the world, a greater number of farms are located in peri-urban areas. For those peri-urban farms, the proximity to urban and suburban populations can provide unique opportunities. For example, the access to urban populations provides farmers with direct markets, such as farmers markets or farm stands, which can result in a higher value for their agricultural products. Research has shown that farms near metro areas are generally smaller, more diverse, and more focused on high-value production; these farms currently make up both a third of the number of all U.S. farms and a third of the value of U.S. farm output. These farmers generally shift to more capital-intensive commodities, and add businesses to their farm that take advantage of the proximity to urban populations.

There are also many challenges to urban agricultural production. Along with everyday agricultural problems, such as pests, urban areas provide unique challenges. For example, there is the potential that cultivation of the food products can take place on contaminated lands or be contaminated by traffic emissions or heavy metals. In addition, the heavy use of chemical fertilizers and pesticides can contaminate local water sources. Although often cities have a large amount of vacant land, land access and tenure are also often challenging. In addition, often the policy environment, whether local or national, is not conducive to urban agriculture as an urban land use, creating a number of political challenges.

Further Reading

Adeyemi, Abiola. *Urban Agriculture: An Abbreviated List of References and Resource Guide 2000*. Beltsville, MD: Alternative Farming Systems Information Center, National Agricultural Library, USDA, 2000.

Brown, Katherine, and Anne Carter. *Urban Agriculture and Community Food Security in the United States: Farming from the City Center to the Urban Fringe*. Venice, CA: Community Food Security Coalition, 2003.

Food and Agriculture Organization of the United Nations. "Issues in Urban Agriculture." *Spotlight*, 1999.

Smit, J., A. Ratt, and J. Nasr. *Urban Agriculture: Food, Jobs, and Sustainable Cities*. New York: United Nations Development Program, 1996.

Van Veenhuizen, Rene, ed. *Cities Farming for the Future: Urban Agriculture for Green and Productive Cities*. RUAF Foundation: International Institute of Rural Reconstruction International Development Research Centre, 2006.

See Also: Farmers Markets; Gardens, Community; Local Food

Lydia Oberholtzer

USDA *Agricultural Research Service*

USDA has provided research support for agriculture for over 130 years, and since 1953, USDA Agriculture Research Service (ARS) has been the Department's primary research organization. Over 2,300 ARS scientists work at more than one hundred locations covering all fifty states. The agency is organized into twenty-two national programs addressing four themes: nutrition, food safety and quality; animal production and protection; natural resources and sustainable agricultural systems; and crop production and protection. Research conducted at ARS differs from that at land grant universities in two important ways. First, its research programs are coordinated by National Program Leaders, helping USDA/ARS scientists to work together to make the biggest impact possible. Second, and more important for organic research, USDA/ARS has five year funding cycles, allowing its scientists to take on longer-term, higher risk projects.

For much of its history, USDA/ARS had little to do with research on organic agriculture but research in organic systems has made considerable gains at the agency over the past decade. In the 1980s and 1990s, as the organic food industry was taking off, ARS researchers who wanted to serve organic farmers had to stretch themselves very thin. They took on organic research projects even though they had research agendas that were already full with conventional agriculture projects. In a 2001 survey of USDA/ARS scientists interested in organic agriculture, one researcher noted that "we have no obstacles except for lack of time, funds, and personnel." This survey helped to create momentum for organic research at USDA/ARS by identifying the types of obstacles that farmers faced, including limited resources, scientific issues, lack of acceptance by the agency, limited cooperators, and regulatory issues. Another milestone in the development of organic research at USDA/ARS was its 2005 stakeholder/scientist workshop in Austin, Texas. The meeting was attended by sixty-three USDA/ARS scientists and included many invited organic farmers, industry representatives and policy makers. One of the most emphatic recommendations to come out of the survey was the need to appoint a national program leader for organic production systems. This suggestion was taken, and the position is occupied by Jeffrey Steiner of the ARS national program staff. Under his guidance, an Organic Research Action plan was introduced in 2005 and updated in the 2008–2012 Action Plan.

Since 2002, USDA/ARS resources allocated to organic production research have increased by 91 percent. As of 2008, there were twenty-three USDA/ARS research units in nineteen states actively conducting research on organic systems. Projects include research on smoothing transitions from conventional to organic farming, biologically based pest and parasite management in animal and plant production systems, and whole-farm strategies for increasing ecosystem services provided by organic production systems. Although organic research has come belatedly to USDA/ARS, the scientific knowledge of organic agriculture produced by the agency is likely to have important impacts on organic production. Because USDA/ARS scientists are located all over the country, yet coordinated at a national level, they are uniquely positioned to conduct research that has relevance at both local and regional levels.

Further Reading

USDA Study Team on Organic Farming. *Report and Recommendations on Organic Farming.* Washington, D.C.: United States Department of Agriculture, 1980.

Lipson, M. *Searching for the "O-Word."* Santa Cruz, CA: Organic Farming Research Foundation, 1997.

Bull, Carolee T. U.S. Federal Organic Research Activity Is Expanding. Crop Management. September 21, 2006. www.cropmanagement.org. DOI:10.1094/CM-2006-0921-18-RV.

Bull, C. T. "Organic Research at the USDA Agricultural Research Service Is Taking Root." *Journal of Vegetable Science* 12 (2006): 5–17.

Steiner, J. USDA/ARS Organic Research Action Plan 2008–2012. 2008. www.ars.usda.gov/research/programs/programs.htm?np_code=216&docid=14994.

See Also: Research; USDA Economic Research Service

Adam Davis

USDA Alternative Farming Systems Information Center

The Alternative Farming Systems Information Center (AFSIC) is a small, topic-oriented information clearinghouse located within the United States Department of Agriculture (USDA) National Agricultural Library in Beltsville, Maryland. AFSIC specializes in library services (locating, accessing, organizing, and distributing information) related to alternative, organic, and sustainable agriculture. The Center serves members of the public, both in the United States and internationally, as well as USDA staff members. Founded in 1985, the center was the first USDA-affiliated organization to focus on and serve the organic farming research community.

In the Food Security Act of 1985 (Farm Bill), Congress recognized that both enhanced agricultural production systems and sound conservation practices were essential to the long-term viability and profitability of U.S. farms. To that end, the bill authorized a study of all literature relating to organic agriculture including information and research relating to legume-crop rotations, use of green manure, animal manures, intercropping and erosion control methods, and biological pest control. In addition, the study called for analysis and dissemination of such literature. The National Agricultural Library, through AFSIC, was designated to fulfill this mandate.

For more than twenty years, staff members have worked to fulfill AFSIC's mission to identify and facilitate access to science-based information in several ways. They compile and distribute timely resource guides and bibliographies related to agricultural research topics and enterprise development, and they build directories and databases. Librarians also provide one-on-one reference services via mail, phone, and e-mail. AFSIC works collaboratively with other information providers to coordinate information technology and services.

Most importantly, AFSIC staff has worked to provide the National Agricultural Library with one of the best collections of alternative and sustainable agriculture-related research materials in the world. The collection of books, journals, reports, videos, databases, and electronic data is searchable via the National Agriculture Library's online catalog and article index, AGRICOLA. If unavailable elsewhere, most materials may be obtained from National Agriculture Library via the Interlibrary Loan program.

AFSIC maintains a comprehensive Web site (http://afsic.nal. usda.gov) that directs users not only to its own information products, but to many other online resources vetted and organized by AFSIC information specialists. Unique research tools include the Organic Roots database—a collection of historical (1880–1943) full-text USDA research bulletins; an annually updated directory, *Educational and Training Opportunities in Sustainable Agriculture*; publications about the history, definitions, regulations, and funding programs pertaining to sustainable and organic agriculture; detailed guides to resources on organic marketing and trade, and community-supported agriculture; and a series of videotaped interviews with organic farming pioneers, including Robert Rodale and Fred Kirschenmann.

AFSIC is supported in part by the USDA Sustainable Agriculture Research and Education (SARE) program, the USDA National Organic Program (NOP), and by a cooperative agreement with the University of Maryland.

See Also: Agriculture, Alternative; Education and Information, Organic; Media

Mary V. Gold

USDA Economic Research Service

The Economic Research Service (ERS) is a nonpartisan research agency within the United States Department of Agriculture. ERS researchers are highly trained economists and social scientists who conduct research on a wide range of economic and policy issues related to agriculture, as well as analyze food and commodity markets, produce policy studies, and develop economic and statistical indicators. ERS's work is used by policy makers, government officials, trade associations, the media, and the research community, and is disseminated to the public through reports, articles, and briefing rooms.

For more than a century, ERS and its predecessor agencies have supported USDA programs with economic data, research, and analysis needed for sound decision-making. Predecessor agencies were the Office of Farm Management (1905–22), and then the Bureau of Agricultural Economics (1922–53). Henry C. Taylor (1873–1969) served as the first chief of the Bureau of Agricultural Economics from 1922 to 1925, and set the stage for much of the work done in ERS today by organizing economic research into one agency and expanding the role of economics for understanding our food and agriculture system. USDA created ERS in 1961.

ERS's work portfolio includes issues related to sustainable agriculture, with a large body of work examining the relationship among natural resources, environmental quality, and agricultural production and consumption. One program of work conducts economic research on the efficiency, effectiveness, and equity of policies and programs directed toward improving the environmental performance of the agricultural sector, including farmland preservation, the Conservation Reserve Program (CRP), and the Environmental Quality Incentive Program (EQUIP). Another program of work focuses on the impact of agricultural production practices on the environment, and examines the critical role of economic and environmental factors in the adoption of management practices and technologies, including the use of conservation tillage, integrated pest management, precision farming, nutrient testing, organic farming, and biotechnology. Research is also conducted into water use and management, including managing water quality in the face of livestock farming.

ERS has an extensive research program on organic agriculture, covering many aspects related to producing and marketing organic food products. Current research related to producing organic food products includes cost of production estimates for organic dairy farms and a comparison of the relative profitability of organic and conventional dairy farms. Estimates of certified organic acreage, the number of certified organic animals, and the number of certified organic farms are provided for most years starting from 1997. Current research into marketing organic products includes an indepth analysis of procurement methods, including the use of contracts, by certified organic manufacturers, processors, and distributors. Other work has considered specific organic sectors, such as dairy, fresh fruits and vegetables, and poultry and eggs. Price premiums for select organic products at the farmgate, wholesale, and retail levels are estimated by ERS economists.

Further Reading

Economic Research Service, USDA. 2008. http://ers.usda.gov.

Economic Research Service, USDA. *Organic Agriculture Briefing Room.* 2008. http://ers.usda.gov/briefing/organic/. The views expressed are those of the author and not necessarily those of the Economic Research Service or the U.S. Department of Agriculture.

See Also: Research; USDA Agricultural Research Service

Carolyn Dimitri

USDA National Organic Program

The National Organic Program (NOP) develops and implements production, processing, and labeling standards for organic products produced and/or distributed in the United States. The NOP is administered through the Agricultural Marketing Service (AMS) of the U.S. Department of Agriculture (USDA). The NOP was established under the 1990 Organic Foods and Production Act because lawmakers wanted to provide consumers with uniform and consistent organic standards. The 1990 law also established the National Organic Standards Board (NOSB), a critical citizen advisory board responsible for recommending standards and materials for organic production. Due to a lengthy rulemaking process, and the large volume of public comments, national organic standards were not implemented until 2002, twelve years after enactment of the law.

PRODUCTION AND HANDLING STANDARDS

Production and processing operations must submit an organic system plan for approval by a certifying agent. The plan must describe all practices and procedures used, including materials; methods to evaluate the effectiveness of the operation; the recordkeeping system used to trace products; and procedures that will keep organic products separate from nonorganic products. In addition to submitting this system plan, organic crop and livestock producers and processors must comply with multiple and very specific management practices, some of which are summarized here but altogether include dozens of pages of regulations posted on the USDA NOP Web site.

For crops to be certified as organic the land must be under organic management practices for a minimum three years prior to the harvest of the crop. Organic farms must establish barriers to prevent contamination from nearby nonorganic crops, which usually means several feet of dedicated land around the perimeter of organic fields there for protective purposes and is not harvested for organic food. Overall crop management must aim to improve soil quality and reduce erosion. Crop rotation, cover crops, and application of plant and animal materials are required. If a producer uses raw animal manure on edible crops it must be applied ninety days in advance of harvest for crops that do not come in contact with soil, or applied 120 days in advance of harvest for crops that are in contact with soil. When compost is used, it must have a carbon to nitrogen ratio between 25:1 and 40:1 and it must reach a temperature between 131°F and 170°F. Sewage sludge is not permitted for use as a soil amendment or fertilizer under the organic program. Organic crop producers must use preventative measures to address pests, weeds, and disease. Such management practices include mulching, mowing, hand weeding, and mechanical cultivation among others. If such measures fail, biological or botanical substances may be used. In rare circumstances synthetic products may be used, but only if these products are recommended by the NOSB and the Secretary of Agriculture after a full scientific review and public comment. Approved synthetics are placed on the National List of approved materials for organic production. All but these few synthetic products on the National List are strictly prohibited.

Most livestock must be under organic management from the last third of gestation or hatching in order to be certified organic. Poultry must be under organic management from the second day of life. All feed rations must be organically produced and feeding of animal slaughter by-products is prohibited. Organic livestock producers must use preventative health care practices. This includes appropriate use of vitamins, minerals, protein, fats, and fiber in feed. It also includes providing livestock with appropriate and sanitary living conditions and room for exercise, movement, and other stress-reducing activities. Vaccines must be used to protect animal health. If preventative measures are not sufficient to maintain animal health, medications included on the National List may be used to treat illnesses. If the medications approved for use on the National List are not sufficient to treat livestock, the producer must seek additional treatment to address animal welfare; however, such animals can no longer be sold as organic. No animal treated with antibiotics can be sold as organic. The use of growth hormones is strictly prohibited, as is the use of approved medications at regular low dose intervals for preventative treatment.

Organic food processors must adhere to a set of standards to maintain the integrity of organic products processed in their facilities. The use of irradiation, genetically engineered processing aids, and synthetic volatile cleaning solvents are prohibited. Processors must take preventative measures to avoid pest infestations; in the event that such measures are insufficient they may use pest control substances from the National List. Additionally, if processors use both organic and nonorganic ingredients, they must take precautions to prevent cross-contamination of the organic products.

CERTIFICATION STANDARDS

Producers and processors seeking certification must submit their organic systems plan to an accredited certifying agent along with an application fee. A site visit is conducted to ensure that the submitted plan accurately reflects the management practices of the operation. The inspecting agent may collect soil, water, plant, or animal samples for verification (e.g., testing for residues of prohibited pesticides). The certifying agent will use the application and information gathered through the site visit to determine whether the operation qualifies for certification. Operations denied certification may reapply at any time, although they must include information regarding any previous denial of certification in the application. Operations that receive certification must update their organic systems plans on an annual basis, noting any changes or modifications to previous practices and undergo an on-site inspection.

ACCREDITATION STANDARDS

The NOP accredits state and private agencies to certify domestic and international producers and handlers. In other words, USDA inspects the inspectors. Certifying agencies must demonstrate to USDA that they have sufficient expertise to determine compliance with organic standards, which they do through record review, organic system plan approvals, site visits, and periodic laboratory testing. Certifying agents must also have recordkeeping procedures in place, maintain client confidentiality, and provide information to the public upon request. Approved agencies are accredited for five year intervals, after which time they must submit an application to renew their status.

Further Reading
Agricultural Marketing Service, USDA. National Organic Program. "Program Overview." www.ams.usda.gov/nop/.

See Also: Certification, Organic; Food Policy Council; National Organic Standards; Organic Labels; Policy, Food; USDA National Organic Standards Board; Appendix 2: Organic Foods Production Act of 1990

Kathleen Merrigan

USDA National Organic Standards Board

The National Organic Standards Board (NOSB) is a fifteen-member citizen board responsible for reviewing materials for organic production and advising the U.S. Secretary of Agriculture on all aspects of the U.S. National Organic Program (NOP). Unlike some federal advisory boards with limited influence on government programs, the NOSB is essential to the operation of the NOP. The centrality of the NOSB within the NOP is the result of widespread concern that existed back in 1990 about political opposition and lack of organic agriculture expertise within the U.S. Department of Agriculture (USDA). When Congress passed the Organic Foods Production Act of 1990 establishing the NOP, bill sponsors described the NOSB as part of the public-private partnership design of the law that would ensure that USDA would always adhere to organic principles.

The fifteen members of the NOSB are appointed by the Secretary of Agriculture to serve staggered five-year terms. By law, the NOSB is to be composed of four farmers, two food processors, one retailer, one scientist, three public interest advocates, three environmentalists, and one USDA accredited certifying agent. The most important role of the NOSB is to

Regulated by the USDA National Organic Program, this USDA Organic Seal means that the product is at least 95% organic. From http://www.ams.usda.gov.

evaluate materials for use in organic production. The basic rule of organic is that if a material is natural, it is allowed for use in production and processing, whereas if it is synthetic, it is prohibited. But as with any rule, there are exceptions, and the NOSB is vested with the authority to determine when those exceptions should be made. Individuals and organizations can petition the USDA to have substances placed on the National List, a list of exceptions to the natural/synthetic rule. The NOSB reviews these petitions and, following a two-thirds vote of the board, a material can be recommended for placement on the National List by the secretary of agriculture. Without a two-thirds vote of the NOSB, the secretary cannot approve synthetic materials for organic production.

The NOSB also advises the secretary on all aspects of the NOP, facilitates communication with the public by conducting meetings and soliciting public comments, provides technical support for the NOP staff, and assists with NOP outreach activities. Board members are divided into committees and task forces to efficiently complete NOSB tasks. NOSB standing committees include Compliance, Accreditation, and Certification; Crops; Handling; Livestock; Materials; and Policy Development. Task forces are established as needed. A task force must include at least one member of the NOSB, but may include several members of the general public. An Organic Pet Food Task Force is working to establish labeling standards for organic pet food, and an Aquatic Animals Task Force is seeking to expand the definition of livestock to include aquatic animals and address the use of fish in livestock feed.

Further Reading
Agricultural Marketing Service, USDA. National Organic Program, National Organic Standards Board. www.ams.usda.gov/NOSB/index.htm.

See Also: Certification, Organic; National Organic Standards; Organic Labels; Policy, Food; USDA National Organic Program

Kathleen Merrigan

V

Vegetarian Diet

A vegetarian diet omits meat. There are several specialized forms of vegetarianism. A vegan diet is completely without animal products; thus, honey, eggs, and milk are rejected. A lacto-ovo vegetarian diet includes eggs and milk, and a pesco-vegetarian diet includes fish. Some people assume that vegetarians only eat vegetables, but this is a mischaracterization. Indeed, vegetarian diets are varied and also include numerous grains, beans, oils, and spices.

There are many motivations for choosing to be vegetarian. Most vegetarians are motivated by religious, ethical, or philosophical reasons. Proponents say that a vegetarian diet offers a more equitable distribution of resources, which could end the famine and malnutrition of millions of people worldwide. At the individual level, there are health benefits to eating a vegetarian diet, with reports showing lower rates of cancers and heart disease. Others are motivated by concern for animals, as fellow intelligent and sentient beings. Animal cruelty, which is visible in the modern industrialized food industry, influences other vegetarians. People for the Ethical

Treatment of Animals (PETA) is an animal rights organization with more than two million members worldwide. Many well-known celebrities have spoken out in favor of a vegetarian lifestyle, including Carrie Underwood, Natalie Portman, Alicia Silverstone, and Paul McCartney, who has said: "It's staggering when you think about it. Vegetarianism takes care of so many things in one shot: ecology, famine, cruelty."

Indeed, there are numerous links between the social and ecological dimensions of food choices that motivate many humanists and environmentalists to choose a vegetarian lifestyle. Vegetarian diets promote sustainability as they reduce energy consumption, land degradation, habitat loss, global warming gases, pollution, and water usage.

The phrase "eating lower on the food chain" implies a greater consumption of plants. Animals are high on the food chain, having themselves grown big and meaty from eating those plants. Vegetarian diets, with ingredients from lower on the food chain, are more ecologically sustainable because they require less energy to produce a given amount of food. Research shows that up to twenty times more fossil fuel is needed to produce a pound of animal protein compared with a pound of vegetable protein. Significant research into this issue of energy conversion from plants to animals indicates that it takes ten times more grain to support a meat-based diet as a vegetarian diet. Indeed, about 80% of the total amount of grain produced in the United States is fed to livestock, not directly to people.

In addition to energy consumption, livestock need lots of land. Livestock production, including both the pasture land they require and the agricultural land in livestock feed crops, take up to 30% of the earth's land area. This pushes out other species as their habitat disappears and greatly reduces biodiversity. Of the deforested land in the Brazilian Amazon, 70% of this area is now in livestock pasture and the other 30% is producing crops for livestock consumption. Reducing the demand for animal protein would slow the current levels of rapid land use change across the planet.

Climate change is another environmental topic that is influenced by meat-based diets. Eighteen percent of global greenhouse gas emissions are caused by livestock. This is a larger percentage than transportation sources. Notably, cattle contribute a significant percentage of the human-induced methane gas, which warms the planet twenty times faster than carbon dioxide. Livestock contribute 9% of the total carbon dioxide emissions globally.

Production of animal protein is water-intensive. It takes about 2,500 gallons of water to produce one pound of beef. That is twenty-six times more water than needed to produce a pound of processed protein food based on soybeans. Livestock production consumes a full one-half of all water used for all purposes in the United States. In regions of the world where water is scarce, the raising of livestock is putting increasing pressure on water supplies.

Water quality is also critical. The production of corn and soybeans for livestock consumption relies heavily on synthetic chemical pesticides and fertilizers. Large-scale meat production (feedlots and processing plants) relies on additional pesticides, disinfectants, and antibiotics. There is concern about the extent to which these chemicals runoff into neighboring waterways, which eventually flow into rivers and impact drinking water sources.

In 2009 it was estimated that twenty million people would die from malnutrition worldwide. If more people chose a vegetarian diet and demand for animal protein decreased, farmers could shift away from growing grains for livestock and instead produce grains and vegetables that would easily feed the entire population of the world. At the same time, deforestation could cease, gases promoting climate change would decrease, water and air pollution would be reduced, and fossil fuel use would decline. The most comprehensive argument in favor of vegetarianism links both ethical and ecological factors.

Further Reading

GoVeg.com. "Meat and the Environment." http://www.goveg.com/environment.asp.

Key, T. G., and Appleby, Davey P. "Health Benefits of a Vegetarian Diet." *Proceedings of the Nutrition Society* 58 (1999): 271–5.

Leitzmann, Claus. "Nutrition Ecology: The Contribution of Vegetarian Diets." *American Journal of Clinical Nutrition* 78 (2003): 657S-9S.

PETA: People for the Ethical Treatment of Animals. "Mission Statement." www.peta.org/about

Pimentel, David, and Marcia Pimentel. *Food, Energy and Society.* Boca Raton, FL: CRC Press, 2007.

Reijnders, Lucas, and Sam Soret. "Quantification of the Environmental Impact of Different Dietary Protein Choices." *American Journal of Clinical Nutrition.* 2003: 664S-8S.

Steinfeld, H. "Livestock's Long Shadow: Environmental Issues and Options." Rome: United Nations, FAO, 2006.

See Also: Animal Welfare; Biodiversity; Climate Change; Environmental Issues; Fossil Fuel Use; Water Quality

Leslie A. Duram

Vermiculture

Worm composting, often referred to as vermicomposting, is not really compost, but rather the manure or "casting" of earth worms. Earthworms flourish on a diet of partially decomposed organic matter. As the worms digest this material, they tunnel through the soil and deposit their manure, which appear as small pellets. Tunneling gives improved drainage and aeration to the soil, while the digestive action of the worms promotes nutrient recycling. The chief composting worms are found in European and North American soils and are prime indicators of the biological health of soil.

Earthworm activity has a profound beneficial effect providing a physical and chemical environment that improves crop and plant growth. Scientific studies at Rodale Institute show that vermicompost can produce superior yields of potatoes and beets. Worms used in "composting" are also used as fish bait. They are also an excellent nutrient supplement for many animals, especially birds.

Besides eating decomposing vegetation, worms are fond of diets including manure. Worms possess strong gizzards that grind their food, promoting the recycling and breakdown of organic materials that increase the soluble nutrients content of the soil. Enhanced nutrient availability in turn stimulates plant and crop development.

As the ground material passes through the worm gullet, it combines with digestive mucus and forms spherical fecal pellets. These fecal pellets not only aid in nutrient availability. One scientist has suggested that earthworm pellets are optimal for improving the physical structure of soil.

As prime indicators of healthy soil conditions, a lack of earthworms generally indicates problems in compaction, drainage, and aeration. Soils that are high in aluminum and acidity are not a conducive environment for earthworms, which predominate in neutral pH soils high in calcium content.

Excessive deep moldboard plowing of soil is very damaging to earth worms as it mechanically disrupts the natural earthworm environment and activity. Also damaging is the use of agricultural pesticides, such as copper and synthetic ammoniated fertilizer, that increase soil acidity.

In Europe, earthworms include *Eisenia foetida* and *Lubricus rubellus*. In the book *Worms Eat My Garbage*, Mary Appelhof explains how home-scale worm rearing became popularized. Simple containers designed for rearing worms are excellent for demonstrating genesis soil organic matter and recycling. For more commercial application, however, products may be produced at larger scale, and materials can be dried to promote shipping, conservation, and reduced costs.

Live earthworms require careful monitoring and control of the environment in which they grow and reproduce.

Charles Darwin was fascinated with the beneficial role of earth worms to the development and productivity of soil. See his book, *The Role of Earth Worms in the Development of Earth Mold*. Clive Roberts of the United Kingdom is often considered the originator of modern wormery and holds patents on certain wormery designs and use. Municipal composting as done in North America has its origin and roots in India. Sir Albert Howard, known as the father of the modern organic method, adapted traditional compost practice. Compost was produced by forming large piles in which carbon- and nitrogen-rich materials are mixed by layering to give a proper carbon and nitrogen ratio of 1:30. These piles, called windrows, provide sufficient mass to insulate themselves, providing a high temperature environment under intense aerobic fermentation. This is a hot and fast process.

Worm composting is also done with a mixture of carbon- and nitrogen-rich materials in windrows. These windrows are much lower to avoid high temperatures that would kill off the beneficial earthworm populations.

In both worm compost and high temperature compost, about three volumes of carbon-rich material is mixed with one volume of nitrogen-rich material. Straw or wood products are commonly used as carbon-rich material; animal manures usually provide the nitrogen-rich component.

High temperatures of municipal compost, 50 to 60°C, can eliminate beneficial microorganisms than flourish under moderate, roomlike temperatures of 30 to 35°C. Worm compost on the other hand do not produce the types of temperatures that pasteurize the compost pile and provide fast compost turnaround.

Worm compost, being rich in soluble nutrients, gives a stimulation of crop and plant growth more akin to raw manures and soluble fertilizers than municipal, high-temperature compost. Because of the lower temperature in its production, worm compost can be an excellent source of beneficial microorganisms. For this reason, vermicompost has been used for making extractions called "compost tea," which are used both as an organic foliar fertilizer and a microbially based tonic.

The chief benefits of vermicompost include soil structure improvement, soil microbial activity enhancement, enhanced soil drainage and aeration, improved soil water holding capacity, enhanced nutrient recycling and availability, stimulation of seed germination and seedling growth, improved root growth, improved health through less soilborne disease.

The chief limitation for use of worm products has been their relatively high production cost. Making these products more economical would allow expanded use of these products. With the advancement of competitive production methods, worm composting could play an expanded role in soil management for more farmers. It has been said that "global worming is the solution to global warming." This catchy saying is based on the amazing capacity of worms to recycle and improve our soils, including its organic matter.

Using worm castings produced under current windrow technology can be used for certified organic production systems with high productivity for small- and medium-scale operations. With improved cost of vermicompost, "global worming" can become a reality.

Further Reading

Appelhof, Mary. *Worms Eat My Garbage*. Kalamazoo, MI: Flower Press, 1997.

Fong, Jen, and Paula Hewitt. Worm Composting Basics. Cornell University. http://compost.css.cornell.edu/worms/basics.html.

Rodale Institute. "Worm Bin Construction Made Simple." http://www.rodaleinstitute .org/20040801/Grube.

Trautmann, N. M., and M. E. Krasny. *Composting in the Classroom*. Dubuque, IA: Kendall/Hunt, 1998. http://compost.css.cornell.edu/CIC.html.

See Also: Agriculture, Organic; Agriculture, Sustainable; Agroecology; Soil Health

Paul Reed Hepperly

Walmart

Walmart is the world's largest retailer. More than 180 million people shop at Walmart stores every week. Over 4,000 stores are located in the United States, and nearly 3,000 stores are located in thirteen other countries around the world.

The first Walmart store opened in 1962 in Rogers, Arkansas, and has since expanded to appeal to diverse markets in the United States and around the world. Walmart Stores, Inc., with headquarters in Bentonville, Arkansas, operates several different prototypes. Walmart discount stores, the earliest form of Walmart stores, sell general merchandise. Walmart Supercenters sell general merchandise, groceries, and often offer special services like vision centers and pharmacies. Neighborhood Markets sell a similar array of products as in Supercenters but are mostly located in large urban areas where zoning restrictions and high property values necessitate smaller store sizes. Sam's Club warehouses sell general merchandise and wholesale items to individuals and businesses that pay an annual membership fee. Walmart.com provides an online shopping option for customers.

Given Walmart's international influence on retailing, a move towards promoting environmentally sustainable practices in 2005 and a wider introduction of organic products in 2006 was greeted by the environmentalist community with both hope and skepticism. Organic products sold at Walmart stores include fresh produce, dairy products, dry goods, and textiles made with organic cotton.

Expanding organic goods availability at Walmart clearly demonstrates promise. Walmart has the capacity to expose a very large consumer base to organic goods and to encourage organic production among suppliers around the world. Walmart's mission to lower prices for consumers extends to organic goods as well. Walmart aims to price organic goods to within 10 percent of conventionally produced goods. It has also vowed to support farmers during transitions from conventional to organic methods and to purchase organic food from local sources when possible.

The main critique leveled against Walmart's foray into the organic market is that the retailer will follow the letter, but not the spirit, of USDA guidelines for organic production. Some fear that, as a retailer that has set industry standards for production and pricing in other goods, Walmart will push for new, lower standards for organic production. Some environmentalists fear that a Walmart-led effort for organic production will not be sustainable, because Walmart's central mission is to provide cheap consumer goods, not to ensure environmental sustainability. Some potentially negative consequences of low-price organics at Walmart are that only organic goods that can be produced cheaply will be supported and that efficiency techniques employed in conventional agriculture, like extensive monocultures or confined animal feeding operations (CAFOs), will be used to mass-produce organic goods. Finally, Walmart has a global supplier network in place so if organic goods from global sources can be produced for less than from domestic sources, more energy will be expended for transportation.

See Also: Brands; Globalization

Julie Weinert

Water Quality

Water quality is affected by myriad factors in agricultural environments. As water cycles through agricultural environments, several processes occur that can alter the chemical composition of water. These alterations can improve or decrease water quality. Many of the negative impacts agriculture has on water quality are exacerbated by human-induced actions.

Agricultural lands are centers of high organic matter density, many of which can dissolve in water, or can be suspended in water and transported. The movement of nutrients and particulates is natural process and one that is important in nutrient cycling. However, the introduction of high concentrations of inorganic fertilizers has led to an increase in concentration of

Buffer pond. Courtesy of Leslie Duram.

some nutrients. In addition, high intensity agriculture has led to increases in soil disturbances, resulting in increased erosion and water turbidity.

The major water quality parameters negatively affected by agricultural waste runoff include increases in pH, nitrogen, phosphorus, salinity, and conductivity, and decreases in dissolved oxygen, which has a negative impact on aquatic species.

Increased salinity is one of the more damaging forms of water impairment that result from agricultural runoff. Highly saline water impairs, and can even be lethal to salt-intolerant plants. The Salton Sea, an inland water body in Southern California fed by agricultural runoff, has salinity as high as 44,000 mg/liter.

Nitrogen runoff from agriculture, usually as the result of the application of inorganic fertilizers to increase production leads to eutrophication, or increased algae growth in stagnant water and subsequently reduced dissolved oxygen content.

Confined animal feedlot operations (CAFOs) are another agricultural source of water contamination. Nitrogen runoff from manure can pose such a threat to water quality that the Clean Water Act regulations were modified to include CAFOs as point-source pollutants, the same category given to factories. High-intensity pig feedlot operations in North Carolina have been linked to increased levels of nitrogen and pesticides in water. Animal waste from CAFOs is usually stored in lagoons with an impermeable layer, but runoff from these can still occur during periods of intense rainfall.

Despite the numerous sources of potential water quality impairment surrounding agriculture, farms need not be liabilities to water quality. Agricultural environments can serve as ecological function hotspots that provide habitat for terrestrial species, migratory birds, and water cycling. In an over-simplified example, farms that use only necessary amounts of nitrogen and phosphorus-based fertilizers and that set aside low-lying areas to serve as natural wetlands could see a natural improvement in water quality as native plants would filter contaminates.

See Also: Environmental Issues; Environmental Protection Agency; Gulf Hypoxia; Pesticides; Precautionary Principle; Vegetarian Diet

Michael Pease

Whole Foods

The first Whole Foods store was opened in 1980 in Austin, Texas. Today, after numerous acquisitions of other alternative food store chains and establishing other new locations, over 270 high-end supermarkets operate in the United States, Canada, and the United Kingdom. The store's mission statement is, "We're highly selective about what we sell, dedicated to stringent Quality Standards, and committed to sustainable agriculture." The Whole Foods brand is built around the idea that people are a part of the natural environment and not superior to nature. Whole Foods recognizes several different stakeholders that must be taken into account for the continued success of the company: customers, employees, investors, the environment, and the good will of the local community.

To accomplish its manifold goals Whole Foods specializes in selling high-quality organic goods. Whole Foods was the first company to be designated as a certified organic grocer by the U.S. Department of Agriculture's National Organic Program (USDA NOP). Whole Foods has recently encountered a few challenges to its success. Increased popularity of organic products has stimulated growth of organics in traditional grocery stores. Now Walmart, Kroger, and Safeway are in direct competition with Whole Foods for environmentally and health conscious shoppers. Organic foods are also highly susceptible to fluctuations in the economy. Like other products, the cost of sustainably produced foods increases when the cost of inputs and fuel increases. However, when consumers have less disposable income, sales of premium-priced organic goods fall more quickly than sales of cheaper, conventionally grown foods.

Several critiques have been leveled against Whole Foods. An overriding issue is whether consumerism and corporate management can be compatible with environmental and social sustainability. The target clientele of

Whole Foods are clearly not lower income families, based on the premium price of goods and the location of most stores in wealthy suburbs. Additionally most stores are located in large cities on the East and West coasts of the United States; it has been argued that stores appeal to elitists that do not represent core American values.

Even though environmentally and socially sustainable practice is a key part of Whole Foods' mission statement, Whole Foods is still a public traded corporation that relies on profit for survival. Whole Foods has been criticized for veering from its focus on wholesome foods by building massive supermarkets that can approach 80,000 square feet and that offer high value-added restaurant and spa services, prepared foods, and three-dollar chocolate bonbons. In addition, the company has taken an anti-union stance, which bothers labor groups. Whole foods has also acquired many of the smaller natural foods stores and smaller health food chains, making consumer groups concerned about a lack of competition.

Whole Foods' commitment to sustainable agriculture has also been questioned. The bulk of products are sourced from large organic distributors rather than from small local farms. One 2006 study showed that the majority of organic beef at Whole Foods was purchased from foreign sources. Even though the food is organic, the resource used for transit of such goods is high, and support for local farms is fairly low.

See Also: Brands; Natural Food; Trader Joe's

Julie Weinert

Wine, Organic

Organic wine results from certified organic practices for both growing winegrapes in the vineyards and making wine. Therefore, the production of organic wine begins with the application of organic grape growing methods.

GROWING ORGANIC WINEGRAPES

Organic viticulture is an evolving science. It involves continual monitoring and adapting specific farming methods to ever-changing environmental conditions in each local vineyard location. Growing winegrapes organically entails applying basic organic principles, including building the health of the soil, elimination of synthetic pesticides and fertilizers, enhancing and conserving biodiversity, using a "systems" approach (a fully integrated orientation), and using cultural and biological methods to prevent pest problems (such as canopy management, water management, cover crops or habitat to attract beneficial insects). Material inputs are applied if needed, only as a last resort. By incorporating these principles, organic viticulture methods generally help protect the health of employees and neighboring communities, protect water and air quality, and produce excellent quality grapes.

Louisiana winery owner whose goal is to make "really good local wine that people can enjoy with our really good local food." (AP Photo/Bill Haber)

SOIL MANAGEMENT, SOIL HEALTH, AND COVER CROPS

An important element of organic vineyard management is building the health of the soil, which includes methods that increase soil organic matter, increase the diversity of soil organisms and biological activity, protect the soil from erosion, and balance the soil to enhance plant growth. Each vineyard site requires different methods to achieve these outcomes. Organic growers aim to increase soil organic matter to about 3 to 6 percent, if possible, because it will offer the following benefits: improved soil tilth, soil structure, and the creation of stable soil aggregates; improved water infiltration and moisture retention; and increased biological activity and biodiversity in the soil macro- and microflora. Thus organic management improves soil fertility.

The main methods for soil management in organic vineyards include compost application and incorporating cover crops between the vine rows. Cover crops are usually seeded in the fall and grow during the winter and spring months. Winegrape growers usually use a mixture of cover crop species. Examples include bell beans, clover, purple vetch, common vetch, winter peas, oats, barley, and mustard. In vineyards planted on more fertile sites, annual legume and small grain mixes are common. Self reseeding annual legumes are often used. They fix nitrogen, and they do not usually compete for moisture with the vines. Cover crops also provide habitat for beneficial insects, mites, and spiders. They prevent erosion, retain moisture, and reduce dust. Often annual grasses are planted to enhance soils and increase biodiversity. Often, the cover crops are mowed during bloom and then tilled into the soil when there is still adequate moisture. Sometimes growers till alternate rows only. This helps to conserve organic matter and reduces tillage. Other growers only mow the cover crops or resident vegetation between vines, using a no-till strategy. Cover crops are grown for soil protection, for attracting beneficial insects, and for absorbing nitrogen. Alyssum, poppies, and other wildflowers may be used both to attract beneficial insects and to add beauty.

Compost is applied in organic vineyards at rate of 1 to 10 tons per acre, depending on the characteristics of the soil. An organic vineyard grower may apply other organically approved soil amendments or minerals if required, as determined by soil testing.

PEST MANAGEMENT

Organic growers should use a "systems" approach for managing pests, including insects, diseases, and weeds. This means that a grower addresses the pest pressures in an integrated approach, monitoring risks, and avoiding dependency on material inputs. In particular, growers are encouraged to analyze, monitor, and establish thresholds. Specifically, they should analyze the life cycle and dynamics of the pests and of their natural enemies, and other aspects of the ecology and the vineyard that influence the incidence of pests and diseases. Next, they should monitor and analyze the pests and beneficials and determine their population levels. Finally, they should establish economic thresholds for each pest, which is the level of the pest population above which the amount of damage to the crop will exceed the cost of controlling the pest. Overall, growers must anticipate pest and disease issues in relation to life cycles, rather than react to problems.

The grower also must demonstrate that cultural, biological, or mechanical methods were attempted before materials are applied, according to the guidelines given by the national organic rules. In other words, growers need to attempt the following types of practices prior to applying organic materials. First, are cultural practices, including crop rotation, good soil management, water management, and other nonchemical preventive methods. (For example, leaf-pulling can reduce mildew and leafhoppers, and changing road surfaces can reduce dust and thereby reduce mites.) Second, are biological control methods: introduce biocontrol agents, augment natural enemies, conserve biological agents, create habitats that attract beneficial insects, or use traps, lures, pheromones, or repellents. Third, vineyards must select resistant grape varieties and adequate placement varieties and avoid varieties that are susceptible to disease in particular areas. If these practices are not effective, then materials listed by the NOP and the Organic Materials Review Institute may be applied.

In organic vineyards, floor vegetation is viewed and treated differently than it is in conventional vineyards. Organic growers generally recognize that having some natural vegetation or weeds in the vineyard is acceptable, and that some kinds of natural vegetation can be desirable for protecting the soil against erosion and for adding organic matter and biodiversity. Some organic growers allow the natural vegetation to grow during part of the season, serving as their cover crop, which is managed and mowed during the growing season. Seeded cover crops also may help suppress noxious weeds but allow helpful native grasses to grow.

However, this does not mean that weeds should be neglected. Weeds still need to be managed and removed in vineyards, particularly under the vines. Without some kind of weed control in vineyards, crop yield and plant vigor will be significantly reduced. In organic systems, weeds are generally managed by mechanical cultivation (tillage), under-the-row mowers, flame weeders (i.e., propane flamers), grazing sheep, and mulches. Mechanical cultivation is the most widely used method of removing weeds in organic vineyards. Cultivation uproots and/or buries the weeds. There are several options of tools or machinery for weed cultivation in vineyards, and most of them are for in-row cultivation.

ORGANIC WINEMAKING, CERTIFICATION FOR ORGANIC WINES, AND MARKET TRENDS

Certification is necessary if a winery wants to use the "USDA Organic" symbol or wants to use the following terms on the label: "100% organic," "organic wine," or "made with organically grown grapes." According to the NOP regulations, 205.301(f)(5), there are two distinct labels for wines that use organic practices: "organic wine," for which no sulfites are added in the winemaking process, and wine "made with organically grown grapes" (or "made with organic grapes"), which requires that added sulfites are below 100 parts per million.

To use these labels, wineries must grow grapes using certified organic viticulture practices, and also need to follow specific organic guidelines in the winemaking process, for sanitation, cleaning, tagging and tracking grapes, and wine-making ingredients. Quality control and consumer relations also must be addressed to qualify for the NOP standards.

Increasing numbers of wineries are complying with these standards in order to develop certified wine brands made with organic grapes, as the market is projected to increase in coming years.

Since the early 1990s, the acreage of organically farmed winegrape vineyards has expanded rapidly in the United States, Italy, France, and also in other winegrowing regions of the world. In California alone, there are nearly ten thousand acres of certified organic vineyards. Although the organic acreage is still a relatively small percentage (3 to 5 percent) of total winegrape acreage, increasing numbers of growers have adopted organic practices, particularly in California. Both large-scale and small-scale organic winegrape vineyards in California provide testimony to the success of organic grape culture in this state.

At the same time, the organic wine market has grown at an accelerated rate since the year 2000, as the overall organic food and beverage market has expanded. Sales of organic wines have risen rapidly in the United States and abroad, just as consumer demand for organic wines (or made with organic grapes) is gaining significantly.

Further Reading

Flaherty, D. L., ed. "Grape Pest Management: Integrated Pest Management." Oakland, CA: University of California ANR Publication 3343, 1992.

McGourty, G. "Wines and Vines." Davis: University of California, Cooperative Extension, 2008.

Thrupp, L. Ann. "Growing Organic Winegrapes: An Introductory Handbook for Growers." Hopland, Mendocino County, CA: Fetzer Vineyards, 2003.

See Also: Integrated Pest Management

Lori Ann Thrupp

Appendix 1:
Report and Recommendations
on Organic Farming, 1980

Bowing to public pressure, the USDA commissioned this first-ever report on the status of organic farming in the country. The report included definitions of key terms in organic agriculture as well as describing the philosophy behind the organic movement. It provided an overview of the relevant topics in organic production, including pest control, soil fertility, nutritional quality, and economic assessments—information that had not been pulled together before that time. Perhaps most important at the time was the inclusion of references for each topic, which propelled research on organic topics well into the next decade. Despite the federal government's publication of this comprehensive and groundbreaking report, critics claim that few actions were actually taken to follow through on the recommendations outlined. The foreword, summary, and conclusions, presented here, indicate the breadth of information encompassed by the report and recommendations.

FOREWORD

We in USDA are receiving increasing numbers of requests for information and advice on organic farming practices. Energy shortages, food safety, and environmental concerns have all contributed to the demand for more comprehensive information on organic farming technology.

Many large-scale producers as well as small farmers and gardeners are showing interest in alternative farming systems. Some of these producers have developed unique systems for soil and crop management, organic recycling, energy conservation, and pest control.

We need to gain a better understanding of these organic farming systems—the extent to which they are practiced in the United States, why they are being used, the technology behind them, and the economic and ecological impacts from their use. We must also identify the kinds of research and education programs that relate to organic farming.

As we strive to develop relevant and productive programs for all of agriculture, we look forward to increasing communication between organic farmers and the U.S. Department of Agriculture.

BOB BERGLAND, Secretary of Agriculture

SUMMARY AND CONCLUSIONS

In April 1979, Dr. Anson R. Bertrand, Director, Science and Education, U.S. Department of Agriculture designated a team of scientists to conduct a study of organic farming in the United States and Europe. Accordingly, the team has assessed the nature and activity of organic farming both here and abroad; investigated the motivations of why farmers shift to organic methods; explored the broad sociopolitical character of the organic movement; assessed the nature of organic technology and management systems; evaluated the level of success of organic farmers and the economic impacts, costs, benefits, and limitations to organic farming; identified research and education programs that would benefit organic farmers; and recommended plans of action for implementation. This report is a condensed version of data and information compiled by the study team. More detailed and documented information will be published later.

In conducting this study, the team relied on a variety of methods and sources to obtain information. These included:

- Selected on-farm case studies of 69 organic farms in 23 States.
- A Rodale Press survey of *The New Farm* magazine readership.
- An extensive review of the literature on organic farming published both here and abroad.
- Interviews and correspondence with knowledgeable organic farming leaders, editors, spokesmen, and practitioners.
- Two study tours of organic farms and research institutes in Europe and Japan.

Public response to this study from both the rural and urban communities has been overwhelming and for the most part highly positive. Thus far approximately 500 letters have been received expressing encouragement and support for the Department's efforts. Many people have generously provided valuable information for the study and innovative ideas on organic methods and techniques. Throughout the study, team members have been invited to speak before various organic producer groups and associations. In all cases, supportive and even enthusiastic receptions were noted. Finally, interviews with team members have been published in numerous newspapers, magazines, and organic newsletters.

It has been most apparent in conducting this study that there is increasing concern about the adverse effects of our U.S. agricultural production system, particularly in regard to the intensive and continuous production of cash grains and the extensive and sometimes excessive use of agricultural chemicals. Among the concerns most often expressed are the following:

1. Sharply increasing costs and uncertain availability of energy and chemical fertilizer, and our heavy reliance on these inputs.
2. Steady decline in soil productivity and tilth from excessive soil erosion and loss of soil organic matter.
3. Degradation of the environment from erosion and sedimentation and from pollution of natural waters by agricultural chemicals.
4. Hazards to human and animal health and to food safety from heavy use of pesticides.
5. Demise of the family farm and localized marketing systems.

Consequently, many feel that a shift to some degree from conventional (that is, chemical-intensive) toward organic farming would alleviate some of these adverse effects, and in the long term would ensure a more stable, sustainable, and profitable agricultural system.

While other definitions exist, for the purpose of this report organic farming is defined as follows:

> Organic farming is a production system which avoids or largely excludes the use of synthetically compounded fertilizers, pesticides, growth regulators, and livestock feed additives. To the maximum extent feasible, organic farming systems rely upon crop rotations, crop residues, animal manures, legumes, green manures, off-farm organic wastes, mechanical cultivation, mineral-bearing rocks, and aspects of biological pest control to maintain soil productivity and tilth, to supply plant nutrients, and to control insects, weeks and other pests.

The concept of the soil as a living system which must be "fed" in a way that does not restrict the activities of beneficial organisms necessary for recycling nutrients and producing humus is central to this definition.

The following is a brief summary of the principal findings of this study:

1. The study team found that the organic movement represents a spectrum of practices, attitudes, and philosophies. On the one hand are those organic practitioners who would not use chemical fertilizers or pesticides under any circumstances. These producers hold rigidly to their purist philosophy. At the other end of the spectrum, organic farmers espouse a more flexible approach. While striving to avoid the use of chemical fertilizers and pesticides, these practitioners do not rule them out entirely. Instead, when absolutely necessary some fertilizers and also herbicides are very selectively and sparingly used as a second line of defense. Nevertheless, these farmers, too, consider themselves to be organic farmers. Failure to recognize that the organic farming movement is distributed over a spectrum can often lead to serious misconceptions. We should not attempt to place all of these organic

practitioners in the same category. For example, we should not lump "organic farmers" and "organic gardeners" together.

2. Organic farming operations are not limited by scale. This study found that while there are many small-scale (10 to 50 acres) organic farms in the northeastern region, there are a significant number of large-scale (more than 100 acres and even up to 1,500 acres) organic farms in the West and Midwest. In most cases, the team members found that these farms, both large and small, were productive, efficient, and well managed. Usually the farmer had acquired a number of years of chemical farming experience before shifting to organic methods.

3. Motivations for shifting from chemical farming to organic farming include concern for protecting soil, human, and animal health from the potential hazards of pesticides; the desire for lower production inputs; concern for the environment and protection of soil resources.

4. Contrary to popular belief, most organic farmers have not regressed to agriculture as it was practiced in the 1930's. While they attempt to avoid or restrict the use of chemical fertilizers and pesticides, organic farmers still use modern farm machinery, recommended crop varieties, certified seed, sound methods of organic waste management, and recommended soil and water conservation practices.

5. Most organic farmers use crop rotations that include legumes and cover crops to provide an adequate supply of nitrogen for moderate to high yields.

6. Animals comprise an essential part of the operation of many organic farms. In a mixed crop/livestock operation, grains and forages are fed on the farm and the manure is returned to the land. Sometimes the manure is composted to conserve nitrogen, and in some cases farmers import both feed and manure from off farm sources.

7. The study team was impressed by the ability of organic farmers to control weeds in crops such as corn, soybeans, and cereals without the use (or with only minimal use) of herbicides. Their success here is attributed to timely tillage and cultivation, delayed planting, and crop rotations. They have also been relatively successful in controlling insect pests.

8. Some organic farmers expressed the feeling that they have been neglected by the U.S. Department of Agriculture and the land-grant universities. They believe that both Extension agents and researchers, for the most part, have little interest in organic methods and that they have no one to turn to for help on technical problems.

9. In some cases where organic fanning is being practiced, it is apparent from a study of the nutrient budget that phosphorus (P) and potassium (K) are being "mined" from either soil minerals or residual fertilizers applied when the land was farmed chemically. While these sources of P and K may sustain high crop yields for some time (depending on soil, climatic, and cropping conditions), it is likely that eventually some organic farmers will have to apply supplemental amounts of these two nutrients.

10. The study revealed that organic farms on the average are somewhat more labor intensive but use less energy than conventional farms. Nevertheless, data are limited and a thorough study of the labor and energy aspects of organic and conventional agriculture is needed.

11. This study showed that the economic return above variable costs was greater for conventional farms (corn and soybeans) than for several crop rotations grown on organic farms. This was largely due to the mix of crops required in the organic system and the large portion of the land that was in legume crops at any one time.

12. There are detrimental aspects of conventional production, such as soil erosion and sedimentation, depleted nutrient reserves, water pollution from runoff of fertilizers and pesticides, and possible decline of soil productivity. If costs of these factors are considered, then cost comparisons between conventional (that is, chemical-intensive) crop production and organic systems may be somewhat different in areas where these problems occur.

In conclusion, the study team found that many of the current methods of soil and crop management practiced by organic farmers are also those which have been cited as best management practices (USDA/EPA joint publication on "Control of Water Pollution from Cropland," Volume I, 1975, U.S. Government Printing Office) for controlling soil erosion, minimizing water pollution, and conserving energy. These include sod based rotations, cover crops, green manure crops, conservation tillage, strip cropping, contouring, and grassed waterways. Moreover, many organic farmers have developed unique and innovative methods of organic recycling and pest control in their crop production sequences. Because of these and other reasons outlined in this report, the team feels strongly that research and education programs should be developed to address the needs and problems of organic farmers. Certainly, much can be learned from a holistic research effort to investigate the organic system of farming, its mechanisms, interactions, principles, and potential benefits to agriculture both at home and abroad.

Source: USDA Study Team on Organic Farming, United States Department of Agriculture, July 1980.

Appendix 2:
Organic Foods Production Act of 1990

(7 U.S.C., including amendments as of January 1, 2004)

In 1990 the U.S. Congress passed the Organic Production Act, which established the National Organic Standards Program within the USDA. The law, excerpted here, set up the guidelines for federal regulations of organic farming for the first time. Although these rules were initiated in 1990, the actual implementation took quite some time, as there was public outcry (in the form of 300,000 public comments) when the initial organic standards were introduced in 1997; critics claimed the initially proposed standards were too weak. Finally, a more stringent rule, which better satisfied organic farmers and organic food advocates, was implemented in 2002.

6501 PURPOSES

It is the purpose of this chapter

(1) to establish national standards governing the marketing of certain agricultural products as organically produced products;

(2) to assure consumers that organically produced products meet a consistent standard; and

(3) to facilitate interstate commerce in fresh and processed food that is organically produced.

6502 DEFINITIONS

As used in this chapter:

(1) Agricultural Product. The term "agricultural product" means any agricultural commodity or product, whether raw or processed, including any commodity or product derived from livestock that is marketed in the United States for human or livestock consumption.

(2) Botanical Pesticides. The term "botanical pesticides" means natural pesticides derived from plants.

(3) Certifying Agent. The term "certifying agent" means the chief executive officer of a State or, in the case of a State that provides for the Statewide election of an official to be responsible solely for the administration of the agricultural operations of a State, such official, and any person (including private entities) who is accredited by the Secretary as a certifying agent for the purpose of certifying a farm or handling operation as a certified organic farm or handling operation in accordance with this chapter.

(4) Certified Organic Farm. The term "certified organic farm" means a farm, or portion of a farm, or site where agricultural products or livestock are produced, that is certified by the certifying agent under this chapter as utilizing a system of organic farming as described by this chapter.

(5) Certified Organic Handling Operation. The term "certified organic handling operation" means any operation, or portion of any handling operation, that is certified by the certifying agent under this chapter as utilizing a system of organic handling as described under this chapter.

(6) Crop Year. The term "crop year" means the normal growing season for a crop as determined by the Secretary.

(7) Governing State Official. The term "governing State official" means the chief executive official of a State or, in the case of a State that provides for the Statewide election of an official to be responsible solely for the administration of the agricultural operations of the State, such official, who administers an organic certification program under this chapter.

(8) Handle. The term "handle" means to sell, process or package agricultural products.

(9) Handler. The term "handler" means any person engaged in the business of handling agricultural products, except such term shall not include final retailers of agricultural products that do not process agricultural products.

(10) Handling Operation. The term "handling operation" means any operation or portion of an operation (except final retailers of agricultural products that do not process agricultural products) that

(A) receives or otherwise acquires agricultural products; and

(B) processes, packages or stores such products.

(11) Livestock. The term "livestock" means any cattle, sheep, goats, swine, poultry, equine animals used for food or in the production of food, fish used for food, wild or domesticated game, or other non-plant life.

(12) National List. The term "National List" means a list of approved and prohibited substances as provided for in section 6517 of this title.

(13) Organic Plan. The term "organic plan" means a plan of management of an organic farming or handling operation that has been agreed to by the producer or handler and the certifying agent and that includes written plans concerning all aspects of agricultural production or handling described in this chapter including crop rotation and other practices as required under this chapter.

(14) Organically Produced. The term "organically produced" means an agricultural product that is produced and handled in accordance with this chapter.

(15) Person. The term "person" means an individual, group of individuals, corporation, association, organization, cooperative, or other entity.

(16) Pesticide. The term "pesticide" means any substance which alone, in chemical combination, or in any formulation with one or more substances, is defined as a pesticide in the Federal Insecticide, Fungicide, and Rodenticide Act (7 U.S.C. 136 et seq.).

(17) Processing. The term "processing" means cooking, baking, heating, drying, mixing, grinding, churning, separating, extracting, cutting, fermenting, eviscerating, preserving, dehydrating, freezing, or otherwise manufacturing, and includes the packaging, canning, jarring, or otherwise enclosing food in a container.

(18) Producer. The term "producer" means a person who engages in the business of growing or producing food or feed.

(19) Secretary. The term "Secretary" means the Secretary of Agriculture.

(20) State Organic Certification Program. The term "State organic certification program" means a program that meets the requirements of section 6506 of this title, is approved by the Secretary, and that is designed to ensure that a product that is sold or labeled as "organically produced" under this chapter is produced and handled using organic methods.

(21) Synthetic. The term "synthetic" means a substance that is formulated or manufactured by a chemical process or by a process that chemically changes a substance extracted from naturally occurring plant, animal, or mineral sources, except that such term shall not apply to substances created by naturally occurring biological processes.

6503 NATIONAL ORGANIC PRODUCTION PROGRAM

(a) In General. The Secretary shall establish an organic certification program for producers and handlers of agricultural products that have been produced using organic methods as provided for in this chapter.

(b) State Program. In establishing the program under subsection (a) of this section, the Secretary shall permit each State to implement a State organic certification program for producers and handlers of agricultural products that have been produced using organic methods as provided for in this chapter.

(c) Wild Seafood.

(1) IN GENERAL—Notwithstanding the requirement of section 2107(a)(1)(A) requiring products be produced only on certified organic farms, the Secretary shall allow, through regulations promulgated after public notice and opportunity for comment, wild seafood to be certified or labeled as organic.

(2) CONSULTATION AND ACCOMMODATION—In carrying out paragraph (1), the Secretary shall

(A) consult with

(i) the Secretary of Commerce;

(ii) the National Organic Standards Board established under section 2119;

(iii) producers, processors, and sellers; and

(iv) other interested members of the public; and

(B) to the maximum extent practicable, accommodate the unique characteristics of the industries in the United States that harvest and process wild seafood.

(d) Consultation. In developing the program under subsection (a) of this section, and the National List under section 6517 of this title, the Secretary shall consult with the National Organic Standards Board established under section 6518 of this title.

(e) Certification. The Secretary shall implement the program established under subsection (a) of this section through certifying agents. Such certifying agents may certify a farm or handling operation that meets the requirements of this chapter and the requirements of the organic certification program of the State (if applicable) as an organically certified farm or handling operation.

6504 NATIONAL STANDARDS FOR ORGANIC PRODUCTION

To be sold or labeled as an organically produced agricultural product under this chapter, an agricultural product shall

(1) have been produced and handled without the use of synthetic chemicals, except as otherwise provided in this chapter;

(2) except as otherwise provided in this chapter and excluding livestock, not be produced on land to which any prohibited substances, including synthetic chemicals, have been applied during the 3 years immediately preceding the harvest of the agricultural products; and

(3) be produced and handled in compliance with an organic plan agreed to by the producer and handler of such product and the certifying agent.

6505 COMPLIANCE REQUIREMENTS

(a) Domestic Products.

(1) In General. On or after October 1, 1993

(A) a person may sell or label an agricultural product as organically produced only if such product is produced and handled in accordance with this chapter; and

(B) no person may affix a label to, or other provide market information concerning, an agricultural product if such label or information implies, directly or indirectly, that such product is produced and handled using organic methods, except in accordance with this chapter.

(2) USDA Standards and Seal. A label affixed, or other market information provided, in accordance with paragraph (1) may indicate that the agricultural product meets Department of Agriculture standards for organic production and may incorporate the Department of Agriculture seal.

(b) Imported Products. Imported agricultural products may be sold or labeled as organically produced if the Secretary determines that such products have been produced and handled under an organic certification program that provides safeguards and guidelines governing the production and handling of such products that are at least equivalent to the requirements of this chapter.

(c) Exemptions for Processed Food. Subsection (a) of this section shall not apply to agricultural products that

(1) contain at least 50 percent organically produced ingredients by weight, excluding water and salt, to the extent that the Secretary, in consultation with the National Organic Standards Board and the Secretary of Health and Human Services, has determined to permit the word "organic" to be used on the principal display panel of such products only for the purpose of describing the organically produced ingredients; or

(2) contain less than 50 percent organically produced ingredients by weight, excluding water and salt, to the extent that the Secretary, in consultation with the National Organic Standards Board and the Secretary of Health and Human Services, has determined to permit the word "organic" to appear on the ingredient listing panel to describe those ingredients that are organically produced in accordance with this chapter.

(d) Small Farmer Exemption. Subsection (a)(1) of this section shall not apply to persons who sell no more than $5,000 annually in value of agricultural products.

6506 GENERAL REQUIREMENTS

(a) In General. A program established under this chapter shall

(1) provide that an agricultural product to be sold or labeled as organically produced must

(A) be produced only on certified organic farms and handled only through certified organic handling operations in accordance with this chapter; and

(B) be produced and handled in accordance with such program;

(2) require that producers and handlers desiring to participate under such program establish an organic plan under section 6513 of this title;

(3) provide for procedures that allow producers and handlers to appeal an adverse administrative determination under this chapter;

(4) require each certified organic farm or each certified organic handling operation to certify to the Secretary, the governing State official (if applicable), and the certifying agent on an annual basis, that such farm or handler has not produced or handled any agricultural product sold or labeled as organically produced except in accordance with this chapter;

(5) provide for annual on-site inspection by the certifying agent of each farm and handling operation that has been certified under this chapter;

(6) require periodic residue testing by certifying agents of agricultural products that have been produced on certified organic farms and handled through certified organic handling operations to determine whether such products contain any pesticide or other nonorganic residue or natural toxicants and to require certifying agents, to the extent that such agents are aware of a violation of applicable laws relating to food safety, to report such violation to the appropriate health agencies;

(7) provide for appropriate and adequate enforcement procedures, as determined by the Secretary to be necessary and consistent with this chapter;

(8) protect against conflict-of-interest as specified under section 6515(h) of this title;

(9) provide for public access to certification documents and laboratory analyses that pertain to certification;

(10) provide for the collection of reasonable fees from producers, certifying agents and handlers who participate in such program; and

(11) require such other terms and conditions as may be determined by the Secretary to be necessary.

(b) Discretionary Requirements. An organic certification program established under this chapter may

(1) provide for the certification of an entire farm or handling operation or specific fields of a farm or parts of a handling operation if

(A) in the case of a farm or field, the area to be certified has distinct, defined boundaries and buffer zones separating the land being operated through the use of organic methods from land that is not being operated through the use of such methods;

(B) the operators of such farm or handling operation maintain records of all organic operations separate from records relating to other operations and make such records available at all times for inspection by the Secretary, the certifying agent, and the governing State official; and

(C) appropriate physical facilities, machinery, and management practices are established to prevent the possibility of a mixing of organic and nonorganic products or a penetration of prohibited chemicals or other substances on the certified area; and

(2) provide for reasonable exemptions from specific requirements of this chapter (except the provisions of section 6511 of this title) with respect to agricultural products produced on certified organic farms if such farms are subject to a Federal or State emergency pest or disease treatment program.

(c) State Program. A State organic certification program approved under this chapter may contain additional guidelines governing the production or handling of products sold or labeled as organically produced in such State as required in section 6507 of this title.

6507 STATE ORGANIC CERTIFICATION PROGRAM

(a) In General. The governing State official may prepare and submit a plan for the establishment of a State organic certification program to the Secretary for approval. A State organic certification program must meet the requirements of this chapter to be approved by the Secretary.

(b) Additional Requirements.

(1) Authority. A State organic certification program established under subsection (a) of this section may contain more restrictive requirements governing the organic certification of farms and handling operations and the production and handling of agricultural products that are to be sold or labeled as organically produced under this chapter than are contained in the program established by the Secretary.

(2) Content. Any additional requirements established under paragraph (1) shall

(A) further the purposes of this chapter;

(B) not be inconsistent with this chapter;

(C) not be discriminatory towards agricultural commodities organically produced in other States in accordance with this chapter; and

(D) not become effective until approved by the Secretary.

(c) Review and Other Determinations.

(1) Subsequent Review. The Secretary shall review State organic certification programs not less than once during each 5-year period following the date of the approval of such programs.

(2) Changes in Program. The governing State official, prior to implementing any substantive change to programs approved under this subsection, shall submit such change to the Secretary for approval.

(3) Time for Determination. The Secretary shall make a determination concerning any plan, proposed change to a program, or a review of a program not later than 6 months after receipt of such plan, such proposed change, or the initiation of such review.

6508 PROHIBITED CROP PRODUCTION PRACTICES AND MATERIALS

(a) Seed, Seedlings and Planting Practices. For a farm to be certified under this chapter, producers on such farm shall not apply materials to, or engage in practices on, seeds or seedlings that are contrary to, or inconsistent with, the applicable organic certification program.

(b) Soil Amendments. For a farm to be certified under this chapter, producers on such farm shall not

(1) use any fertilizers containing synthetic ingredients or any commercially blended fertilizers containing materials prohibited under this chapter or under the applicable State organic certification program; or

(2) use as a source of nitrogen: phosphorous, lime, potash, or any materials that are inconsistent with the applicable organic certification program.

(c) Crop Management. For a farm to be certified under this chapter, producers on such farm shall not

(1) use natural poisons such as arsenic or lead salts that have long-term effects and persist in the environment, as determined by the applicable governing State official or the Secretary;

(2) use plastic mulches, unless such mulches are removed at the end of each growing or harvest season; or

(3) use transplants that are treated with any synthetic or prohibited material.

6509 ANIMAL PRODUCTION PRACTICES AND MATERIALS

(a) In General. Any livestock that is to be slaughtered and sold or labeled as organically produced shall be raised in accordance with this chapter.

(b) Breeder Stock. Breeder stock may be purchased from any source if such stock is not in the last third of gestation.

(c) Practices. For a farm to be certified under this chapter as an organic farm with respect to the livestock produced by such farm, producers on such farm

(1) shall feed such livestock organically produced feed that meets the requirements of this chapter;

(2) shall not use the following feed

(A) plastic pellets for roughage;

(B) manure refeeding; or

(C) feed formulas containing urea; and

(3) shall not use growth promoters and hormones on such livestock, whether implanted, ingested, or injected, including antibiotics and synthetic trace elements used to stimulate growth or production of such livestock.

(d) Health Care.

(1) Prohibited Practices. For a farm to be certified under this chapter as an organic farm with respect to the livestock produced by such farm, producers on such farm shall not

(A) use subtherapeutic doses of antibiotics;

(B) use synthetic internal paraciticides on a routine basis; or

(C) administer medication, other than vaccinations, in the absence of illness.

(2) Standards. The National Organic Standards Board shall recommend to the Secretary standards in addition to those in paragraph (1) for the care of livestock to ensure that such livestock is organically produced.

(e) Additional Guidelines.

(1) Poultry. With the exception of day old poultry, all poultry from which meat or eggs will be sold or labeled as organically produced shall be raised and handled in accordance with this chapter prior to and during the period in which such meat or eggs are sold.

(2) Dairy Livestock. A dairy animal from which milk or milk products will be sold or labeled as organically produced shall be raised and handled in accordance with this chapter for not less than the 12-month period immediately prior to the sale of such milk and milk products.

(f) Livestock Identification.

(1) In General. For a farm to be certified under this chapter as an organic farm with respect to the livestock produced by such farm, producers on such farm shall keep adequate records and maintain a detailed,

verifiable audit trail so that each animal (or in the case of poultry, each flock) can be traced back to such farm.

(2) Records. In order to carry our paragraph (1), each producer shall keep accurate records on each animal (or in the case of poultry, each flock) including

(A) amounts and sources of all medications administered; and

(B) all feeds and feed supplements bought and fed.

(g) Notice and Public Comment. The Secretary shall hold public hearings and shall develop detailed regulations, with notice and public comment, to guide the implementation of the standards for livestock products provided under this section.

6510 HANDLING

(a) In General. For a handling operation to be certified under this chapter, each person on such handling operation shall not, with respect to any agricultural product covered by this chapter

(1) add any synthetic ingredient during the processing or any post harvest handling of the product;

(2) add any ingredient known to contain levels of nitrates, heavy metals, or toxic residues in excess of those permitted by the applicable organic certification program;

(3) add any sulfites, except in the production of wine, nitrates, or nitrites;

(4) add any ingredients that are not organically produced in accordance with this chapter and the applicable organic certification program, unless such ingredients are included on the National List and represent not more than 5 percent of the weight of the total finished product (excluding salt and water);

(5) use any packaging materials, storage containers or bins that contain synthetic fungicides, preservatives, or fumigants;

(6) use any bag or container that had previously been in contact with any substance in such a manner as to compromise the organic quality of such product; or

(7) use, in such product water that does not meet all Safe Drinking Water Act [42 U.S.C.A. 300f et seq.] requirements.

(b) Meat. For a farm or handling operation to be organically certified under this chapter, producers on such farm or persons on such handling operation shall ensure that organically produced meat does not come in contact with nonorganically produced meat.

6511 ADDITIONAL GUIDELINES

(a) In General. The Secretary, the applicable governing State official, and the certifying agent shall utilize a system of residue testing to test products sold or labeled as organically produced under this chapter to assist in the enforcement of this title.

(b) Pre-Harvest Testing. The Secretary, the applicable governing State official, or the certifying agent may require preharvest tissue testing of any crop grown on soil suspected of harboring contaminants.

(c) Compliance Review.

(1) Inspection. If the Secretary, the applicable governing State official, or the certifying agent determines that an agricultural product sold or labeled as organically produced under this chapter contains any detectable pesticide or other non-organic residue or prohibited natural substance the Secretary, the applicable governing State official, or the certifying agent shall conduct an investigation to determine if the organic certification program has been violated, and may require the producer or handler of such product to prove that any prohibited substance was not applied to such product.

(2) Removal of Organic Label. If, as determined by the Secretary, the applicable governing State official, or the certifying agent, the investigation conducted under paragraph (1) indicates that the residue is

(A) the result of intentional application of a prohibited substance; or

(B) present at levels that are greater than unavoidable residual environmental contamination as prescribed by the Secretary of the applicable governing State official in consultation with the appropriate environmental regulatory agencies; such agricultural product shall not be sold or labeled as organically produced under this chapter.

(d) Recordkeeping Requirements. Producers who operate a certified organic farm or handling operation under this chapter shall maintain records for 5 years concerning the production or handling of agricultural products sold or labeled as organically produced under this chapter, including

(1) a detailed history of substances applied to fields or agricultural products; and

(2) the names and addresses of persons who applied such substances, the dates, the rate, and method of application of such substances.

6512 OTHER PRODUCTION AND HANDLING PRACTICES

If a production or handling practice is not prohibited or otherwise restricted under this chapter, such practice shall be permitted unless it is determined that such practice would be inconsistent with the applicable organic certification program.

6513 ORGANIC PLAN

(a) In General. A producer or handler seeking certification under this chapter will submit an organic plan to the certifying agent and the State organic certification program (if applicable), and such plan shall be reviewed by the certifying agent who shall determine if such plan meets the requirements of the programs.

(b) Crop Production Farm Plan.

(1) Soil Fertility. An organic plan shall contain provisions designed to foster soil fertility, primarily through the management of the organic content of the soil through proper tillage, crop rotation, and manuring.

(2) Manuring.

(A) Inclusion in Organic Plan. An organic plan shall contain terms and conditions that regulate the application of manure to crops.

(B) Application of Manure. Such organic plan may provide for the application of raw manure only to

(i) any green manure crop;

(ii) any perennial crop;

(iii) any crop not for human consumption; and

(iv) any crop for human consumption, if such crop is harvested after a reasonable period of time determined by the certifying agent to ensure the safety of such crop, after the most recent application of raw manure, but in no event shall such period be less than 60 days after such application.

(C) Contamination by Manure. Such organic plan shall prohibit raw manure from being applied to any crop in a way that significantly contributes to water contamination by nitrates or bacteria.

(c) Livestock Plan. An organic livestock plan shall contain provisions designed to foster the organic production of livestock consistent with the purposes of this chapter.

(d) Mixed Crop Livestock Production. An organic plan may encompass both the crop production and livestock production requirements in subsections (b) and (c) of this section if both activities are conducted by the same producer.

(e) Handling Plan. An organic handling plan shall contain provisions designed to ensure that agricultural products that are sold or labeled as organically produced are produced and handled in a manner that is consistent with the purposes of this chapter.

(f) Management of Wild Crops. An organic plan for the harvesting of wild crops shall

(1) designate the area from which the wild crop will be gathered or harvested;

(2) include a 3 year history of the management of the area showing that no prohibited substances have been applied;

(3) include a plan for the harvesting or gathering of the wild crops assuring that such harvesting or gathering will not be destructive to the environment and will sustain the growth and production of the wild crop; and

(4) include provisions that no prohibited substances will be applied by the producer.

(g) Limitation on Content of Plan. An organic plan shall not include any production or handling practices that are inconsistent with this chapter.

6514 ACCREDITATION PROGRAM

(a) In General. The Secretary shall establish and implement a program to accredit a governing State official, and any private person, that meets the requirements of this section as a certifying agent for the purpose of certifying a farm or handling operation as a certified organic farm or handling operation.

(b) Requirements. To be accredited as a certifying agent under this section, a governing State official or private person shall

(1) prepare and submit, to the Secretary, an application for such accreditation;

(2) have sufficient expertise in organic farming and handling techniques as determined by the Secretary; and

(3) comply with the requirements of this section and section 6515 of this title.

(c) Duration of Designation. An accreditation made under this section shall be for a period of not to exceed 5 years, as determined appropriate by the Secretary, and may be renewed.

6515 REQUIREMENTS OF CERTIFYING AGENTS

(a) Ability to Implement Requirements. To be accredited as a certifying agent under section 6514 of this title, a governing State official or a person shall be able to fully implement the applicable organic certification program established under this chapter.

(b) Inspectors. Any certifying agent shall employ a sufficient number of inspectors to implement the applicable organic certification program established under this chapter, as determined by the Secretary.

(c) Recordkeeping.

(1) Maintenance of Records. Any certifying agent shall maintain all records concerning its activities under this chapter for a period of not less than 10 years.

(2) Access for Secretary. Any certifying agent shall allow representatives of the Secretary and the governing State official access to any and all records concerning the certifying agent's activities under this chapter.

(3) Transference of Records. If any private person that was certified under this chapter is dissolved or loses its accreditation, all records or copies of records concerning such person's activities under this chapter shall be transferred to the Secretary and made available to the applicable governing State official.

(d) Agreement. Any certifying agent shall enter into an agreement with the Secretary under which such agent shall

(1) agree to carry out the provisions of this chapter; and

(2) agree to such other terms and conditions as the Secretary determines appropriate.

(e) Private Certifying Agent Agreement. Any certifying agent that is a private person shall, in additional to the agreement required in subsection (d) of this section

(1) agree to hold the Secretary harmless for any failure on the part of the certifying agent to carry out the provisions of this chapter; and

(2) furnish reasonable security, in an amount determined by the Secretary, for the purpose of protecting the rights of participants in the applicable organic certification program established under this chapter.

(f) Compliance with Program. Any certifying agent shall fully comply with the terms and conditions of the applicable organic certification program implemented under this chapter.

(g) Confidentiality. Except as provided in section 6506 (a)(9) of this title, any certifying agent shall maintain strict confidentiality with respect to its clients under the applicable organic certification program and may not disclose to third parties (with the exception of the Secretary or the applicable governing State official) any business related information concerning such client obtained while implementing this chapter.

(h) Conflict of Interest. Any certifying agent shall not

(1) carry out any inspections of any operation in which such certifying agent, or employee of such certifying agent has, or has had, a commercial interest, including the provision of consultancy services;

(2) accept payment, gifts, or favors of any kind from the business inspected other than prescribed fees; or

(3) provide advice concerning organic practices or techniques for a fee, other than fees established under such program.

(i) Administrator. A certifying agent that is a private person shall nominate the individual who controls the day-to-day operation of the agent.

(j) Loss of Accreditation.

(1) Noncompliance. If the Secretary or the governing State official (if applicable) determines that a certifying agent is not properly adhering to the provisions of this chapter, the Secretary or such governing State official may suspend such certifying agent's accreditation.

(2) Effect on Certified Operations. If the accreditation of a certifying agent is suspended under paragraph (1), the Secretary or the governing State official (if applicable) shall promptly determine whether

farming or handling operations certified by certifying such agent may retain their organic certification.

6516 PEER REVIEW OF CERTIFYING AGENTS

(a) Peer Review. In determining whether to approve an application for accreditation submitted under section 6514 of this title, the Secretary shall consider a report concerning such applicant that shall be prepared by a peer review panel established under subsection (b) of this section.

(b) Peer Review Panel. To assist the Secretary in evaluating applications under section 6514 of this title, the Secretary may establish a panel of not less than three persons who have expertise in organic farming and handling methods, to evaluate the State governing official or private person that is seeking accreditation as a certifying agent under such section. Not less than two members of such panel shall be persons who are not employees of the Department of Agriculture or of the applicable State government.

6517 NATIONAL LIST

(a) In General. The Secretary shall establish a National List of approved and prohibited substances that shall be included in the standards for organic production and handling established under this chapter in order for such products to be sold or labeled as organically produced under this chapter.

(b) Content of List. The list established under subsection (a) of this section shall contain an itemization, by specific use or application, of each synthetic substance permitted under subsection (c) (1) of this section or each natural substance prohibited under subsection (c)(2) of this section.

(c) Guidelines for Prohibitions or Exemptions.

(1) Exemption for Prohibited Substances. The National List may provide for the use of substances in an organic farming or handling operation that are otherwise prohibited under this chapter only if

(A) the Secretary determines, in consultation with the Secretary of Health and Human Services and the Administrator of the Environmental Protection Agency, that the use of such substances

(i) would not be harmful to human health or the environment;

(ii) is necessary to the production or handling of the agricultural product because of unavailability of wholly natural substitute products; and

(iii) is consistent with organic farming and handling;

(B) the substance

(i) is used in production and contains an active synthetic ingredient in the following categories: copper and sulfur compounds; toxins derived from bacteria; pheromones, soaps, horticultural oils, fish emulsions, treated seed, vitamins and minerals; livestock paraciticides and medicines and production aids including netting, tree wraps and seals, insect traps, sticky barriers, row covers, and equipment cleansers;

(ii) is used in production and contains synthetic inert ingredients that are not classified by the Administrator of the Environmental Protection Agency as inerts of toxicological concern; or

(iii) is used in handling and is non-synthetic but is not organically produced; and

(C) the specific exemption is developed using the procedures described in subsection (d) of this section.

(2) Prohibition on the use of Specific Natural Substances. The National List may prohibit the use of specific natural substances in an organic farming or handling operation that are otherwise allowed under this chapter only if

(A) the Secretary determines, in consultation with the Secretary of Health and Human Services and the Administrator of the Environmental Protection Agency, that the use of such substances

(i) would be harmful to human health or the environment; and

(ii) is inconsistent with organic farming or handling, and the purposes of this chapter; and

(B) the specific prohibition is developed using the procedures specified in subsection (d) of this section.

(d) Procedure for Establishing National List.

(1) In General. The National List established by the Secretary shall be based upon a proposed national list or proposed amendments to the National List developed by the National Organic Standards Board.

(2) No Additions. The Secretary may not include exemptions for the use of specific synthetic substances in the National List other than those exemptions contained in the Proposed National List or Proposed Amendments to the National List.

(3) Prohibited Substances. In no instance shall the National List include any substance, the presence of which in food has been prohibited by Federal regulatory action.

(4) Notice and Comment. Before establishing the National List or before making any amendments to the National List, the Secretary shall publish the Proposed National List or any Proposed amendments to the National List in the Federal Register and seek public comment on such proposals. The Secretary shall include in such Notice any changes to such proposed list or amendments recommended by the Secretary.

(5) Publication of National List. After evaluating all comments received concerning the Proposed National List or Proposed Amend-

ments to the National List, the Secretary shall publish the final National List in the Federal Register, along with a discussion of comments received.

(e) Sunset Provision. No exemptions or prohibition contained in the National List shall be valid unless the National Organic Standards Board has reviewed such exemption or prohibition as provided in this section within 5 years of such exemption or prohibition being adopted or reviewed and the Secretary has renewed such exemption or prohibition.

6518 NATIONAL ORGANIC STANDARDS BOARD

(a) In General. The Secretary shall establish a National Organic Standards Board (in accordance with the Federal Advisory Committee Act (5 U.S.C. app. 2 et seq.)) (hereafter referred to in this section as the "Board") to assist in the development of standards for substances to be used in organic production and to advise the Secretary on any other aspects of the implementation of this chapter.

(b) Composition of Board. The Board shall be composed of 15 members, of which

(1) four shall be individuals who own or operate an organic farming operation;

(2) two shall be individuals who own or operate an organic handling operation;

(3) one shall be an individual who owns or operates a retail establishment with significant trade in organic products;

(4) three shall be individuals with expertise in areas of environmental protection and resource conservation;

(5) three shall be individuals who represent public interest or consumer interest groups;

(6) one shall be an individual with expertise in the fields of toxicology, ecology, or biochemistry; and

(7) one shall be an individual who is a certifying agent as identified under section 6515 of this title.

(c) Appointment. No later than 180 days after November 28, 1990, the Secretary shall appoint the members of the Board under paragraph (1) through (6) of subsection (b) of this section (and under subsection (b)(7) of this section at an appropriate date after the certification of individuals as certifying agents under section 6515 of this title) from nominations received from organic certifying organizations, States, and other interested persons and organizations.

(d) Term. A member of the Board shall serve for a term of 5 years, except that the Secretary shall appoint the original members of the Board for staggered terms. A member cannot serve consecutive terms unless such member served an original term that was less than 5 years.

(e) Meetings. The Secretary shall convene a meeting of the Board not later than 60 days after the appointment of its members and shall convene subsequent meetings on a periodic basis.

(f) Compensation and Expenses. A member of the Board shall serve without compensation. While away from their homes or regular places of business on the business of the Board, members of the Board may be allowed travel expenses, including per diem in lieu of subsistence, as is authorized under section 5703 of Title 5 for persons employed intermittently in the Government service.

(g) Chairperson. The Board shall select a Chairperson for the Board.

(h) Quorum. A majority of the members of the Board shall constitute a quorum for the purpose of conducting business.

(i) Decisive Votes. Two-thirds of the votes cast at a meeting of the Board at which a quorum is present shall be decisive of any motion.

(j) Other Terms and Conditions. The Secretary shall authorize the Board to hire a staff director and shall detail staff of the Department of Agriculture or allow for the hiring of staff and may, subject to necessary appropriations, pay necessary expenses incurred by such Board in carrying out the provisions of this chapter, as determined appropriate by the Secretary.

(k) Responsibilities of the Board.

(1) In General. The Board shall provide recommendations to the Secretary regarding the implementation of this chapter.

(2) National List. The Board shall develop the proposed National List or proposed amendments to the National List for submission to the Secretary in accordance with section 6517 of this title.

(3) Technical Advisory Panels. The Board shall convene technical advisory panels to provide scientific evaluation of the materials considered for inclusion in the National List. Such panels may include experts in agronomy, entomology, health sciences and other relevant disciplines.

(4) Special Review of Botanical Pesticides. The Board shall, prior to the establishment of the National List, review all botanical pesticides used in agricultural production and consider whether any such botanical pesticides should be included in the list of prohibited natural substances.

(5) Product Residue Testing. The Board shall advise the Secretary concerning the testing of organically produced agricultural products for residues caused by unavoidable residual environmental contamination.

(6) Emergency Spray Programs. The Board shall advise the Secretary concerning rules for exemptions from specific requirements of this chapter (except the provisions of section 6511 of this title) with respect to agricultural products produced on certified organic farms if such farms are subject to a Federal or State emergency pest or disease treatment program.

(l) Requirements. In establishing the proposed National List or proposed amendments to the National List, the Board shall

(1) review available information from the Environmental Protection Agency, the National Institute of Environmental Health Studies, and such other sources as appropriate, concerning the potential for adverse human and environmental effects of substances considered for inclusion in the proposed National List;

(2) work with manufacturers of substances considered for inclusion in the proposed National List to obtain a complete list of ingredients and determine whether such substances contain inert materials that are synthetically produced; and

(3) submit to the Secretary, along with the proposed National List or any proposed amendments to such list, the results of the Board's evaluation and the evaluation of the technical advisory panel of all substances considered for inclusion in the National List.

(m) Evaluation. In evaluating substances considered for inclusion in the proposed National List or proposed amendment to the National List, the Board shall consider

(1) the potential of such substances for detrimental chemical interactions with other materials used in organic farming systems;

(2) the toxicity and mode of action of the substance and of its breakdown products or any contaminants, and their persistence and areas of concentration in the environment;

(3) the probability of environmental contamination during manufacture, use, misuse or disposal of such substance;

(4) the effect of the substance on human health;

(5) the effects of the substance on biological and chemical interactions in the agroecosystem, including the physiological effects of the substance on soil organisms (including the salt index and solubility of the soil), crops and livestock;

(6) the alternatives to using the substance in terms of practices or other available materials; and

(7) its compatibility with a system of sustainable agriculture.

(n) Petitions. The Board shall establish procedures under which persons may petition the Board for the purpose of evaluating substances for inclusion on the National List.

(o) Confidentiality. Any confidential business information obtained by the Board in carrying out this section shall not be released to the public.

6519 VIOLATIONS OF CHAPTER

(a) Misuse of Label. Any person who knowingly sells or labels a product as organic, except in accordance with this chapter, shall be subject to a civil penalty of not more than $10,000.

(b) False Statement. Any person who makes a false Statement under this chapter to the Secretary, a governing State official, or a certifying agent shall be subject to the provisions of section 1001 of Title 18.

(c) Ineligibility.

(1) In General. Except as provided in paragraph (2), any person who

(A) makes a false Statement;

(B) attempts to have a label indicating that an agricultural product is organically produced affixed to such product that such person knows, or should have reason to know, to have been produced or handled in a manner that is not in accordance with this chapter; or

(C) otherwise violates the purposes of the applicable organic certification program as determined by the Secretary;

after notice and an opportunity to be heard, shall not be eligible, for a period of 5 years from the date of such occurrence, to receive certification under this chapter with respect to any farm or handling operation in which such person has an interest.

(2) Waiver. Notwithstanding paragraph (1), the Secretary may reduce or eliminate the period of ineligibility referred to in such paragraph if the Secretary determines that such modification or waiver is in the best interests of the applicable organic certification program established under this chapter.

(d) Reporting of Violations. A certifying agent shall immediately report any violations of this chapter to the Secretary or the governing State official (if applicable).

(e) Violations by Certifying Agent. A certifying agent that is a private person that violates the provisions of this chapter or that falsely or negligently certifies any farming or handling operation that does not meet the terms and conditions of the applicable organic certification program as an organic operation, as determined by the Secretary or the governing State official (if applicable) shall, after notice and an opportunity to be heard

(1) lose its accreditation as a certifying agent under this chapter; and

(2) be ineligible to be accredited as a certifying agent under this chapter for a period of not less than 3 years subsequent to the date of such determination.

(f) Effect of Other Laws. Nothing in this chapter shall alter the authority of the Secretary under the Federal Meat Inspection Act (21 U.S.C. 601 et seq.), the Poultry Products Inspection Act (21 U.S.C. 451 et seq.), and the Egg Products Inspection Act (21 U.S.C. 1031 et seq.) concerning meat, poultry and egg products, nor any of the authorities of the Secretary of Health and Human Services under the Federal Food, Drug and Cosmetic Act (21 U.S.C. 301 et seq.), nor the authority of the Administrator of the Environmental Protection Agency under the Federal Insecticide, Fungicide and Rodenticide Act (7 U.S.C. 136 et seq.).

6520 ADMINISTRATIVE APPEAL

(a) Expedited Appeals Procedure. The Secretary shall establish an expedited administrative appeals procedure under which persons may appeal an action of the Secretary, the applicable governing State official, or a certifying agent under this chapter that
(1) adversely affects such person; or
(2) is inconsistent with the organic certification program established under this chapter.
(b) Appeal of Final Decision. A final decision of the Secretary under subsection (a) of this section may be appealed to the United States District Court for the District in which such person is located.

6521 ADMINISTRATION

(a) Regulations. Not later than 540 days after the date of enactment of this title, the Secretary shall issue proposed regulations to carry out this chapter.
(b) Assistance to State.
(1) Technical and Other Assistance. The Secretary shall provide technical, administrative, and Extension Service assistance to assist States in the implementation of an organic certification program under this chapter.
(2) Financial Assistance. The Secretary may provide financial assistance to any State that implements an organic certification program under this chapter.

6522 AUTHORIZATION OF APPROPRIATIONS

There are authorized to be appropriated for each fiscal year such sums as may be necessary to carry out this chapter.

Source: Title XXI of 1990 Farm Bill: "Food Agricultural Conservation and Trade Act of 1990," Pub. L. No. 101-624, 104 Stat. 3359 (Nov. 28, 1990).

Appendix 3: U.S. Organic Farming Emerges in the 1990s: Adoption of Certified Systems, 2001

This report was written at a time of rapid growth in organic agricultural production in the United States, but before USDA National Organic Standards were implemented in 2002, significantly changing the playing field. It documents the national trends in the types of crops and geography of the crops grown and the demand for and pricing of organic products, labeling issues, and certification procedures. Specific data on adoption of organic agricultural methods are described within the categories of field crops, specialty crops, and livestock. In addition, information on certification and organic agricultural organizations is provided. Much of this material had never been published before—certainly not in one USDA document. The abstract and summary are presented here to provide a brief overview of the findings.

ABSTRACT

Farmers have been developing organic farming systems in the United States for decades. State and private institutions also began emerging during this period to set organic farming standards and provide third-party verification of label claims, and legislation requiring national standards was passed in the 1990s. More U.S. producers are considering organic farming systems in order to lower input costs, conserve nonrenewable resources, capture high-value markets, and boost farm income. Organic farming systems rely on practices such as cultural and biological pest management, and virtually prohibit synthetic chemicals in crop production and antibiotics or hormones in livestock production. This report updates U.S. Department of Agriculture (USDA) estimates of land farmed with organic practices during 1992–94 with 1997 estimates, and provides new State- and crop-level detail.

SUMMARY

The amount of U.S. farmland managed under certified organic farming systems expanded substantially during the 1990s, as did consumer demand for organically grown food. More U.S. producers are considering organic farming systems in order to lower input costs and conserve nonrenewable resources, as well as to capture high-value markets and improve farm income. Organic farming systems foster the cycling of resources, rely on practices such as cultural and biological pest management, and virtually prohibit synthetic chemicals in crop production and antibiotics or hormones in livestock production.

Regulation of the organic food industry—which began several decades ago when private organizations began developing standards and third-party certification to support organic farming and thwart consumer fraud—also evolved rapidly during the 1990s. By the mid-1990s, over half the States had laws or rules regulating the production and marketing of organically grown food and fiber, and Congress had passed legislation requiring that national standards be set. The U.S. Department of Agriculture (USDA) published the final rule implementing this legislation in December 2000, and by late-2002, all except the smallest organic growers will have to be certified by a State or private agency accredited under these national standards.

This report updates USDA estimates of land farmed with certified organic practices during 1992–94 with 1997 estimates, and provides new State- and crop-level detail. The procedures used in this report are similar to those used in the first study of certified acreage: data from State and private certifiers were collected and analyzed, uncertified production was excluded, and double-certified acreage was excluded whenever possible. Forty organic certification organizations—12 State and 28 private—conducted third-party certification of organic production in 1997.

Certified organic farming systems were used on 1.5 million acres of cropland and pasture in 49 States in 1997. U.S. certified organic cropland more than doubled between 1992 and 1997, and two organic livestock sectors—eggs and dairy—grew even faster. However, the organic sector is still of modest size because of the low starting base. Also, the percentage increase in the number of certified organic growers between 1995 and 1997 was much less than the increase in farmland certified. Many existing organic farmers expanded their operations, and a number of new large-scale operations became certified.

The structure of the organic sector differs substantially from the agriculture industry as a whole, with fruits, vegetables and other high-value specialty crops making up a much larger proportion of this sector. About 2 percent of top specialty crops—lettuce, carrots, grapes, and apples—were grown under certified organic farming systems in 1997, while only 0.1 percent of the top U.S. field crops—corn and soybeans—were certified organic. Also, organic field crop producers grow a greater diversity of crops than their conventional counterparts because of the key role that crop rotation plays in organic pest and nutrient management. At least 1 percent of the oat, spelt (.7 percent), millet, buckwheat (.0 percent), rye, flax, and dry pea crops were under organic management. While some large-scale organic farms emerged during the 1990s, small farms producing "mixed vegetable" crops for marketing direct to consumers and restaurants still made up a large segment of the organic sector in 1997.

Government policy in the United States has focused primarily on developing national certification standards to assure consumers that organic commodities meet a consistent standard and to facilitate interstate and international commerce. However, several States have begun subsidizing conversion to organic farming systems as a way to capture the environmental benefits of these systems. Potential benefits from organic farming systems include improved soil tilth and productivity, lower energy use, and reduced use of pesticides, which can cause acute and chronic illness in humans as well as damage to fish and wildlife. Most European countries have been providing direct financial support for conversion since the late 1980s, with conversion levels much higher than in the United States.

Obstacles to adoption include large managerial costs and risks of shifting to a new way of farming, limited awareness of organic farming systems, lack of marketing and technical infrastructure, and inability to capture marketing economies. State and private certifier fees for inspections, pesticide residue testing, and other services represent an added expense for organic producers. And farmers cannot command certified organic price premiums during the 3-year required conversion period before crops and livestock can be certified as organic, although a few organic buyers offer a smaller premium for crops from land that is under conversion.

Strong market signals for organically produced agricultural goods, along with growing public and private support for organic farming systems, make it likely that organic production will remain a fast-growing segment of U.S. agriculture. Although government involvement in the United States has focused primarily on developing national certification standards, USDA has recently begun several small organic programs, including export promotion, farming systems trials, and a pilot program to provide financial assistance for certification costs.

Source: Catherine R. Greene, U.S. Department of Agriculture, Economic Research Service, Resource Economics Division, Agriculture Information Bulletin No. 770. www.ers.usda.gov/Publications/AIB770/.

Appendix 4:
Recent Growth Patterns in the
U.S. Organic Foods Market, 2002

This important USDA document presents national trends in the sales of organic products in the United States. Most of these data are not readily available and have not been assembled in one document previously. Its broad coverage is particularly interesting in that it describes various segments of the organic market, including livestock, grains, vegetables, fiber, and herbs. In addition, the report provides an overview of the various USDA agencies that are involved with organic production. Here is a brief excerpt that includes the abstract and summary of key findings. At the time this report was published, market growth in organic products was substantial, but federal organic standards had not yet been fully implemented. Thus, this report presents a key snapshot in the evolution of organic foods in the United States. All studies conducted after this point can draw from this baseline study to investigate the effects of policy changes after the USDA National Organic Standards were put into place in late 2002.

ABSTRACT

Organic farming is one of the fastest growing segments of U.S. agriculture. As consumer interest continues to gather momentum, many U.S. producers, manufacturers, distributors, and retailers are specializing in growing, processing, and marketing an ever-widening array of organic agricultural and food products. This report summarizes growth patterns in the U.S. organic sector in recent years, by market category, and describes various research, regulatory, and other ongoing programs on organic agriculture in the U.S. Department of Agriculture.

Keywords: organic agriculture, organic farming systems, organic marketing, organic marketing channels, certified organic acreage and livestock, price premiums, national organic rules, specialty agriculture, high-value crops, USDA research.

SUMMARY

Burgeoning consumer interest in organically grown foods has opened new market opportunities for producers and is leading to a transformation in the organic foods industry. Once a niche product sold in a limited number of retail outlets, organic foods are currently sold in a wide variety of venues including farmers markets, natural product supermarkets, conventional supermarkets, and club stores. Since the early 1990s, certified organic acreage has increased as producers strive to meet increasing demand for organic agricultural and food products in the United States. The dramatic growth of the industry spurred Federal policy to facilitate organic product marketing, and is leading to new government activities in research and education on organic farming systems.

This report summarizes growth patterns in the U.S. organic sector in recent years, by market category, and traces the marketing channels for major organic commodity groups. The report describes various research, regulatory, and other ongoing programs on organic agriculture in the U.S. Department of Agriculture.

- The U.S. organic food industry crossed a threshold in 2000: for the first time, more organic food was purchased in conventional supermarkets than in any other venue.
- Growth in retail sales has equaled 20 percent or more annually since 1990. Organic products are now available in nearly 20,000 natural foods stores, and are sold in 73 percent of all conventional grocery stores.
- According to the most recent USDA estimates, U.S. certified organic cropland doubled between 1992 and 1997, to 1.3 million acres.
- The new U.S. Department of Agriculture standards for organic food, slated to be fully implemented by October 2002, are expected to facilitate further growth in the organic foods industry.
- Fresh produce is the top-selling organic category, followed by nondairy beverages, breads and grains, packaged foods (frozen and dried prepared foods, baby food, soups, and desserts), and dairy products. During the 1990s, organic dairy was the most rapidly growing segment, with sales up over 500 percent between 1994 and 1999.
- Nine USDA agencies have expanded research, regulatory, and other programs on organic agriculture.

- The main regulatory program is the creation, implementation, and administration of the USDA organic standard. Other programs include crop insurance for organic farmers, information provision, and promotion of organic exports.
- USDA also funds projects for international market development and for natural resource conservation. Funding is also extended to projects assisting adoption of organic practices and exploration of new farming systems, methods, and educational opportunities.
- USDA research includes agronomic studies on soil management, biological control of pests and weeds, livestock issues, and post-harvest fruit treatment. Economic research focuses on tracking growth in the organic sector, demand for organic products, and organic farmers' risk management strategies.

Source: Carolyn Dimitri and Catherine Greene, U.S. Department of Agriculture, Economic Research Service, Market and Trade Economics Division and Resource Economics Division. Agriculture Information Bulletin Number 777. www.ers.usda.gov/Publications/AIB777/.

Selected Bibliography

Adam, Katherine. Direct Marketing. Fayette, AR: Appropriate Technology Transfer for Rural Areas, National Center for Appropriate Technology. 1999. http://attra .ncat.org/attra-pub/directmkt.html.

Allen, Patricia. *Together at the Table: Sustainability and Subsistence in the American Agrifood System.* University Park: Pennsylvania State University Press, 2004.

Altieri, Miguel A. *Agroecology: The Science of Sustainable Agriculture,* 2nd ed. Boulder, CO: Westview Press, 1995.

Bacon, C. M., V. E. Méndez, S. Gliessman, D. Goodman, and J. A. Fox, eds. *Confronting the Coffee Crisis: Fair Trade, Sustainable Livelihoods, and Ecosystems in Mexico and Central America.* Cambridge, MA: MIT Press, 2008.

Balfour, Lady Eve. *The Living Soil.* London: Faber and Faber, 1943.

Barratt-Brown, Michael. *Fair Trade: Reform and Realities in the International Trading System.* London: Zed, 1993.

Becker, Geoffrey. *Humane Treatment of Farm Animals: Overview and Issues.* CRS Report for Congress. Washington, D.C.: Congressional Research Service, 2008.

Bell, Michael. *Farming for Us All: Practical Agriculture and the Cultivation of Sustainability.* University Park: Pennsylvania State University Press, 2004.

Benbrook, Charles M. "Minimizing Pesticide Dietary Exposure through the Consumption of Organic Food." Organic Center State of Science Review. 2004. www.organic-center.org/reportfiles/PESTICIDE_SSR.pdf.

Berry, Wendell. *The Unsettling of America: Culture and Agriculture.* San Francisco: Sierra Club Books, 1977.

Berry, Wendell. *What Are People For?* New York: North Point Press, 1990.

Berry, Wendell. *The Art of the Commonplace: The Agrarian Essays of Wendell Berry.* Berkeley, CA: Counterpoint, 2003.

Beus, Curtis E., and Riley E. Dunlop. "Conventional Versus Alternative Agriculture: The Paradigmatic Roots of the Debate." *Rural Sociology* 55(4) 1990: 590–616.

Bird, Elizabeth Ann R., Gordon L. Bultena, and John C. Gardner. *Planting the Future: Developing an Agriculture that Sustains Land and Community.* Ames: Iowa State University Press, 1995.

Brown, Katherine, and Anne Carter. *Urban Agriculture and Community Food Security in the United States: Farming from the City Center to the Urban Fringe.* Venice, CA: Community Food Security Coalition, 2003.

Brown, Lester. *Building a Sustainable Society.* New York: W. W. Norton, 1981.

Buchmann, Stephen, and Gary Nabhan. *The Forgotten Pollinators.* Washington, D.C.: Island Press, 1997.

Butler, S. J., J. A. Vickery, and K. Norris. "Farmland Biodiversity and the Footprint of Agriculture." *Science* 315 (2007): 381–84.

Byczynski, Lynn, ed. Growing for Market. 2008. www.growingformarket.com.

Canavari, M., and K. D. Olson, eds. *Organic Food. Consumers' Choices and Farmers' Opportunities.* New York: Springer, 2007.

Center for Food Safety. "Genetically Engineered Crops." 2008. www.centerforfood safety.org/geneticall2.cfm.

Center for Food Safety. "Not Ready for Prime Time: FDA's Flawed Approach to Assessing the Safety of Food from Animal Clones." 2007. www.centerforfood-safety.org/Policy.cfm.

Centers for Disease Control and Prevention. "Environmental Hazards and Health Effects: Concentrated Animal Feeding Operations (CAFOs)." 2004. www.cdc.gov/cafos/about.htm.

Ching, Lim Li. "GM Crops Increase Pesticide Use." Institute of Science in Society. 2007. www.i-sis.org.uk/GMCIPU.php.

Colborn, Theo, Dianne Dumanoki, and John Peterson Myers. *Our Stolen Future.* New York: Plume, 1997.

Community Food Security Coalition. "North American Food Policy Council." www.foodsecurity.org/FPC/.

Conford, Philip. *Origins of the Organic Movement.* Edinburgh, UK: Floris Books, 2001.

Cook, Christopher. *Diet for a Dead Planet: How the Food Industry Is Killing Us.* New York: The New Press, 2004.

Cooperative Grocer. "Food Cooperative Directory." www.cooperativegrocer .coop/coops.

Danbom, David. *Born in the Country: A History of Rural America.* Baltimore: John Hopkins University Press, 2006.

DeCarlo, Jacqueline. *Fair Trade: A Beginner's Guide.* Oxford, UK: One World, 2007.

Dimitri, Carolyn, and Catherine Greene. *Recent Growth Patterns in U.S. Organic Foods Market.* Agricultural Information Bulletin 777. Washington, D.C.: USDA Economic Research Service, 2002.

Dimitri, Carolyn, and Nessa J. Richman. *Organic Food Markets in Transition.* Policy Studies Report No. 14. Henry A. Wallace Center for Agriculture and Environmental Policy, 2000.

Drake University Agricultural Law Center. "State and Local Food Policy Councils." www.statefoodpolicy.org.

Duram, Leslie A. *Good Growing: Why Organic Farming Works.* Lincoln: University of Nebraska Press, 2005.

Environmental Defense Fund. "Seafood Selector."www.edf.org/seafood.

Environmental Working Group. "Farming and Agriculture." www.ewg.org/farming.

Environmental Working Group. "Shopper's Guide to Pesticides in Produce." 2009. www.foodnews.org.

Fossel, Peter V. *Organic Farming: Everything You Need to Know.* Minneapolis, MN: Voyageur Press, 2007.

Fromartz, Samuel. *Organic, Inc.: Natural Foods and How They Grew.* New York: Harcourt, 2006.

Funes, Fernando, Luis García, Martin Bourque, Nilda Pérez, and Peter Rosset, eds. *Sustainable Agriculture and Resistance: Transforming Food Production in Cuba.* Oakland, CA: Food First Books, 2002.

Gerdes, L. I. *Genetic Engineering. Opposing Viewpoints Series.* Farmington Hills, MI: Greenhaven Press, 2004.

Gliessman, Stephen R. *Agroecology: The Ecology of Sustainable Food Systems*, 2nd ed. Boca Raton, FL: CRC Press, 2006.

Gould, Kenneth A., David N. Pellow, and Allan Schnaiberg. *The Treadmill of Production.* Boulder, CO: Paradigm Press, 2008.

Greene, Catherine. *U.S. Organic Farming in the 1990's: Adoption of Certified Systems.* Agriculture Information Bulletin No. 770. Washington, D.C.: USDA Economic Research Service, 2001.

Greene, Catherine. *Land in U.S. Organic Production.* USDA Datasets. USDA Economic Research Service. 2004. www.ers.usda.gov/Data/organic/.

Gurian-Sherman, D. *Holes in the Biotech Safety Net: FDA Policy Does Not Assure the Safety of Genetically Engineered Foods.* Washington, D.C.: Center for Science in the Public Interest, 2003.

Guthman, Julie. *Agrarian Dreams: The Paradox of Organic Farming in California.* Berkeley: University of California Press, 2004.

Halweil, Brian. *Eat Here: Homegrown Pleasures in a Global Supermarket.* New York: W. W. Norton, 2004.

Hannaford, Stephen G. *Market Domination! The Impact of Industry Consolidation on Competition, Innovation, and Consumer Choice.* Westport, CT: Praeger, 2007.

Hightower, Jim. *Hard Tomatoes, Hard Times: The Failure of the Land Grant College Complex.* Cambridge, MA: Schenkman, 1973.

Hill, Holly. "Food Miles: Background and Marketing." National Sustainable Agriculture Information Service. National Center for Appropriate Technology. 2008. www.attra.ncat.org/attra-pub/PDF/foodmiles.pdf.

Hinrichs, Clare, and Thomas Lyson. *Remaking the American Food System: Strategies for Sustainability.* Lincoln: University of Nebraska Press, 2008.

Horne, James, and Maura McDermott. *The Next Green Revolution: Essential Steps to a Healthy, Sustainable Agriculture.* New York: Haworth Press, 2001.

Howard, Sir Albert. *An Agricultural Testament.* Emmaus, PA: Rodale Press, 1943.

Howard, Sir Albert. *Farming and Gardening for Health or Disease (Soil and Health).* London: Faber and Faber, 1945.

Howard, Philip H. "Organic Industry Structure." www.msu.edu/~howardp/organicindustry.html.

Human Genome Project. "Genetically Modified Foods and Organisms." 2008. www.ornl.gov/sci/techresources/Human_Genome/elsi/gmfood.shtml.

Ikerd, John. *Small Farms Are Real Farms: Sustaining People through Agriculture.* Austin, TX: Acres USA, 2007.

Ikerd, John. *Crisis and Opportunity: Sustainability in American Agriculture.* Lincoln: University of Nebraska Press, 2008.

Intergovernmental Panel on Climate Change. "Summary for Policymakers." In *Climate Change 2007: The Physical Science Basis. Contribution of Working Group I to the Fourth Assessment Report of the Intergovernmental Panel on Climate Change*, edited by S. Solomon, D. Qin, M. Manning, Z. Chen, M. Marquis, K. B. Averyt, M. Tignor, and H. L. Miller. New York: Cambridge University Press, 2007.

Johnson, J. Richard, and Timothy Gower. *The Sugar Fix: The High Fructose Fallout that Is Making You Fat and Sick.* New York: Rodale, 2008.

Jackson, Dana L., and Laura L. Jackson. *The Farm as Natural Habitat: Reconnecting Food Systems with Ecosystems*. Washington, D.C.: Island Press, 2002.

Jaffee, Daniel. *Brewing Justice: Fair Trade Coffee, Sustainability and Survival*. Berkeley: University of California Press, 2007.

Jasanoff, S. *Designs on Nature: Science and Democracy in Europe and the United States*. Princeton, NJ: Princeton University Press, 2008.

Kimbrell, Andrew, ed. *Fatal Harvest: The Tragedy of Industrial Agriculture*. Washington, D.C.: Island Press, 2002.

Kingsolver, Barbara, with Steven L. Hopp and Camille Kingsolver. *Animal, Vegetable, Miracle*. New York: HarperCollins, 2007.

Kirschenmann, Fred. "Ecological Morality: A New Ethic for Agriculture." In *Agroecosystems Analysis*, edited by D. Rickerl and C. Francis, chap. 11. Madison, WI: American Society of Agronomy, Crop Science Society of America, and the Soil Science Society of America, 2004.

Kloppenburg, Jack. *First the Seed: The Political Economy of Plant Biotechnology*, 2nd ed. Madison: University of Wisconsin Press, 2004.

Koepf, Herbert H., Bo D. Pettersson, and Wolfgang Shaumann. *Bio-dynamic Agriculture: An Introduction*. Spring Valley, NY: Anthroposophic Press, 1976.

Lawson, Laura. *City Bountiful: A Century of Community Gardening in America*. Berkeley: University of California Press, 2005.

Leclair, M. S. "Fighting the Tide: Alternative Trade Organizations in the Era of Global Free Trade." *World Development* 30(6) 2002: 949–58.

Lipson, M. *Searching for the "O-Word."* Santa Cruz, CA: Organic Farming Research Foundation, 1997. www.ofrf.org.

Local Harvest. "Community Supported Agriculture. Real Food, Real Farmers, Real Community." 2008. www.localharvest.org.

Lockeretz, William, Georgia Shearer, and Daniel H. Kohl. "Organic Farming in the Corn Belt." *Science* 211 (1981): 540–47.

Losey, John, Laura Rayor, and Maureen E. Carter. "Transgenic Pollen Harms Monarch Larvae." *Nature* 399 (1999): 214–15.

Losey, John E., and Vaughan, Mace. "The Economic Value of Ecological Services Provided by Insects." *Bioscience* 56 (2006): 311–23.

Lu, C., K. Toepel, R. Irish, R. Fenske, D. Barr, and R. Bravo. "Organic Diets Significantly Lower Children's Dietary Exposure to Organophosphate Pesticides." *Environmental Health Perspectives* 114(2) 2006: 260–63.

Lyson, T. A. *Civic Agriculture: Reconnecting Farm, Food and Community*. Medford, MA: Tufts University Press, 2004.

Mäder, Paul, Andreas Fließbach, David Dubois, Lucie Gunst, Padruot Fried, and Urs Niggli. "Soil Fertility and Biodiversity in Organic Farming." *Science* 296 (5573) May 31, 2002: 1694–7.

Magdoff, Fred, John Bellamy Foster, and Frederick Buttel, eds. *Hungry for Profit: The Agribusiness Threat to Farmers, Food and the Environment*. New York: Monthly Review Press, 2000.

Mies, Maria, and Vandana Shiva. *Ecofeminism*. London: Zed Books, 1993.

Mitchell, Don. *The Lie of the Land: Migrant Workers and the California Landscape*. Minneapolis: University of Minnesota Press, 1996.

Monterey Bay Aquarium. "Seafood Watch." www.seafoodwatch.org.

National Oceanic and Atmospheric Administration. "FishWatch: U.S. Seafood Facts." www.nmfs.noaa.gov/fishwatch/.

National Research Council. *The Use of Drugs in Food Animals: Benefits and Risks*. Washington, D.C.: National Academies Press, 1999.

National Research Council. *Nutrient Control Actions for Improving Water Quality in the Mississippi River Basin and Northern Gulf of Mexico*. National Academies Press, 2008.

National Sustainable Agriculture Information Service. "Organic Farming." Updated February 2009. http://attra.ncat.org/organic.html.

Nestle, Marion. *Food Politics: How the Food Industry Influences Nutrition and Health.* Berkeley: University of California Press, 2002.

Novacek, M. J., ed. *The Biodiversity Crisis: Losing What Counts.* New York: New Press, 2001.

Organic Center. "Organic Essentials: Pocket Guide for Reducing Pesticide Dietary Exposure." In *Simplifying the Pesticide Risk Equation: The Organic Option*, Organic Center State of Science Review, March 2008. www.organic-center.org/reportfiles/TOC_Pocket_Guide.pdf.

Organic Trade Association. "Organic Facts." www.ota.com/organic.

Pederson, Donald B.,and Keith G. Meyer. *Agricultural Law in a Nutshell.* St. Paul, MN: West Publishing, 1995.

Perfecto, Ivette, Robert A. Rice, Russell Greenberg, Martha E. Van der Voort. "Shade Coffee: A Disappearing Refuge for Biodiversity." *BioScience* 46(8) 1996: 598–609.

Petrini, Carlo. *Slow Food Nation: Why Our Food Should Be Good, Clean, and Fair.* New York: Rizzoli Ex Libris, 2007.

Pimentel, D., P. Hepperly, J. Hanson, D. Douds, and R. Seidel. "Environment, Energy, and Economic Comparisons of Organic and Conventional Farming Systems." *Bioscience* 55(7) 2005: 573–82.

Pimentel, David, and Marcia Pimentel. *Food, Energy and Society,* 3rd ed. Boca Raton, FL: CRC Press, 2008.

Pimm, S. L., G. J. Russell, J. L. Gittleman, and T. M. Brooks. "The Future of Biodiversity." *Science* 269 (1995): 347–50.

Pirog, Rich,Timothy Van Pelt, Kamyar Enshayan, and Ellen Cook. "Food Fuel, and Freeways: An Iowa Perspective on How Far Food Travels, Fuel Usage, and Greenhouse Gas Emissions." Iowa State University, Leopold Center for Sustainable Agriculture, 2001. www.leopold.iastate.edu/pubs/staff/ppp/food_mil.pdf.

Pollan, Michael. *The Omnivore's Dilemma: A Natural History of Four Meals.* New York: Penguin, 2006.

Pollan, Michael. *In Defense of Food: An Eater's Manifesto.* New York: Penguin, 2008.

Reganold, John P., Jerry D. Glover, Preston K. Andrews, and Herbert R. Hinman. "Sustainability of Three Apple Production Systems." *Nature* 410(6831) 2001: 926–30.

Reilly, Jean K. "Moonshine, Part 2: A Blind Sampling of 20 Wines Shows that Biodynamics Works. But How? (This, by the Way, Is Why We Went into Journalism.)" *Fortune.* August 23, 2004.

Robinson, Guy. *Geographies of Agriculture: Globalisation, Restructuring, and Sustainability.* Essex: Pearson Education, 2004.

Robinson, J. M., and J. A. Hartenfeld. *The Farmers' Market Book: Growing Food, Cultivating Community.* Bloomington: Indiana University Press, 2007.

Rodale. "The Organic Transition Course." www.tritrainingcenter.org/course/.

Rodale, J. I. *Pay Dirt: Farming and Gardening with Composts.* Emmaus, PA: Rodale Press, 2005.

Ruddiman W. F. *Plows, Plagues, and Petroleum: How Humans Took Control of Climate,* Princeton, NJ: Princeton University Press, 2005.

Schlosser, Eric. *Fast Food Nation: The Dark Side of the All-American Meal.* Boston: Houghton Mifflin Company, 2001.

Shiva, Vandana. *Stolen Harvest: The Hijacking of the Global Food Supply.* Cambridge, MA: South End Press, 2000.

Shiva, Vandana, and Gitanjali Bedi. *Sustainable Agriculture and Food Security: The Impact of Globalization.* New Delhi: Sage, 2002.

Simon, Michelle. *Appetite for Profit: How the Food Industry Undermines Our Health and How to Fight Back.* New York: Nation Books, 2006.

Sligh, Michael, and Carolyn Christman. *Who Owns Organic? The Global Status, Prospects and Challenges of a Changing Organic Market.* Pittsboro, NC: Rural Advancement Foundation International-USA, 2003.

Smith, Alisa, and J. B. MacKinnon. *Plenty: One Man, One Woman, and a Raucous Year of Eating Locally.* New York: Harmony, 2007.

Sooby, Jane. *State of the States. Organic Farming Systems Research at Land Grant Institutions 2000–2001.* Santa Cruz, CA: Organic Farming Research Foundation, 2001. www.ofrf.org.

Stanhill, G."The Comparative Productivity of Organic Agriculture." *Agriculture, Ecosystems, and Environment* 30(1–2) 1990: 1–26.

Steiner, Rudolf. *Agriculture,* 3rd ed. London: Rudolf Steiner Press, 1974.

Steingraber, Sandra. *Living Downstream.* Vintage Books: New York, 1997.

Strange, Marty. *Family Farming: The New Economic Vision,* 2nd ed. Lincoln: University of Nebraska Press, 2008.

Tokar, Brian, ed. *Redesigning Life: The Worldwide Challenge to Genetic Engineering.* London: Zed Books, 2001.

United Nations Environmental Programme. "Organic Agriculture and Food Security in Africa" United Nations Conference on Trade and Development. UNEP-UNCTAD Capacity-building Task Force on Trade, Environment and Development. New York and Geneva: United Nations, 2008.

United States Department of Agriculture. *Soils and Men: Yearbook of Agriculture.* Washington, D.C.: USDA, 1938.

USDA Agricultural Marketing Service. "National Organic Program." www.ams.usda.gov/nop/.

USDA Agricultural Marketing Service. "Wholesale and Farmers' Markets." www.ams.usda.gov/farmersmarkets.

USDA Alternative Farming Systems Information Center, National Agricultural Library. "Community Supported Agriculture." 2008. www.nal.usda.gov/afsic/pubs/csa/csa.shtml.

USDA Alternative Farming Systems Information Center, National Agricultural Library. "Direct Marketing, Alternative Marketing and Business Practices." http://afsic.nal.usda.gov.

USDA Economic Research Service. "Biotechnology and Crops." 2008. www.ers.usda.gov/Data/BiotechCrops.

USDA Economic Research Service. "Organic Agriculture Briefing Room." 2008. www.ers.usda.gov/Briefing/Organic/.

USDA Study Team on Organic Farming. *Report and Recommendations on Organic Farming.* Washington, D.C.: USDA, 1980. www.nal.usda.gov/afsic/pubs/pubsindex.shtml.

USDA Sustainable Agriculture Research and Education. "Transitioning to Organic Production." www.sare.org/publications/organic.

U.S. Environmental Protection Agency. "Mississippi River Gulf of Mexico Watershed Nutrient Task Force." 2008. www.epa.gov/msbasin/.

U.S. Environmental Protection Agency. "Pesticide Issues." www.epa.gov/pesticides.

U.S. Food and Drug Administration. "Animal Cloning: A Risk Assessment–FINAL, January 15, 2008." 2008. www.fda.gov/cvm/CloneRiskAssessment_Final.htm.

U.S. Geological Survey. "Investigating the Environmental Effects of Agriculture Practices on Natural Resources." 2007. http://pubs.usgs.gov/fs/2007/3001.

U.S. Geological Survey. "Synthesis of U.S. Geological Survey Science for the Chesapeake Bay Ecosystem and Implications for Environmental Management." Circular 1316. 2008.

Vaughan, M., M. Shepherd, C. Kremen, and S. Black. *Farming for Bees: Guidelines for Providing Native Bee Habitat on Farms,* 2nd ed. 2007. Portland, OR: Xerces Society for Invertebrate Conservation, 2007.

Weaver, William. *Vegetable Gardening: A Master Gardener's Guide to Planting, Growing, Seed Saving, and Cultural History.* New York: Henry Holt, 1997.

Wilson, E. O. *The Diversity of Life.* New York: W.W. Norton, 1992.

WEB SITES

Agribusiness Accountability Initiative. www.agribusinessaccountability.org.

American Community Gardening Association. www.communitygarden.org.

American Farmland Trust. www.aft.org.

Center for Rural Affairs. www.cfra.org.

Demeter International. www.demeter.net.

Demeter USA. www.demeter-usa.org.

Environmental Defense Fund. www.edf.org.

Farm to School. www.farmtoschool.org.

Food and Agriculture Organization of the United Nations. www.fao.org.

Food Routes Network. www.foodroutes.org.

National Agricultural Law Center. www.nationalaglawcenter.org.

National Cooperative Grocers Association. www.ncga.coop.

Organic Consumers Association. www.organicconsumers.org.

Organic Farming Research Foundation. www.ofrf.org.

Organic Trade Association. www.ota.com.

People for the Ethical Treatment of Animals. www.goveg.com.

Scientific Congress on Organic Agricultural Research, Organic Farming Research Foundation. http://ofrf.org/scoar/index.html.

Seafood Choices Alliance. www.seafoodchoices.org.

Soil Association, U.K. www.soilassociation.org.

Sustainable Agriculture Coalition. www.sustainableagriculturecoalition.org.

U.S. Food and Drug Administration. www.fda.gov.

Union of Concerned Scientists. www.ucsusa.org.

USDA Agricultural Marketing Service. www.ams.usda.gov.

USDA Economic Research Service. www.ers.usda.gov.

World-wide Opportunities on Organic Farms. www.wwoofusa.org.

About the Editor and Contributors

Samuel Adu-Prah is a lecturer and geographic information system lab director in the Department of Geography and Environmental Resources in Southern Illinois University, Carbondale. He conducts research in geographic data mining and geovisualization in understanding environmental and public health data.

Dan Anderson is a research specialist in agriculture in the natural resources and environmental sciences department at the University of Illinois, Urbana-Champaign. His research area is organic farming and local food systems.

Christopher M. Bacon is an environmental social scientist. His work focuses on fair trade, knowledge production, and sustainable community development in the Americas. He is a Ciriacy-Wintrup Postdoctoral Fellow in the geography department at the University of California, Berkeley.

Shauna M. Bloom is a doctoral candidate in the Department of Geography at the University of Guelph. Her research interests revolve around environmental and resource management issues, particularly issues relating to community, food, and agriculture. Her doctoral research addresses producer engagement in local food systems and organizational capacity of local food promotion groups.

Steven Bossu is an undergraduate research assistant at the University of Illinois, Urbana-Champaign.

Jerry Bradley is media supervisor at the Neighborhood Coop Grocery in Carbondale, Illinois, and a local food advocate, blogger, and freelance journalist

Lynn Byczynski is the editor of *Growing for Market* (www.growingformarket.com), a periodical about local food and sustainable farming. She has also been a market farmer for more than twenty years.

Kate Clancy is a food systems consultant focusing on the development of regional food systems, research to support sustainable agriculture policy, and land-use policies related to small and midsize farm viability.

Cynthia Abbott Cone is professor emerita of anthropology at Hamline University. Her research interests include alternative agricultural strategies and images of landscape among American Indian cultures.

Adam Davis is an ecologist with the USDA-ARS Invasive Weed Management Unit in Urbana, Illinois. His research focuses on the ecological management of weeds in field crops and nonarable areas.

Carolyn Dimitri is a senior economist with USDA's Economic Research Service, and her research centers on marketing organic products.

Holly A. S. Dolliver is an assistant professor of soil science and geology at the University of Wisconsin, River Falls. Her research focuses on quantifying the influence of soil management practices on antibiotic, carbon, and nutrient transport in agricultural systems.

Leslie A. Duram is professor of geography at Southern Illinois University, Carbondale, and the author of *Good Growing: Why Organic Farming Works* (2005) and numerous research articles. She served on the Illinois Local Food and Farm Task Force and is a Fulbright Scholar to Ireland.

Tim Fitzgerald is a marine scientist in Environmental Defense Fund's Oceans Program. His research focuses on conservation and human health issues in the U.S. seafood market.

Lauren Flowers is an undergraduate research assistant at the University of Illinois, Urbana-Champaign.

Wendy Fulwider was raised on a sixty-cow Wisconsin dairy farm. As an Organic Valley/CROPP Cooperative animal husbandry specialist, she works to improve animal well-being.

Fernando Funes is national coordinator of the Agroecological Programme of the Cuban Association of Agricultural and Forestry Technics (ACTAF), Havana, Cuba.

Fernando R. Funes-Monzote is researcher of Indio HatueyPastures and Forages Research Statión, University of Matanzas, Cuba

Mari Gallagher is president of both Mari Gallagher Research & Consulting Group and the nonprofit National Center for Public Research. Her experience as a practitioner and researcher has been focused on food access, public health, and commercial markets.

Brian J. Gareau is visiting assistant professor in the Department of Agricultural Economics and Rural Sociology at the Pennsylvania State University, and his research focuses on the linkages between globalization, science, and politics in global environmental governance, particularly in the methyl bromide phase-out of the Montreal Protocol.

Tara Pisani Gareau is post-doctoral researcher of agroecology at the Pennsylvania State University, State College. Her research interests include the value of noncrop vegetation in the conservation of insect natural enemies and restoration of biological control services and the sociopsychological factors that influence growers' conservation behavior.

Joel Gehrig is a research assistant at the University of Illinois, Urbana-Champaign.

Holly Givens is public affairs advisor at the Organic Trade Association. Her focus is helping people better understand organic agriculture and products.

Mary V. Gold is retired from the USDA National Agricultural Library where she was a specialist at the Alternative Farming Systems Information Center.

Ramu Govindasamy is chair and associate professor of the Department of Agricultural, Food and Resource Economics at Rutgers-The State University of New Jersey. He is also the associate director of research at the Rutgers Food Policy Institute and extension specialist at Rutgers Cooperative Extension. His research focus is food consumer perceptions, attitude and demand, especially organic produce and ethnic foods.

Jonathan M. Gray is an associate professor in the Department of Speech Communication, Southern Illinois University, Carbondale. His research interests include performing nature, queer identity and ecological advocacy, visual argumentation, and environmental rhetoric.

Julie Guthman is an associate professor of community studies at the University of California at Santa Cruz. Her research is about various efforts and social movements to transform the way food is produced, distributed, and consumed and the effect these have on social and environmental justice.

Paul Reed Hepperly is the director of research at the Rodale Institute in Kutztown, Pennsylvania. His research has focused on the long-term analysis of farming systems and food and agricultural issues. He has measured the effect of farming practices on carbon sequestration and energy use and how this relates to greenhouse gas issues.

Annette M. Hormann is a doctoral student in environmental resources and policy at Southern Illinois University, Carbondale. Her research focuses on environmental endocrine disruptors; looking for correlations between the herbicides break-down components and incidence of thyroid disease in local farming communities.

Philip H. Howard is assistant professor of community, food, and agriculture at Michigan State University. He conducts research on food system changes and their social impacts, focusing on retailers and ecolabels.

Thomas W. Hutcheson is regulatory and policy manager for the Organic Trade Association. He manages domestic regulatory and legislative relations and represents the interests of the organic trade, from farmers to retailers.

John Ikerd is professor emeritus of agricultural economics, University of Missouri, Columbia. His academic and professional focus is sustainable agriculture with an emphasis on economic and rural development issues.

Barry Krissoff is deputy division director for research in the Market and Trade Economics Division, Economic Research Service, U.S. Department of Agriculture. His research focus is on agricultural trade and trade policy issues, particularly as they relate to fruits and vegetables.

Fred Kuchler is a senior economist in the Food Economics Division, Economic Research Service, U.S. Department of Agriculture. His research focus is on consumer behavior, particularly with respect to food purchasing patterns.

Audrey Law is a graduate student in crop science at the University of Kentucky. Her research focused on agricultural systems in vegetable production and how they affect soil chemical and biological properties, and how the soil microbial community affects plant growth.

E. F. Legner is professor emeritus at the University of California and his research focuses on biological pest control, ontogeny, and reproduction of parasitic insects.

Mark Lipson is the senior policy analyst for the Organic Farming Research Foundation. His primary focus is cultivating organic research programs in the public sector. His seminal work *Searching for the "O-Word"* documented the absence of publicly funded organic research and provided the platform for OFRF's advocacy success.

Luanne Lohr is a professor of agricultural and applied economics at the University of Georgia.

Eric Mader is the pollinator outreach coordinator for the Xerces Society. He works to raise awareness of about native pollinator conservation techniques among growers and government agencies.

Jesse Marsh is Program Officer at World Wildlife Fund in the San Francisco Bay Area.

John Masiunas is an associate professor in the Department of Natural Resources and Environmental Sciences at the University of Illinois. His research focuses on ecology and management of weedy plants in sustainable and organic food cropping systems.

Kathleen Merrigan is the Deputy Secretary of Agriculture for the USDA. She formerly headed the UDSA's Agriculture Marketing Service which oversaw the National Organic Program.

The National Cooperative Grocers Association (NCGA) is a business services cooperative for natural food co-ops located throughout the United States.

Martha L. Noble is a senior policy associate with the National Sustainable Agriculture Coalition in Washington, D.C. She is also a vice-chair of the Agricultural Management Committee of the American Bar Association's Section of Environment, Energy, and Resources, the Board of Directors of the Clean Water Network and the EPA's Farm, Ranch and Rural Communities Advisory Committee.

Jennifer Obadia is a doctoral candidate studying urban agriculture at Tufts University.

Lydia Oberholtzer is a researcher with Penn State University through a cooperative agreement with USDA's Economic Research Service. She conducts research on organic agriculture and retailing, agricultural marketing for small and mid-sized farms, local food systems, and issues related to farmland preservation and farm viability.

Michael Pease is an environmental consultant and adjunct lecturer in the Department of Geography at Central Washington University. His research is in water policy and natural resource management.

Sharon Peterson is a registered dietitian and an assistant professor at Southern Illinois University, specializing in community nutrition issues. Her overall research focus is factors that influence human food choices and community-based lifestyle intervention for children with risk factors for type 2 diabetes.

David Pimentel is a professor in the College of Agriculture and Life Sciences at Cornell University, in the Department of Entomology and the Department of Ecology and Evolutionary Biology. His research focus is on energy, population, and environment.

Harrison Mauzy Pittman is director of the National Agricultural Law Center at the University of Arkansas School of Law. His research focuses on a number of areas including the impact of globalization on federal farm policy, animal welfare, the National Organic Program, and environmental law

Meredith Redlin is an associate professor of rural sociology at South Dakota State University. Her research interests are sustainable community development and race and minority studies in rural areas.

Philip D. Reid is Louise C. Harrington Professor Emeritus at Smith College, Northampton, Massachusetts. He was instrumental in developing the technique of tissue printing to locate specific enzymes in plant tissues.

Taylor Reid is a doctoral student in community food and agriculture at Michigan State University. His research focuses on organic standards and first-generation organic farmers.

Jim Riddle is the University of Minnesota's organic outreach coordinator. He is former chair of the USDA's National Organic Standards Board, founding chair of the Independent Organic Inspectors Association, and co-author of the IFOAM/IOIA International Organic Inspection Manual.

Robin Jane Roff does research and publishes on the political economy of agricultural biotechnology in North America and the effect of neoliberalization on alternative food activism.

Justin T. Schoof is an assistant professor of geography and environmental resources at Southern Illinois University. His research focuses on regional climate change and statistical applications in climatology.

George Siemon is the CEO and a founding farmer of Organic Valley Family of Farms.

Sylvia Smith is an assistant professor in the Department of Animal Science, Food and Nutrition at Southern Illinois University. Her principal area of research is food tourism and local foodways.

Catherine H. Strohbehn is adjunct associate professor in hotel, restaurant, and institution management and extension specialist at Iowa State University. Her research investigates local food use by institutions and commercial foodservice operations, as well as food safety and sanitation issues.

Angie Tagtow is a Food & Society Policy Fellow—Connecting Soil to Food to Health, a registered dietitian who owns an environmental nutrition consulting business, Environmental Nutrition Solutions, and a national speaker on food issues.

Brise Tencer is the Washington Representative for the Food and Environment Program of the Union of Concerned Scientists (UCS) where she advocates for legislation and policy positions on issues pertaining to organic and sustainable agriculture. Prior to joining UCS, Brise served as Legislative Coordinator and Acting Policy Program Director for the Organic Farming Research Foundation. Brise worked on a crop diversification project with Peace Corps Guatemala. She has a masters degree in International Environmental Policy and is certified in Conflict Mediation.

Matthew D. Therrell is an assistant professor of geography at Southern Illinois University. His interests include studying the impacts of climate variability and global change on the environment and society. His research expertise is dendroclimatology.

Lori Ann Thrupp is manager of sustainability and organic development at Fetzer Vineyards., where she provides technical assistance and information for growers, wineries, and the broader community.

Julie Weinert is a lecturer in the Department of Geography and Environmental Resources at Southern Illinois University, Carbondale, and her research interests are in ecotourism, Latin America, and feminist geography.

Richard Weinzierl is a professor and extension entomologist at the University of Illinois. His research focuses on the management of insect pests in conventional and organic production of fruits and vegetables.

Evan Weissman is a graduate student in geography at Syracuse University, where his research focuses on urban agriculture and community gardens in the United States.

Rick Welsh is an associate professor of sociology at Clarkson University in Potsdam, New York. His research focuses on social and technological change in development with emphases on agri-food systems, science, technology and society and environmental sociology. He is the coeditor of *Food and the Mid-level Farm: Renewing an Agriculture of the Middle*.

Mark A. Williams is an associate professor in the Department of Horticulture at the University of Kentucky. His research and educational efforts focus on organic horticultural crop production. He is also the director of the sustainable agriculture undergraduate degree program.

Camilla Yandoc Ables is an associate program officer at the National Academy of Sciences Board on Agriculture and Natural Resources. Her research expertise is in biological control of weeds and methyl bromide alternatives.

Index

Note: the letter "t" following a page number indicates a table. Boldfaced page numbers indicate main articles.